HANDBOOK OF SKETCH-BASED MODELING FOR ENGINEERING DESIGN

A Practical Approach with REFER

HANDBOOK OF SKETCH-BASED MODELING FOR ENGINEERING DESIGN

A Practical Approach with REFER

Pedro Company
Jaume I University, Spain

Jorge D Camba
Purdue University, USA

World Scientific

NEW JERSEY • LONDON • SINGAPORE • BEIJING • SHANGHAI • HONG KONG • TAIPEI • CHENNAI • TOKYO

Published by

World Scientific Publishing Co. Pte. Ltd.
5 Toh Tuck Link, Singapore 596224
USA office: 27 Warren Street, Suite 401-402, Hackensack, NJ 07601
UK office: 57 Shelton Street, Covent Garden, London WC2H 9HE

Library of Congress Cataloging-in-Publication Data
Names: Company, Pedro author | Camba, Jorge D. author
Title: Handbook of sketch-based modeling for engineering design : a practical approach with REFER / author Pedro Company, Jaume I University, Spain, Jorge D. Camba, Purdue University, USA.
Description: New Jersey : World Scientific, [2026] | Includes bibliographical references and index.
Identifiers: LCCN 2025047260 | ISBN 9789819825653 hardcover | ISBN 9789819825660 ebook for institutions | ISBN 9789819825677 ebook for individuals
Subjects: LCSH: REFER (Computer program) | Three-dimensional modeling--Computer programs | Computer-aided design--Computer programs | Engineering design--Data processing | Artificial intelligence--Engineering applications | LCGFT: Handbooks and manuals
Classification: LCC T385 .C54866 2026
LC record available at https://lccn.loc.gov/2025047260

British Library Cataloguing-in-Publication Data
A catalogue record for this book is available from the British Library.

For any available supplementary material, please visit
https://www.worldscientific.com/worldscibooks/10.1142/14656#t=suppl

Desk Editors: Nambirajan Karuppiah/Veronica Lee

Typeset by Stallion Press
Email: enquiries@stallionpress.com

To my amazing wife María José,
for her love, patience, and support.

Pedro

To Elizabeth and Emma,
the brightest part of every day.

Jorge

Preface

The problem of sketch-based modeling (SBM) can be addressed through the combined use of what may be termed the classical approach and the artificial intelligence (AI) approach. These paradigms, rather than being seen as competing, ought to be leveraged in a complementary, synergistic manner.

The classical approach brings with it a wealth of accumulated expertise, knowledge that is essential for refining the strategies currently driving many AI-based methods. Unfortunately, such integration has not been the prevailing trend. In SBM, the focus has shifted from an almost obsessive pursuit of computational efficiency (what might be described as "micro-efficiency") to a reliance on massive computing resources, often at the expense of established methodological insights that could substantially alleviate the extraordinary computational demands of modern approaches.

The case of Dijkstra's algorithm illustrates this shift. For decades, sustained research efforts improved the computational efficiency of pathfinding in graphs ("minimum distance between nodes," etc.), even though the original method was already efficient when it was first conceived more than 60 years ago. In parallel, SBM research demonstrated that engineering sketches, long recognized as amenable to graph-based representations, could be analyzed using tools such as Dijkstra's algorithm to extract cues about the underlying three-dimensional geometry. This enabled the incremental recovery of three-dimensional coordinates by traversing the nodes of the associated graph during the inflation of procedural models. In contrast, many contemporary AI-based methods seek to recover comparable information while disregarding the accumulated knowledge concerning the underlying graph structures of the sketches.

A first paradox of the current situation is that, whereas even minor inefficiencies in real-time computational cost were once considered unacceptable, the increasing growth of deferred computational cost (whether in cloud-based execution or model training) now appears to face neither clear limits nor significant criticism. This paradigm reflects a prevailing doctrine of "as long as I do not have to wait, the cost does not matter." The field has thus moved from carefully weighing the implications of even the smallest variations in methods of computation and data storage to a general acceptance of the need for massive storage and processing capabilities as the default solution to every problem.

A second paradox is that existing knowledge, i.e., knowledge that does not need to be generated anew and that is highly specific to the problems under consideration (for example, the conversion of sketches into CAD models), is often disregarded. Problems are instead approached as "black boxes," thereby overlooking the substantial body of expertise already accumulated on many of these issues. The common justification for ignoring such knowledge is that it has not enabled solutions to large-scale problems (we do not know how to scale it) nor to all possible variants of the problem (we do not know how to generalize it). Little attention is given to the fact that AI-based approaches tend to assume spontaneous scalability (the belief that "although training is conducted on specific cases, generalization will occur automatically") and to conflate the part with the whole (the belief that, although training is based on biased examples, the process of training itself will eliminate the bias).

In the specific case of SBM, our proposal for overcoming these challenges and fostering synergy begins with a closer examination of a method for converting sketches of geometric parts into procedural CAD models. The method consists of two main stages; each subdivided into several sub-stages (or sub-tasks). The two primary stages involve: (1) detecting features in the sketch, and (2) combining those features into a procedural model tree. The feature detection stage can be extended in two directions: first, to develop the ability to recognize a wide variety of feature types, and second, to "deconstruct" any drawing to reveal features that may be partially or entirely obscured by others.

As described throughout this book, reasonable success has been achieved in detecting a limited set of features (e.g., ribs, dividers, rails, slots, steps, and pockets). A viable method has also been implemented for deconstructing the drawing by successively removing the detected features, thereby revealing the core of the represented part. It is important to emphasize the difficulty of this task, as all these operations are performed on a two-dimensional drawing, before the corresponding 3D model is available.

The task of combining features requires not only determining how the features represented in the sketch are related to one another but also defining their precise location. At the most basic level, the task can be reduced to identifying whether features are "additive" or "subtractive." The ability to intersect features is also useful and is already well established in commercial CAD applications. Together, these constitute the so-called Boolean operations for combining solids, which form the foundation of Constructive Solid Geometry (CSG). In many cases, identifying a feature type inherently determines how it can be combined (for example, a rail is subtractive, whereas a slot is additive). Nevertheless, the detection of more sophisticated feature combinations in 2D drawings remains an open problem.

The problem of locating features in space can be addressed efficiently through the definition of datums and datum systems. In modern CAD, two extreme approaches can be applied: defining datums explicitly prior to modeling or defining them "on the fly" during the modeling process. In practice, a hybrid approach is typically employed, in which the most important (or critical) references are established in advance, while others are defined as needed.

In SBM approaches aimed at generating CSG-based models without the use of AI, the ability to detect datum planes in the model has been achieved only to a limited extent (e.g., auxiliary planes, such as bounding faces, and certain types of symmetry planes). This task is further complicated by the fact that it is performed starting from an imperfect drawing, i.e., a sketch that may contain geometric inaccuracies.

Therefore, classical approaches allow for the detection of a limited set of features, as well as determining their relative positions and the manner in which they should be combined. In short, they provide the ability to generate procedural model trees for simple polyhedral geometric shapes. However, classical methods remain neither scalable nor generalizable. The problem is inherently complex, not only because it requires extracting three-dimensional geometric information from a two-dimensional drawing, but also because it seeks to obtain geometrically precise information from an imprecise representation. Interestingly, humans, and designers and engineers in particular, are able to extract this information with relative ease, in part because they do not rely solely on mathematical or geometric reasoning but also employ visual perception and technical knowledge.

Classical SBM methods have thus far achieved limited success when addressing the problem from a mathematical or geometric perspective. No general methods have been found that are applicable to all geometric shapes. Similarly, progress in the perceptual approach has been insufficient. Replicating human perception is extremely complex, first because many of the mechanisms underlying perception are not yet well understood, and second because computational attempts to emulate perception rely on heuristic methods applicable only to specific cases. Finally, most methods also fail to incorporate design intent information, that is, the functional purpose of the reconstructed object, a fundamental aspect in CAD. Nevertheless, it would not be prudent to disregard classical methods entirely simply because they are not generalizable or scalable. These methods can provide valuable insights into how SBM problems can be decomposed and efficiently addressed using AI-based tools.

Building on this insight, we propose training AI to recognize known features in well-defined cases. Procedural CAD models can be used to extract 2D drawings, which can then be annotated by leveraging the information already available in their model trees. For example, one can identify "rib" operations in the tree and mark the corresponding elements in the 2D drawing (e.g., by color, layer assignment, or textual labels). Once annotated in this manner, the drawings can serve as training data for AI models. Moreover, the approach allows presenting the AI not only with drawings that include a given feature, but also with versions in which that feature has been removed, thereby training the system both on detecting and deconstructing features. Drawings can also be distorted into "pseudo-sketches," to train the model to detect features even under imperfect representations, deliberately introducing alterations without bias and emulating the highly variable nature of human sketches.

In parallel with feature detection, existing computational knowledge about graph structures can also be leveraged to guide AI learning. The goal is not to replicate with AI models the task of finding the shortest path in a graph (an efficient solution already exists in Dijkstra's algorithm). Rather, the challenge is to capitalize on established insights, such as the role of orthogonal paths in the graphs of polyhedral parts. By doing so, an

AI model can be trained to use these paths to extract information relevant for deciding how best to "inflate" a drawing into a complete 3D model.

A similar approach can be applied to datums. An AI model can be trained to detect, on 2D projections of procedural 3D models, the datums used in their construction (which are readily available in their model trees). Artificial imperfections can be introduced to better approximate datum detection in 2D sketches. Finally, AI models can be trained to combine features based on the detected datums, producing a procedural model tree that reproduces the original CAD model from which the "view" was extracted.

We are confident in this proposal, yet we recognize that it is not without numerous challenges. Perhaps the most significant is that, in this final stage, the issues associated with the low quality of many models currently available in databases will become more apparent. Training AI on low-quality models cannot be expected to yield high-quality reconstructions. For this reason, we consider it critical to have the means to monitor and ensure the quality of the models used for training.

Despite the many challenges (and perhaps precisely because of them) we hope that this book will guide readers in leveraging the knowledge accumulated through classical SBM methods to inform AI-based approaches, beginning with problem decomposition and applying methods that have already been shown to be viable, albeit limited, in non-AI contexts.

Acknowledgments

This handbook compiles over two decades of the authors' knowledge and experience in Sketch-Based Modeling (SBM) for engineering design. Rather than providing a theoretical description of the problem, it focuses on demonstrating different methods and algorithms through interactive implementations. This approach allows readers and users to experiment firsthand, making the learning process more practical and engaging.

The book is connected to REFER; a software application developed over the years to experiment with and test various SBM methods and approaches. To make both the book and REFER more useful to readers, the authors have gathered, revised, and updated the source code.

This process demanded countless hours devoted to identifying and fixing the many bugs that inevitably accumulate over time in a research software application. Despite our best efforts, some issues are likely to remain. Nevertheless, we decided to release the current version, confident that, errors aside, it will still be valuable to readers interested in the evolution and current state of SBM in engineering design.

The authors have made every effort in the preparation of this book to ensure the accuracy of the information. However, the information contained in this book is sold without warranty, either expressed or implied. Any projections concerning the future of SBM that are interwoven throughout the explanations of the functioning of the different methods are purely the authors' opinions.

Naturally, this work builds upon the foundation of prior collective efforts. The chronological sequence of contributions to earlier versions of the source code is outlined in the "About REFER" section, along with references to relevant publications resulting from these efforts. The authors would like to acknowledge with special thanks the contributions of Juan Vicente Andreu, Julian Conesa, Ana Piquer, Ferran Naya, Peter Varley, and Raquel Plumed, as well as the fruitful collaborations with Manuel Contero, to the different versions of the software which have enriched REFER's ever-evolving code. It has been an honor and a pleasure to collaborate with all of them. Without their respective contributions, REFER could not have reached this seventh version, which is now available to readers as part of this handbook.

About the Authors

Pedro Company is a Professor in the Department of Mechanical Engineering and Construction at Universitat Jaume I (UJI) in Castellon, Spain, where he also served as Head of Technology from 1995 to 2005. Dr. Company graduated from UPV with degrees in Industrial Engineering (BS) and Mechanical Engineering (PHD). He has authored several textbooks on geometric design, standardized technical drawing, industrial drafting, computer-aided design, and computer graphics. Dr. Company's research interests center around computer-aided design, computer graphics, and sketch-based modeling, with more than 200 research papers and communications published in these areas. His research, both in the areas of parametric CAD quality of sketch-based modeling, is multidisciplinary and requires the development of methods and procedures specific to communication, computer science, and engineering design and their application to mechanical and product development problems.

Jorge D. Camba is an Associate Professor in the School of Engineering Technology at Purdue University in West Lafayette, IN. Before joining Purdue, he was a faculty member in the Dwight Look College of Engineering at Texas A&M University and, most recently, at the University of Houston, TX. He holds degrees in Computer Science (BS, MS), Technology (MS), and Systems and Engineering Management (PhD). Dr. Camba is the author of 8 books and more than 100 peer-reviewed publications. His research focuses on parametric model complexity and optimization, design reusability, knowledge management, and the model-based enterprise.

Contents

List of Figures

Chapter 1

Generalities

1.1 Introduction

This handbook serves as an interactive state-of-the-art reference for Sketch-Based Modeling (SBM) in engineering design, offering both a practical, hands-on introduction to SBM and a user guide for the REFER software application.

REFER (from the Catalan "Repetir o tornar a fer una cosa que s'ha desfet," meaning "to repeat or redo something that has been undone") is a research-oriented software application designed to offer a state-of-the-art perspective on SBM. It serves both as a modular framework and an experimental laboratory for developing, exploring, and validating SBM algorithms and techniques. We distinguish between two primary branches of SBM: the artistic branch, which aims to support intuitive 3D content creation from hand-drawn inputs; and the engineering branch, which focuses on generating precise 3D models suitable for integration into CAD/CAM/CAE workflows. REFER is specifically oriented toward engineering, offering tools and functionalities that support the accurate interpretation and reconstruction of technical sketches for engineering design purposes.

REFER is not intended as commercial software. The program is free and open source. You can redistribute it and/or modify it under the terms of the GNU General Public License, as published by the Free Software Foundation. The program is intended for academic and research purposes only. It is distributed WITHOUT ANY WARRANTY, not even the implied warranty for MERCHANTABILITY or FITNESS FOR A PARTICULAR PURPOSE.

REFER is an incomplete software system under permanent evolution and released as a work-in-progress. It is provided as-is, with the intent of sharing a functional architecture and a set of core algorithms that can serve as a foundation for further development. Our goal is to support the research community by offering a modular platform that others can build upon, refine, and extend – contributing to the advancement of SBM in engineering design. The book is structured in the following seven chapters:

(1) Generalities.
(2) Input.
(3) Reconstruction.
(4) Display.
(5) Output.
(6) Implementation.
(7) Final remarks.

Chapter 1, entitled "Generalities," includes a section with information about the evolution of REFER, along with guidance on getting REFER installed and how to get started. It also contains an introductory section on the fundamentals of SBM. As mentioned earlier, this work is not simply a manual for a specific application, it also serves as a broader review of SBM concepts and methodologies.

The book is structured into chapters and sections that follow the logical stages involved in transforming a 2D sketch into a 3D model. Each chapter includes explanations of how and why the application behaves the way it does (or does not), with visuals to clarify behavior. Larger topics are broken down into sub-chapters for ease of navigation. Chapter 6 is especially relevant for readers who wish to explore the source code or contribute to improving and extending the application.

Readers who plan to use this book as a general review of SBM may skip Chapters 1 (except for the SBM overview) and 6, as well as detailed software-specific content located at the end of other chapters. Instead, they are encouraged to focus on the introductory sections of each chapter, which provide rich context and a state-of-the-art overview of the SBM process. Chapter 7 discusses current challenges and future directions in SBM.

1.2 About REFER

REFER is a Sketch-Based Modeling application under continuous development by the research group REGEO. The current version is REFER v7.0. The latest version of the application and its source code are available in REFER's Github repository (https://github.com/REFER-SBM/REFER/).

The version history is as follows:

REFER v1.0 was developed as part of the following capstone project by Andreu (2000). The software began as a C implementation (in a Windows environment) of the optimization approach for line-drawing reconstruction proposed by Lipson and Shpitalni (1996). Major improvements included:

(1) Open code.
(2) Interactive interface.
(3) Expandable database.

REFER v2.0 was a result of Conesa's PhD Dissertation [Conesa, 2001]. Major improvements included:

(1) Upgraded detection of regularities.
(2) Implementation and tuning of two variants of Simulated Annealing algorithms (Conesa *et al.*, 2000).
(3) Implementation of five direct inflation approaches (Conesa *et al.*, 1999):
    (i) Axonometric
    (ii) Levels
    (iii) Cavalier
    (iv) Marill's MSDA
    (v) Random

REFER v3.0 was released as a result of Piquer's PhD Dissertation (Piquer, 2003). Major improvements included:

(1) Improvements in face detection (new approach derived from Courter and Brewer (1986).
(2) Bilateral symmetry detection in 2D line-drawings (Piquer *et al.*, 2003).
(3) 2D offline refinement through optimization.

REFER v4.0 was an unpublished update developed in 2004. While searching for a smart sketch system, Naya *et al.* (2003) added sketching capabilities through the CALI Library (Fonseca *et al.*, 2002). Major improvements in versions 3 and 4 were published by Company *et al.* (2004).

REFER v5 was a major update that started in 2009:

(1) v5.1. Update to C++ and MFC, plus new database.
(2) v5.2. Code structured by classes and new execution flow.
(3) v5.3. A new algorithm for finding faces in wireframes (Varley and Company, 2010).
(4) v5.4. Adaptation of corner detection approach by Xiong and LaViola (2010).
(5) v5.5. Feature detection (Company *et al.*, 2010).
(6) v5.6. Improvements in detection of cues (Company and Varley, 2012).
(7) v5.7. Sketch editor and vectorization module (Company *et al.*, 2013).
(8) v5.8. Improvements in features detection (Plumed *et al.*, 2013).

REFER v6 was a result of Plumed's PhD Dissertation (Plumed, 2014). Major improvements included:

(1) v6.1. New vectorization code (Company *et al.*, 2015b).
(2) v6.2. Detection of vanishing points and grouping of parallel edges (Company *et al.*, 2014).
(3) v6.3. Improvements on perceptually-based detection of features (Company *et al.*, 2012; Plumed *et al.*, 2013).
(4) v6.4. Improvements on mirror symmetry detection in 2D line-drawings (Plumed *et al.*, 2016).
(5) v6.5. New method to obtain a 3D CSG models from a 2D engineering wireframe sketch which depicts a polyhedral shape (Plumed *et al.*, 2014). Major improvements in version 6.5 were published by Plumed *et al.* (2014) and Plumed *et al.* (2022).

REFER v7 included a major update to the help file, providing a clearer explanation of the application's new approach, which emphasizes the reconstruction of procedural models. Major improvements in version 7.0 were published by Camba *et al.* (2022).

1.3 Sketch-Based Modeling

1.3.1 *Introduction to sketch-based modeling*

Sketch-Based Modeling is a specialized area within the broader field of Artificial Vision (or Computer Vision). In SBM, the input consists of both geometric figures drawn in 2D and standardized symbols commonly used in engineering drawings. The output is typically a 3D geometric model, similar to those used in Computer-Aided Design/Computer-Aided Manufacturing/Computer-Aided Engineering (CAD/CAM/CAE) applications.

The motivation behind SBM stems from a key gap in current design workflows. Despite the technical advances in CAD systems, user interfaces are not well-suited for the early stages of product development. During these initial phases, engineers and designers rely heavily on free-hand sketches to express and refine their ideas visually. Computers, however, are unable to effectively assist in this stage because they cannot fully interpret or leverage the fluid and intuitive nature of sketching. As a result, conceptual design remains largely disconnected from the digital CAD environment, with CAD models typically created only after the conceptual stage has been completed. Bridging this gap remains a major objective in SBM research, which aims to create systems that support creativity while streamlining the transition from early ideas to formalized models.

The first attempt to automatically infer 3D shapes from 2D line drawings is attributed to Lawrence Roberts (1963). Early efforts in the field were further documented by Sugihara (1986), while a more comprehensive review was provided years later by Martin Cooper (2008). Initial research in this area concentrated primarily on the vectorization of blueprints and sketches. As the field evolved, the focus expanded from simple vectorization to the more complex challenge of reconstructing the 3D shapes implied by 2D drawings. This shift introduced new challenges, most notably the issue of "realizability," the requirement that the reconstructed object must correspond to a physically plausible three-dimensional shape. This critical aspect of geometric reconstruction has also been highlighted in more recent studies (Azariadis *et al.*, 2013). It is important to distinguish reconstruction from retrieval (also referred to as recovery or recognition), a related but fundamentally different problem, which is briefly discussed in the Reconstruction section of this handbook.

It was later widely recognized that Sketch-Based Modeling is not merely a numerical process governed by deterministic geometric rules. In practice, the geometric information contained in sketches is often incomplete, ambiguous, or distorted. As a result, successful modeling requires the computer not only to interpret explicit lines and points, but also to "perceive" the overall structure and intent behind the 2D figures—using "perception" in the psychological and cognitive sense (Bonnici *et al.*, 2019). Addressing these challenges moves SBM beyond traditional geometry and into the broader field of Artificial Perception, which focuses on enabling machines to process and interpret sensory information in ways that resemble human understanding.

In addition, the likely presence of standardized symbols such as dimensions, tolerances, and other geometric constraints, introduces an additional layer of complexity. Beyond simply recognizing lines and shapes, the system must also interpret elements of the graphical language traditionally used in engineering drawings. Properly understanding these annotations is essential, as they carry critical information about the design intent. Moreover, interpreting these symbols enables the integration of constraints and specifications into the resulting 3D model, enhancing the model's completeness and usefulness in downstream applications. At the time of this writing, annotations are not supported by REFER (they are not processed from the input or included in output models). It is anticipated that future developments will support the integration of Product Manufacturing Information (PMI).

The evolution of keywords in fields related to SBM is illustrated in Figure 1.1.

A recent state-of-the-art review by Bonnici *et al.* (2019) emphasized that "Sketching is used for a variety of purposes, from the expression of ideas on two-dimensional (2D) physical media, to object creation, manipulation, or deformation in three-dimensional (3D) immersive environments." While artists and graphic designers are potential users of SBM systems, their objectives and workflows differ significantly from those of engineers and product designers. In particular, the nature of the input and the expected output in artistic applications often emphasize creative freedom and visual aesthetics, whereas engineering applications prioritize precision, constraints, and manufacturability.

Two distinct but related branches of SBM have evolved: one aimed primarily at facilitating 3D content creation for artistic and entertainment purposes, and another focused on the integration of engineering designs into CAD/CAM/CAE workflows. Cook and Agah's work (2009) provides

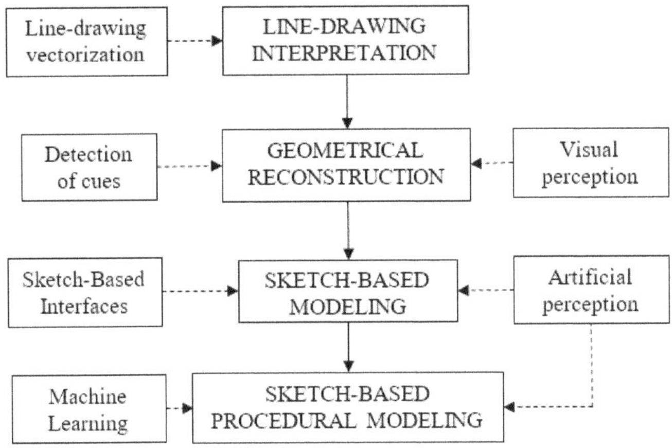

Figure 1.1. Evolution of sketch-based modeling.

a helpful introduction to the artistic and creative branch of SBM. Likewise, current advances in 3D content creation based on sketching techniques were recently examined by Bhattacharjee and Chaudhuri (2020).

An updated review of SBM within the context of engineering design was provided by Camba *et al.* (2022). The authors also outline the two distinct branches of SBM research: one oriented toward "artistic" purposes and the other governed by "engineering" principles. The review emphasizes that, while significant advances have been achieved in the automatic conversion of artistic sketches into 3D models (particularly for creative industries and entertainment), progress in the generation of CAD models that feature controllable, semantic, and parametric solid geometry remains limited. The production of engineering-grade models that can seamlessly integrate into CAD/CAM/CAE environments continues to pose substantial technical challenges and, as of today, has not led to substantial contributions.

It is important to note that, depending on whether the emphasis is placed on the goal, the process, or the interface, other closely related fields can also be considered, which may either benefit from developments in SBM or contribute to such advancements. Three of these related fields are listed as follows:

(1) Pen-based input—also referred to as Pen-driven Computing or simply Pen Computing—is a user interface method where a pen on a sensitive screen (such as a tablet) replaces traditional input devices like the

keyboard and mouse. In this paradigm, typing is replaced by handwriting, while navigating menus is largely handled through gestures. Consequently, the core computational challenges associated with pen-based systems involve the recognition and interpretation of both handwriting and gestures. Several commercial libraries such as Microsoft's Tablet PC Platform SDK (Jarret and Su, 2002) have been developed to address these challenges. The most recent contributions in the field have shifted from traditional tablets to the use of advanced hand-tracking sensors (Kim and Bae, 2016) that allow users to interact with the model directly in three-dimensional space without the need for physical contact with a device. Furthermore, it has been proposed that combining tablet-based input with augmented reality (AR) environments could open new possibilities for sketching, blending the precision of pen input with the immersive capabilities of AR. This hybrid approach holds the potential to enable more natural, flexible, and intuitive sketching experiences in design and engineering applications (Xin *et al.*, 2008).

(2) Varley *et al.* (2004) provided a concise yet insightful review of the state-of-the-art in the machine interpretation of line drawings, summarizing key developments, challenges, and emerging trends in the field.

(3) Van Dijk (1995) and Company *et al.* (2009) offered insights into how SBM can be leveraged as a tool to enhance Computer-Aided Sketching during the early conceptual stages of product design. Their research emphasizes the potential of SBM to bridge the gap between freehand sketching, which is often used by designers to quickly visualize ideas, and the precise, structured design processes facilitated by CAD systems.

With the advent of modern machine learning techniques, the initial, simplistic attempts to "read" sketches are being superseded by sophisticated methods rooted in neural networks. These advanced approaches aim to emulate human deductive reasoning by training on extensive datasets of sketches with known interpretations. Lun *et al.* (2017) argue that "shape interpretation is fundamentally a learning problem; otherwise, it is highly under-constrained." This perspective drives their proposal for a learned, view-based representation designed to generate sculptured 3D shapes directly from sketches. A state-of-the-art review by Bonnici *et al.* (2019) highlights how AI is reshaping the field, noting the shift towards leveraging deep learning and neural networks to improve the accuracy, flexibility, and usability of sketch interpretation systems. This evolving approach promises to enhance the capability of SBM systems to process more complex, abstract sketches and better align with human cognitive processes.

To put these new advances in context, we begin by reviewing the fundamentals of Artificial Intelligence (AI) and the fundamentals of multi-agent systems, followed by an exploration of how SBM and AI intersect. In a literal sense, REFER does not implement formal AI techniques, but its architecture does exhibit typical AI characteristics. Therefore, we recommend examining the reconstruction flow to gain insight into the hierarchical organization of the reconstruction process and the complex synergy between the required local and global perspectives.

1.3.2 *Artificial perception*

The idea that producing 3D shapes from 2D drawings involves more than just geometry has been recognized since the early days of SBM research. The pioneering work by Perkins, as part of Project Zero, examined how people perceive drawings that represent objects. Perkins identified the key geometric relationships that must be preserved and outlined the circumstances under which certain relationships can be disregarded (Perkins, 1968, 1971). This research has been instrumental in advancing the understanding of how sketches can be interpreted beyond simple geometric constructs. Notably, cubic corners have remained a powerful tool in SBM, serving as an "isolated contribution" of perception to the modeling process (Company *et al.*, 2001; Lee and Fang, 2011).

The realization that SBM is as much about perception as it is about geometry truly gained momentum when two key factors converged. First, sketches began to replace line drawings as the primary input method, offering a more intuitive and flexible means for users to convey ideas. Second, researchers began to recognize the inherent imperfections in sketches (previously dismissed as "minor artifacts" or secondary issues) not as flaws, but as essential and idiosyncratic elements of sketch-based communication (Piquer, 2003). This shift in perspective laid the groundwork for a deeper understanding of how human perception shapes the interpretation and transformation of 2D sketches into 3D models.

Understanding the role of perception in both natural and artificial interpretation of sketches is critical. However, this aspect alone does not capture the full complexity of artificial perception. While significant advances have been made, our understanding of the actual process, particularly in the context of machine interpretation, remains incomplete. In this regard, artificial perception should not be limited to simply replicating human visual perception. In fact, heuristic methods, while inspired by human perception, often take alternative approaches that may offer novel insights that can further inform our understanding. In other words, while artificial perception is primarily grounded in the science of human perception, the relationship between the two fields is reciprocal: each field influences and shapes the other. This interplay allows for a more nuanced exploration of both natural and artificial processes, ultimately enhancing the accuracy and efficiency of systems that aim to interpret visual information.

Throughout this handbook, the key contributions of visual perception to SBM are outlined. It will become clear that perception plays a pivotal role in the identification and interpretation of cues. Notable works, such as those by Suwa and Tversky (1997), the encyclopedic book by Palmer (1999), and the "algorithmically oriented" approach proposed by Hoffman (2000), provide key contributions to advancing the artificial detection of cues in sketch-based systems. These works collectively drive forward the development of systems that can "perceive" sketches in ways that are similar to human interpretation. However, the principles of human perception are not confined to just one phase of SBM. Instead, perception is interwoven throughout the entire SBM process, influencing how sketches are understood at every step. Understanding and integrating these perceptual principles are crucial for creating more accurate and intuitive SBM systems.

The recent work by Hertzmann (2024) holds particular significance. Hertzmann's primary objective is to explain how viewers perceive 3D shapes and spatial relationships in images, considering both linear and nonlinear projections. To this end, the author introduced a two-stage model for human 3D vision: "In pictures, the first stage perceives a 3D shape for the current gaze. The second stage continually integrates some of the per-gaze information into an overall interpretation of a picture." Hertzmann's framework emphasizes that local perception of objects (gaze vision) is more profoundly influenced by perspective rules, while the global perception of the scene as a whole operates on a more integrated level. Since the goal of SBM is not the analysis of complex scene images but rather the automatic processing of sketches, distinguishing between objects and scenes may not be directly applicable. Instead, the idea that perception unfolds in two stages—a "local" stage followed by a "global" stage—is highly relevant. This model aligns with how sketch-based systems can interpret individual elements of a drawing before integrating them into a coherent whole.

The practical implication of this insight is that the concurrence of individual algorithms, each designed to artificially perceive different information items from sketches, is a more effective approach than relying on the currently dominant sequential flows. Nevertheless, sequential flows remain popular due to the lack of viable and efficient concurrent procedures, which also applies to vectorization schedules and refining 2D schedules. According to Liu and Bessmeltsev (2025), the core challenges of sketch processing include "determining sketch topology, detecting and disambiguating intersections, junctions, and occlusions, segmenting background from the foreground, robustly defining tangents, finding centerlines and correspondences." In our experience, treating these core challenges as independent from one another is a necessary simplification in the absence of algorithmic methods capable of providing mutual feedback.

1.3.3 *Fundamentals of artificial intelligence*

AI is a field of study that focuses on developing systems capable of performing tasks that replicate certain cognitive abilities of human intelligence. These tasks, traditionally considered difficult to automate, include image recognition, natural language processing, decision-making, and more.

On a theoretical level, a distinction is often made between "weak" AI, which is designed to perform specific tasks or solve particular problems, and "strong" AI, a hypothetical AI with the cognitive abilities and consciousness comparable (or even superior) to a human. In practice, all existing AI is considered weak, as current systems lack "consciousness," awareness of other minds, or the ability to experience emotions.

AI systems may include one or more "rational agents" that autonomously make decisions and take actions to achieve specific goals, based on both past and current inputs. Since a single agent is often unable to handle complex problems independently, AI systems typically incorporate multiple cooperative agents that work together towards a common objective. Two main approaches are commonly identified within AI: symbolic AI and machine learning.

Symbolic AI, also known as conventional AI or expert systems, involves two key components: (1) selecting a set of logical rules that represent knowledge and allow reasoning about a specific domain, and (2) enabling a computer to apply these rules to make decisions. In symbolic AI, expert rules dictate how a system should behave when performing tasks such as medical diagnosis or transportation route planning. Although symbolic AI excels at tasks requiring complex reasoning, it struggles to capture the intricacies and unpredictability of the real world.

According to the classification proposed by Arend Hintze (2016), machines that implement symbolic AI are considered reactive, i.e., they respond to stimuli without utilizing past experiences to inform future actions. Nevertheless, the rules they follow encapsulate the accumulated knowledge and expertise of their human creators. In contrast, proactive machines possess transient memory, allowing them to adapt their behavior based on prior interactions.

Machine Learning (ML) represents a different paradigm that focuses on the creation of systems that can learn and adapt from data without being explicitly programmed for each individual task. Through exposure to large datasets, machine learning algorithms identify patterns and relationships that enable them to make informed decisions, continuously improving their performance over time. Unlike symbolic AI, machine learning systems do not rely on predefined logical rules; instead, they "learn" by generalizing from examples. This characteristic makes them particularly well-suited for tasks such as image recognition, natural language processing, and predictive analytics, where explicit rule-writing would be impractical or impossible. In general, machine learning involves three primary stages:

(1) *Training*: exposing the model to data so it can learn patterns.
(2) *Validation*: tuning the model using separate data to ensure it generalizes well.
(3) *Deployment*: applying the trained model to real-world tasks.

Since SBM is inherently a complex cognitive problem, applying AI techniques presents significant challenges. In particular, compiling sufficiently large and representative databases of engineering sketches—essential for training AI systems—is a difficult and resource-intensive task. As a result, the application of AI to SBM is not straightforward. This topic is explored in greater depth in subsequent subsections.

1.3.4 *Fundamentals of multi-agent systems*

An agent is a computer system capable of independently determining and executing actions on behalf of its user or owner (Jennings *et al.*, 1998; Sycara, 1998). To address complex problems, the collaboration of multiple agents becomes necessary (Deloach *et al.*, 2001). According to Luck *et al.* (2005), this collaboration can be structured by first capturing the system's goals through a hierarchical organization of functional requirements and then identifying the specific roles needed to fulfill those goals. Each role, in turn, gives rise to a set of associated tasks, which are implemented as cooperative agents organized into "layers."

The study of multi-agent systems falls within the broader field of Distributed Artificial Intelligence (DAI). In a multi-agent system, autonomous, heterogeneous, and independent agents apply intelligence to solve discrete parts of a larger problem. Through interaction and coordination, these agents collaboratively work toward a complete solution. As Stone articulates, "Multi-Agent Systems (MAS) is the subfield of AI that aims to provide both principles for the construction of complex systems involving multiple agents and mechanisms for the coordination of independent agents' behaviors" (Stone, 1998).

When designing agent-based systems, it is important to carefully select agents with appropriate characteristics for the task at hand. In the context of SBM, it is evident that a multi-agent system would offer more advantages than a single-agent approach. However, it remains an open question whether all agents in a hypothetical SBM system should be proactive, or if reactive agents might also play a valuable role, particularly in capturing low-level semantic information. Before extending these approaches to dynamic sketching environments, more experience and experimentation in controlled or fixed environments are needed.

In REFER, the flow is formally sequential. However, it is organized with a layered structure that could readily support adaptation into a formal multi-agent system. This design choice reflects the intent to maintain modularity and potential for future scalability. Nevertheless, as Cardoso and Ferrando observe, "there is little incentive for developers to switch to current Agent Programming Languages (APLs), as the behaviors that can be easily programmed are sufficiently simple to be implementable in mainstream languages with only a small overhead in coding time" (Cardoso and Ferrando, 2021). In this context, REFER's sequential yet layered structure strikes a balance between practical development constraints and the architectural flexibility needed for future enhancements. The layers in REFER are evident in the detection process, where low-level semantic cues are detected first. These cues then serve as the basis for detecting higher-level semantic cues which, in turn, facilitate the detection of features.

More recently, deep learning approaches have demonstrated the capability to simultaneously manage both the sequence of stages required to reconstruct a 3D CAD model from a 2D sketch and the predominantly (though not exclusively) hierarchical relationships between local and global information processed at each stage. These methods provide an alternative to traditional sequential flows by learning complex dependencies directly from data, allowing for a more integrated and flexible interpretation of sketches throughout the reconstruction process.

1.3.5 *Fundamentals of machine learning*

ML is a subset of AI focused on developing systems capable of making decisions based on data. The process typically involves three stages: training, validation, and deployment.

During the training phase, ML algorithms are used to create models (often referred to as "machines") that can make predictions or decisions without requiring explicit instructions from the developer for each possible scenario. This stage depends heavily on large datasets, as the model must learn from a wide range of examples that reflect the types of problems it will later face. Once trained, the model undergoes validation to evaluate the accuracy, generalizability, and confidence of its predictions or decisions. This step ensures that the model performs reliably on data it has not previously encountered. Finally, the model is deployed into real-world applications, where it continues to generate predictions or support decision-making based on new incoming data.

Machine learning has evolved into four distinct approaches, each with its own techniques and algorithms: supervised learning, unsupervised learning, reinforcement learning, and deep learning.

In supervised learning, "datasets" are composed of both problem descriptions (inputs) and their corresponding solutions (outputs or labels). The data used during training is labeled, meaning that each example explicitly associates the input information with the correct outcome. This structure allows the algorithm to learn how to distinguish between the input features and the desired output results.

Throughout the training phase, the algorithm analyzes the labeled data to identify patterns and build a model capable of correctly labeling new, similar samples. During the validation phase, the model is evaluated using new, unseen data to measure how well it generalizes beyond the training set. The model's predictions are compared against the true labels, and a variety of metrics (e.g., accuracy, precision, or recall) are used to assess performance. An example of a supervised learning algorithm is K-nearest neighbors, which classifies new instances based on the labels of the most similar examples in the training data.

In unsupervised learning, the algorithm works with unlabeled data, aiming to identify underlying patterns, structures, or clusters without predefined categories or outcomes. Because the algorithm must independently discover meaningful groupings or features, unsupervised learning typically requires larger datasets to ensure reliable pattern detection.

Despite the absence of labeled examples, unsupervised learning can often provide greater flexibility and stronger decision-making capabilities in complex or poorly understood domains compared to supervised approaches. The performance of unsupervised learning algorithms is generally evaluated using cost functions that measure the quality of the discovered structures, such as intra-cluster similarity or reconstruction error. Many unsupervised learning techniques leverage neural networks, particularly in more advanced methods like autoencoders or self-organizing maps. Classic examples of unsupervised learning algorithms include K-means clustering, Hierarchical Cluster Analysis (HCA), and Principal Component Analysis (PCA).

Reinforcement learning is inspired by the trial-and-error learning process, where an agent interacts with its environment by performing actions and receiving feedback in the form of rewards or penalties. Favorable actions that lead to positive outcomes are reinforced, while unfavorable actions are discouraged, allowing the agent to gradually learn an optimal strategy through experience. Over time, the system improves its decision-making ability by maximizing cumulative rewards. A well-known example of a reinforcement learning algorithm is Q-learning, which enables an agent to learn the value of actions in different states without requiring a model of the environment.

Deep learning is a subset of machine learning that leverages artificial neural networks with multiple layers to process complex data (Jain, 2010; LeCun *et al.*, 2015; Schmidhuber, 2015). The term "deep" refers to the multiple layers of the network, where each layer processes information at different levels of abstraction. These layers can be organized either sequentially, in a "cascading" structure, where the output from one layer becomes the input for the next, or hierarchically, where higher-level features are derived from lower-level ones, enabling the network to learn increasingly complex representations of the data. Deep learning models are particularly effective in tasks such as image recognition, natural language processing, and autonomous systems due to their ability to automatically learn features from raw data without the need for manual feature engineering.

A recent study by Reisfeld and Sharf (2023) delved into the learning of sketch-to-shape relationships by examining how simple human strokes can effectively capture high-level 2D shape semantics. Their neural network demonstrated the ability to retrieve 2D objects from basic one-stroke sketches with impressive accuracy, showcasing the potential of sketch-based input in design and modeling. Koch *et al.* (2019) further emphasized that the combination of large data collections and deep learning algorithms has transformed various areas of computer science, particularly in tasks such as image and speech recognition. However, this transformation has not yet fully extended to three-dimensional geometric models. Acquiring or creating high-quality 3D models requires a substantial amount of data and resources.

Similarly, the task of gathering large sets of 2D sketches for training remains a significant obstacle. An innovative solution to this problem was proposed by Hahnlein (2022) and Hahnlein *et al.* (2022), who developed an algorithm capable of synthesizing human-like concept sketches directly from CAD models. This approach offers a promising direction for overcoming the data limitations in training models for 3D shape recognition and sketch interpretation. It was used to develop Free2CAD (Li *et al.*, 2022).

A recent improvement on this line is due to Sun *et al.* (2025), who built on the idea of analyzing sketches first, and then using grouping information to reconstruct CAD modeling sequences. The authors claim two important aspects were improved: "identify diverse CAD feature reconstruction like fillet and chamfer with better segmentation performance," and "reconstruct these challenging modified features and generate feasible operation orderings without requiring the input stroke order as a prerequisite."

These results are promising in terms of replacing heuristics dependent on complex multi-parameter settings. However, there does not yet seem to be a satisfactory answer to address the implicit bias from the examples used to train the models. Some sketches generated by different authors appear to be copies of the same "master drawing." However, it is clear that different people making personal interpretations of the same drawing is not the same as different people expressing their own designs through drawings. In addition, most of the current training examples are clean and perfect sketches—"cleaned," or "completed" in the terms of Sun *et al.* (2025)—while the most important obstacle in SBM is known to be the management of imperfections which, despite being potentially large, do not usually prevent humans from correctly interpreting the sketches.

Achieving effective synergies between SBM and AI is a complex and challenging process that does not guarantee immediate success. Nonetheless, the integration of local and global vision during the "perception" of geometric information in sketches has been successfully implemented in REFER through a combination of detection techniques followed by refinement procedures. These steps are crucial for handling the nuances and imperfections inherent in sketches, making them interpretable by a software system. This approach could potentially be replicated and enhanced using deep learning techniques, which have demonstrated a remarkable ability to process and learn from data in a more sophisticated manner. A clear example of this process is the 2D Refine stage, which takes place after the vectorization process.

1.3.6 *Sketch-based modeling and artificial intelligence*

The current version of REFER does not implement any formal AI techniques for generating 3D models or interpreting 2D sketches. Nevertheless, recent advances in AI applied to Sketch-Based Modeling (SBM) have yielded promising results, particularly in the procedural generation of CAD models directly from sketches (Li *et al.*, 2022; Hähnlein, 2022). This section discusses the potential synergies between traditional SBM approaches and contemporary AI techniques, and the opportunities that a suitable—though not yet fully defined—integration of these disciplines might offer.

SBM is a complex problem. As Erik J. Larson (2021) observed in the context of natural language, "natural language understanding proved vastly more difficult and complex than anyone had imagined." A similar level of complexity is evident in SBM, which also confronts ambiguity, where a single visual element may admit multiple interpretations, and strong contextual dependencies, where the meaning of one element is influenced by surrounding elements in the sketch.

In some domains, interpreting drawings may rely on subjective perception or artistic intuition. In contrast, engineering design demands a precise and unambiguous decoding of technical intent. The challenges faced in natural language processing (NLP) are, to a great extent, analogous to those in SBM. Just as phonemes and morphemes form the foundation of language, primitive geometric elements serve as the building blocks of sketches. The linear sequence of words in a sentence is replaced by the spatial arrangement of lines, arcs, and shapes in a 2D layout. Despite the presence of ambiguity and contextual influence, this "graphical language" must be interpreted with the same precision and systematic reasoning that we apply to natural language.

There is also a prevalent misconception surrounding AI, particularly the belief that intelligence can inevitably be generated through facts and rules alone. Hubert L. Dreyfus addressed this issue in the introduction to his seminal book: "It is now clear to all but a few diehards that this attempt to produce general intelligence has failed. This failure does not mean that this sort of AI is impossible; no one has been able to come up with such a negative proof. Rather, it has turned out that, for the time being at least, the research program based on the assumption that human beings produce intelligence using facts and rules has reached to a dead end, and there is no reason to think it could ever succeed" (Dreyfus, 1992).

Dreyfus' observations underscore a critical inflection point in the evolution of AI research. While the early hope of building intelligent systems purely from predefined facts and rules has largely fallen short, it does not imply that true artificial intelligence is unattainable. Rather, it highlights the limitations of a specific paradigm. The shortcomings of rule-based systems have redirected attention toward alternative approaches, particularly those that learn from data or aim to emulate human cognition more closely, as more promising pathways for achieving robust AI.

In the pursuit of computer-based solutions to complex problems, it is common to draw analogies between computational processes and human cognition. In some cases, the aspiration is to develop general AI capable of mimicking human cognitive abilities. However, one of the primary challenges in this endeavor lies in our incomplete understanding of both the biological mechanisms of the brain and the psychological foundations of the human mind. Despite these uncertainties, such analogies remain valuable: they inform the development of computational models and inspire practical strategies for tackling difficult challenges. From this perspective, AI should not be seen as a replacement for existing methods in SBM, but rather as a complementary tool, one that can extend and enrich ongoing efforts.

Three distinct types of reasoning can be used to produce inferences: deductive, inductive, and abductive (Minnameier, 2017), as illustrated in Figure 1.2.

Deductive reasoning involves drawing highly probable (though not necessarily certain) conclusions based on the analysis of available information. Its reliability increases significantly when grounded in established theories. For instance, the formula $F = m{\cdot}a$ enables one to accurately deduce the force of a car crash, given its mass and acceleration. In contrast, in the absence of a theoretical foundation, deductive conclusions drawn from limited or circumstantial clues tend to be weak. However, the availability of large volumes of data can substantially strengthen such reasoning. This is a key advantage of modern AI systems, particularly those based on data-driven approaches (Chinesta *et al.*, 2020, 2023). As Larson (2021) recently noted, deductive reasoning is relatively straightforward for both humans and machines, since it involves deriving specific conclusions from general, pre-established rules. This logical structure makes it one of the more accessible forms of cognition to computational implementation.

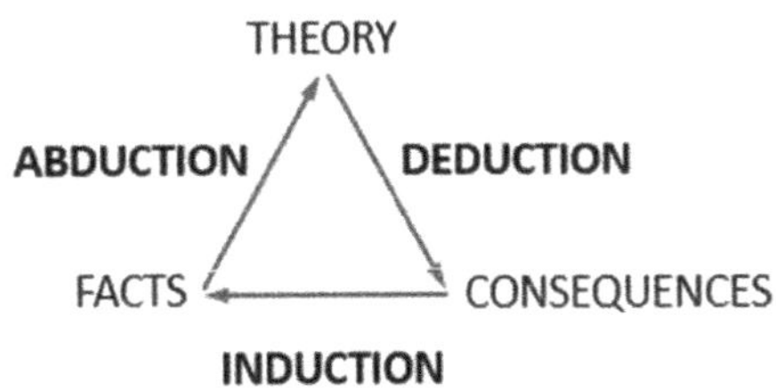

Figure 1.2. Types of reasoning.

Deductive reasoning has already found effective applications in SBM contexts. However, the diversity and complexity of the facts and rules involved in interpreting and modeling sketches make it difficult to rely on deduction alone. As discussed earlier, generating 3D geometry from 2D sketches is not merely a geometric computation problem. It depends both on geometric and perceptual cues.

One underexplored dimension in this process is the use of design principles to guide interpretation. While some researchers have applied informal heuristics, a formal incorporation of design principles as interpretive cues is a relatively recent proposal (Hähnlein, 2022). Future advancements in SBM should aim to systematically integrate all three types of cues—geometric, perceptual, and design-based—into a unified modeling framework.

Unlike deductive reasoning, inductive reasoning is significantly more complex, as it involves extrapolating from observed data to infer the underlying causes or general principles. It moves from the specific to the general, proposing hypotheses based on the observation of particular instances. The strength of an inductive inference depends not only on the quantity and quality of the data but also on the experience or intuition underlying the "leap" in reasoning. While deduction is typically described as top-down, induction is a bottom-up approach.

In the context of SBM and AI, ML (especially through neural networks) replaces manually encoded rules with data-driven pattern recognition. These systems learn to map inputs to outputs by adjusting internal parameters based on large datasets of paired examples. As such, ML is fundamentally inductive. Nevertheless, as Larson (2021) cautioned, "induction works on games, not life." Games operate within fixed, deterministic boundaries, whereas the real world, including engineering contexts, often presents unstructured and unpredictable challenges.

Engineering sketches, which serve as a medium for conveying design intent, frequently feature irregular, novel, or unconventional shapes. This inherent variability limits the effectiveness of purely inductive approaches, as interpreting unfamiliar sketches often requires abductive reasoning. In this regard, the challenge remains significant even for advanced ML systems. A possible strategy is to decompose complex sketches into simpler, more interpretable subcomponents. This decomposition could reduce reliance on inductive inference by enabling the reuse of known deductions and minimizing the scope of abductive leaps. However, it is important to recognize that the whole is often greater than the sum of its parts; design meaning often emerges from the holistic composition. Therefore, inferring intent from isolated components may not sufficiently capture the richness or coherence of the complete design.

A neural network can be defined as a large, massively interconnected network of simple processors, where each processor has a numerical activation value that changes in response to the activity of the processors to which it is connected and the particular weights assigned to those connections. Parallel processing enables neural networks to generate solutions efficiently by handling multiple computations simultaneously. The operation is limited to feeding data into the subset of processors designated as inputs and collecting the resulting output from the corresponding output processors.

In the context of SBM, it may be beneficial to explore the deployment of specialized neural networks organized in a cascading architecture. In such a structure, an initial network would focus on detecting low-level cues from the input sketches. The outputs of this stage would then serve as the inputs to a second network, responsible for identifying and interpreting higher-level features based on the detected cues. Finally, a third network would take the recognized features and attempt to "assemble" them into coherent and meaningful 3D shapes or models. This sequential, modular approach could offer several advantages, including improved interpretability, greater flexibility in handling diverse sketch styles, and the potential to isolate and address errors at specific processing stages. A similar methodology was recently demonstrated by Yin *et al.* (2022) in the context of refining 2D vertices.

Large, high-quality input-output datasets play a critical role in training AI systems, both in terms of the volume and the accuracy of the data. In many fields, approaches based on straightforward statistical analysis of extensive datasets have proven capable of capturing useful regularities and producing inductive solutions that are "good enough to be useful," as illustrated by the advances in Machine Translation (Dreyfus, 1992). However, assembling comparable datasets for engineering design presents a unique challenge. Unlike more open domains, engineering sketches and models often involve trade secrets and sensitive intellectual property, which is often difficult to access and share.

As discussed in the section on fundamentals of machine learning, one promising alternative is the synthesis of human-like concept sketches derived from CAD models (Hähnlein, 2022). Some recent datasets provide useful foundations for this approach: ABC (Koch *et al.*, 2019) (focused on B-Rep models), SketchGraphs (Seff *et al.*, 2020), Computer-Aided Design as Language (Ganin *et al.*, 2021) (both primarily addressing 2D profiles), and Vitruvion (Seff *et al.*, 2022). Ensuring the quality of the CAD models used in this synthesis process, however, remains a significant challenge (Company *et al.*, 2015a; Gonzalez-Lluch *et al.*, 2017), as the reliability of the resulting sketches—and, by extension, the effectiveness of the training datasets—depends heavily on the underlying quality of the source models.

In the final section of this handbook, we review the different models employed within REFER and identify those that appear most promising for guiding future efforts, particularly with respect to enhancing inferencing capabilities. These models are intended to support limited inductive reasoning. The development of abductive reasoning capabilities is not currently planned for future versions of REFER. This decision rests on the premise that, within data-driven learning paradigms, meaningful inferences must already be implicitly present in the training datasets, rather than being derived autonomously by the system.

1.4 Getting REFER Installed

REFER can be downloaded from http://www.regeo.uji.es/software.htm. The installation package includes a setup assistant that automatically installs the executable software, the required configuration files, and a collection of sample projects.

REFER can also be compiled and linked from its source code. The source code is written in C++[a] and provided as a Microsoft Visual Studio Solution that leverages Microsoft Foundation Classes (MFC)[b] to build the interactive environment. It can be compiled and linked using Microsoft Visual Studio 2008 (©Microsoft Corporation 2007). The solution can be exported to Microsoft Visual Studio Community 2015 or 2019. The warnings generated during the migration process can be safely ignored.[c]

---

[a] For readers interested in learning C++, we recommend (Eckel, 2000).

[b] Microsoft Foundation Class Library (MFC) is a C++ object-oriented library for developing desktop applications for Windows. For those interested in learning MFC, we recommend (Jones, 2000).

[c] The application uses two different approaches to produce a vectorized line drawing. One of these approaches is a third-party contribution (Fonseca *et al.*, 2002). To compile the

In both cases, the terms of use are similar (but not identical) to those of copyleft licenses (specifically, the source code is not bundled with the executable). You can study, use, distribute and modify the code, as long as the software and any derivatives remain free and open source. You cannot claim ownership of the software or pass it off as your own original work. If you use it for academic purposes, please cite this handbook.

Please note that REFER is not designed as a commercial application but as a testbed of ideas. As such, it is not optimized for delivering the quickest solution. Instead, it requires users to interact with various dialogs to configure the application parameters and find the most suitable solution for each example.

1.5 Getting Started with REFER

1.5.1 *Getting started*

After installing REFER, double click *Refer.exe* to launch the application. Once the program starts, the main window (or parent window) will automatically open, as shown in Figure 1.3.[d]

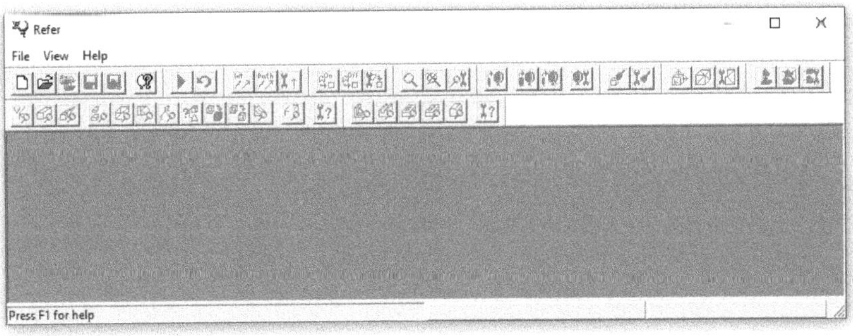

Figure 1.3. REFER's main window.

application with an alternative to Microsoft Visual Studio 2008, the CALI library must be rebuilt using Microsoft Visual Studio Community 2015 or 2019, or similar compiler and linker: Open and run the file *Recognizer.sln* located in the RECOGNIZER folder. Once the resulting library (*Recognizer.lib*) is generated, copy it to the "lib" folder, where the *freeglut* lib is located.

[d]The first time REFER is launched, the *About* dialog will be automatically displayed.

The parent window contains a set of menus and different types of child windows. Use the main menu to begin the input process. Select *File-New* to start the creation of a new hand-drawn sketch. Alternatively, select *File-Open* to use a previously created REFER file. Once the desired sketch or line drawing is created, the system can interactively produce the corresponding 3D model through the reconstruction flow. To exit the application, click the *File-Exit* button.

1.5.2 *Menus*

REFER implements a WIMP-style[e] graphical user interface. Although the ultimate goal is to provide a complete pen-based interface, REFER is intended as a testbed, a platform to test ideas and refine algorithms that will ultimately provide the necessary know-how to develop stand-alone, robust, and user-friendly pen-based applications.

Two types of menus are available in REFER, which are equivalent and can be used interchangeably:

(1) The menu bar is located on the top of the window below the title bar (see Figure 1.4). The commands in this menu are context-sensitive, i.e., only valid commands are available depending on the current state of the execution flow. Since the process is sequential, child tasks will not be accessible until their parent tasks have been executed and valid output data has been generated.

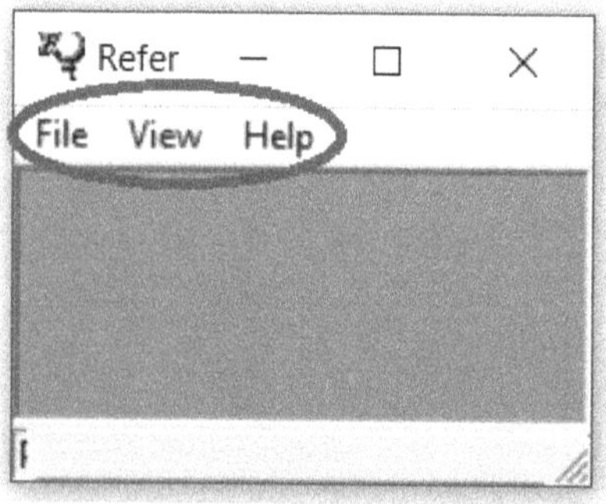

Figure 1.4. Menu bar in REFER.

[e]"Windows, Icons, Menus, Pointer" in human–computer interaction.

(2) Icons organized in toolbars located in the upper or left side of the corresponding parent or child window[f] (see Figure 1.5). Note that toolbars do not include all the available commands. Some specific actions can only be accessed through the menu bar.

Toolbars can be toggled on or off[g] by clicking the *View* option in the menu bar (see Figure 1.6).

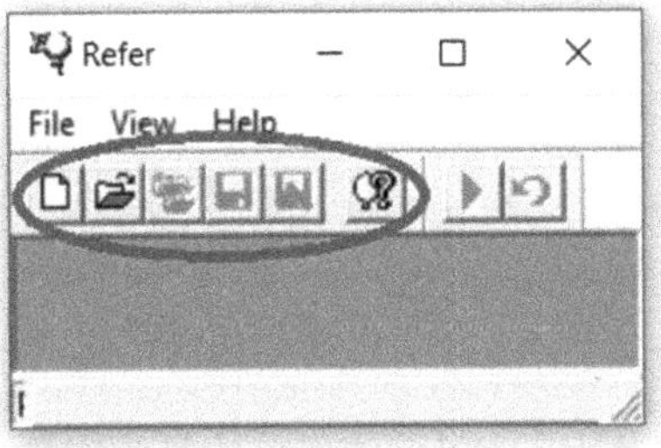

Figure 1.5. Icons and toolbars in REFER.

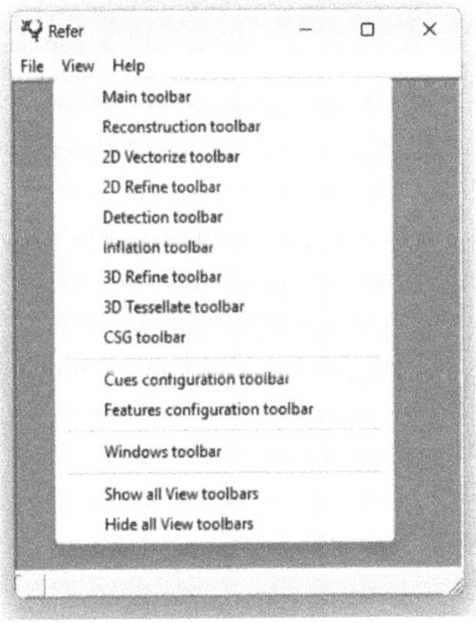

Figure 1.6. View option in the REFER's menu bar.

[f]Toolbars in REFER are dockable. Simply click and drag the toolbar frame to move the toolbar to a different location.

[g]Toolbars linked to child windows are only enabled when their corresponding windows are open.

1.5.3 *Windows*

1.5.3.1 *Child windows*

The information processed by REFER is displayed in specific types of child windows located within the parent window. These child windows are automatically activated and selected based on the current stage of the process and the type of information that is being displayed.

Child windows open automatically when needed. They can also be opened or closed using the *Windows* toolbar, which also includes additional commands to automatically arrange the windows. Note that child windows are only available after a file has been loaded (see Figure 1.7). Alternatively, child windows can be controlled by clicking *View* in the Menu Bar (see Figure 1.8).

Any attempt to close the 2D Sketch window will trigger a warning message, informing the user that the current document will also be closed, as shown in Figure 1.9.

Figure 1.7. Windows toolbar in REFER.

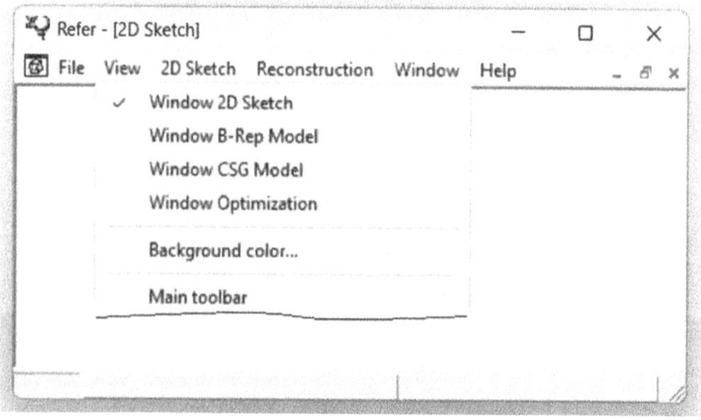

Figure 1.8. *View* option in REFER's menu bar.

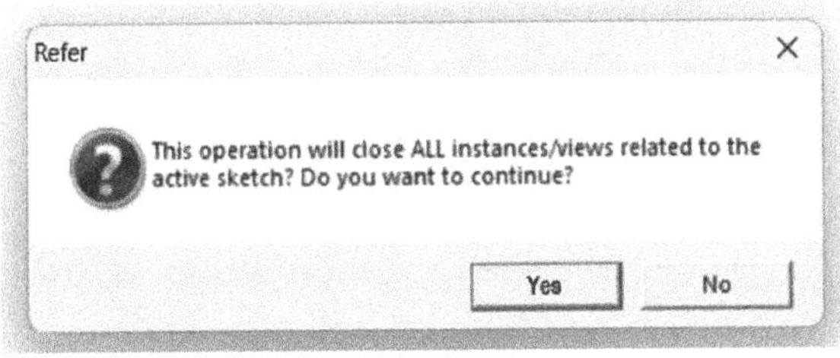

Figure 1.9. Warning message in REFER when closing the 2D sketch window.

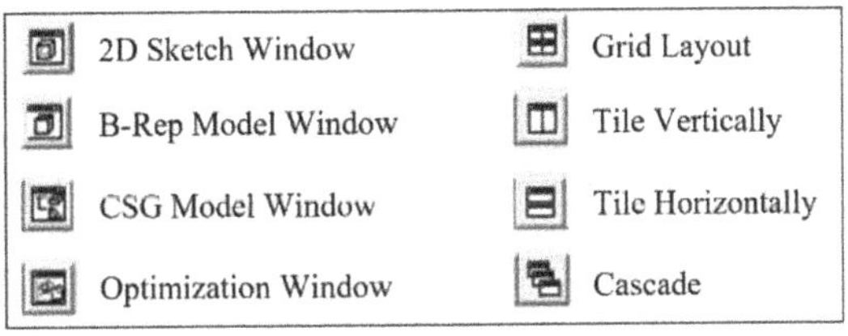

Figure 1.10. *Windows* toolbar commands.

1.5.3.2 *Windows toolbar*

The *Windows* toolbar controls the child windows and includes four commands to automatically arrange these child windows on the screen. The commands are shown in Figure 1.10.

When the 2D Sketch, 3D Model, and Optimization windows are open at the same time, and the CSG window is closed, the Grid layout command will position the 2D Sketch window on the left side of the main window and arrange the other two windows vertically on the right side.

1.5.3.3 *2D sketch window*

The 2D Sketch window opens automatically when a file is loaded. If it is a new file, the window displays an empty sketching area. Otherwise, the contents of the file will be displayed in the sketch window.

The 2D Sketch window is intended for:

(1) Sketching new drawings.
(2) Editing the current sketch.

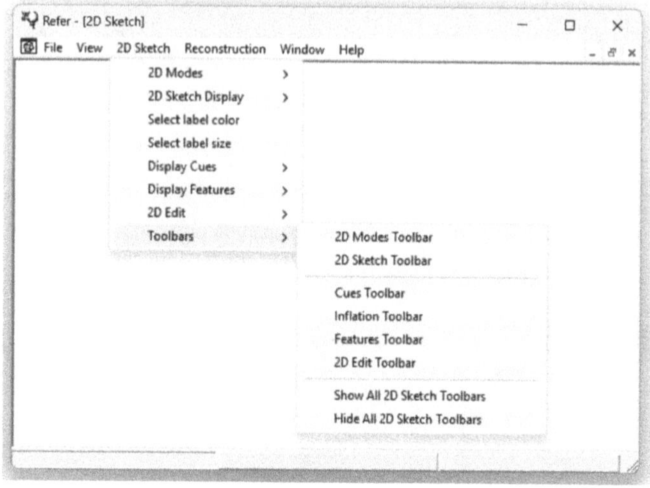

Figure 1.11. Toolbars linked to the 2D Sketch window.

(3) Displaying vectorized line-drawings.[h]
(4) Displaying information related to the different cues and features detected in the sketch/drawing.[i]

The contents displayed in the sketch window can be controlled via the *Display 2D Sketch* toolbar. Different 2D input modes are available to operate inside the 2D Sketch window. The user can switch between these modes using the *2D Modes Toolbar*. The *2D Sketch* window is linked to various toolbars that control its operation and the contents being displayed within the window, as shown in Figure 1.11.

1.5.3.4 *B-rep model window*

The *B-Rep Model* window is designed to display the explicit 3D model generated after the reconstruction of the input sketch. The model is

---

[h]Visualization tools for vectorized line-drawings can be found in the *Display 2D Sketch* toolbar.

[i]Visualization tools for cues and features can be found in the *Display Cues* toolbar and *Display Features* toolbar.

rendered using Open GL (Version 1.2.2.0). The user can interact with the model in two ways:

(1) Changing visualization parameters through the mouse.
  (a) The point of view can be changed interactively by orbiting around the model. To do so, move the mouse while holding the left button.[j]
  (b) Scrolling the wheel up and down is used to zoom in and out.
(2) Model attributes can be changed by using the menu options available in the *B-Rep Model* toolbar.

The *B-Rep Model* window is associated with a specific toolbar that controls the content that is being displayed, as shown in Figure 1.12.

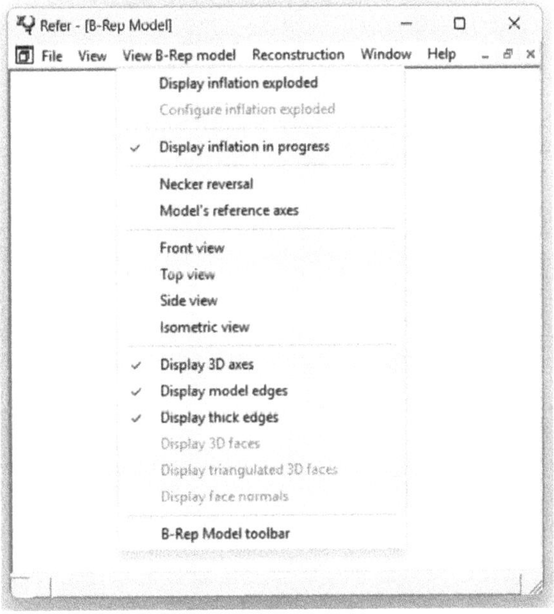

Figure 1.12. View B-Rep model menu.

[j]To rotate the model, the mouse functions as a virtual trackball. This feature is based on the original trackball implementation developed by Gavin Bell for Silicon Graphics in November 1988 (http://www.visgraf.impa.br/Projects/3dp/doc/html2/trackball_8h.html). Releasing the mouse button while moving the model will cause the model to continue rotating until it is stopped. The rotation can be stopped by clicking the left mouse button a second time.

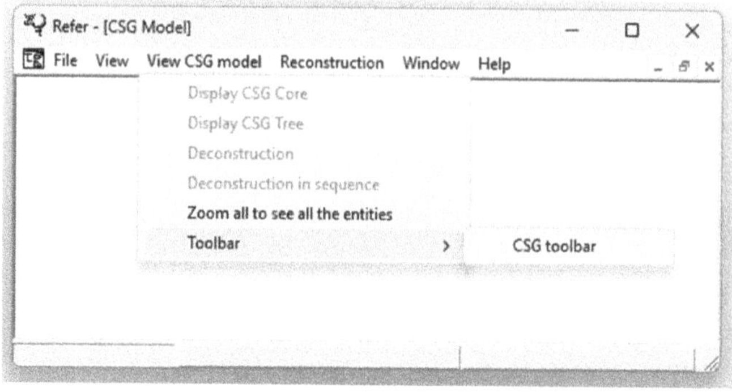

Figure 1.13. View CSG model menu.

1.5.3.5 *CSG model window*

The Constructive Solid Geometry (CSG) model window is used to visualize the reconstruction process based on CSG. The CSG window displays various visual elements associated with the CSG-based reconstruction process. The CSG window is linked to a dedicated toolbar that controls the information being displayed (see Figure 1.13).

1.5.3.6 *Optimization window*

The *Optimization* window is designed to display information related to the optimization-based process of inflation, including details of the ongoing adjustments and refinements. The information displayed is controlled by the *Optimization Process* toolbar. The *Optimization* window is linked to a toolbar which controls the information being displayed (see Figure 1.14).

1.5.4 *Main menu*

1.5.4.1 *Main toolbar*

The Main Toolbar is located in the upper left corner of the parent window, as its controls affect the entire application. The toolbar contains the following commands, as shown in Figure 1.15. The commands are also available under *File* in the menu bar, which also provides access to the *General Settings* (not available from the toolbar).

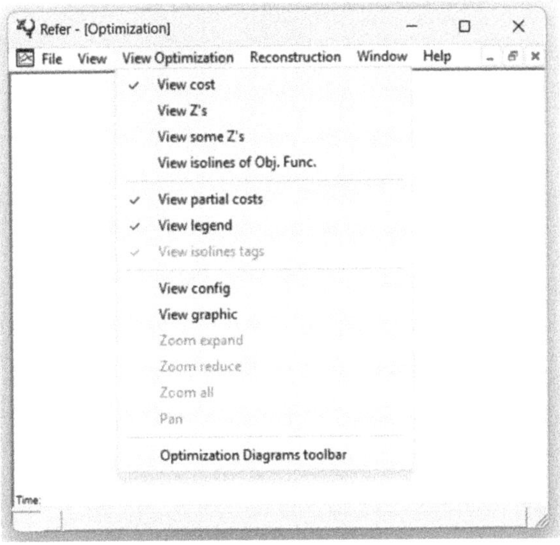

Figure 1.14. View Optimization menu.

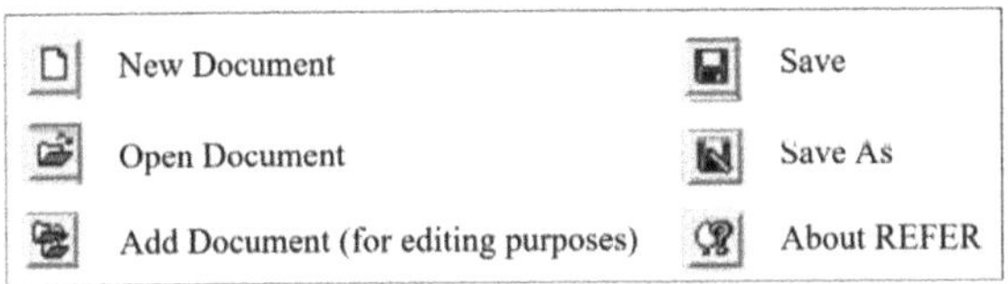

Figure 1.15. Main toolbar commands.

1.5.4.2 *General settings*

The menu bar provides access to the General Settings (not available from the toolbar) which allows users to customize the default execution flow of the application, so the system can be adapted to the style of each user and reduce the need for interaction. The number of "Most Recently Used (MRU)" files provided by the menu can also be configured through this dialog, as well as the type of entities used to output IGES files. The *General Settings* dialog box is shown in Figure 1.16.

The application must be relaunched for some changes to take effect. General settings are saved in the *REFER.ini* file. This file is automatically generated the first time REFER is executed and updated every time a user

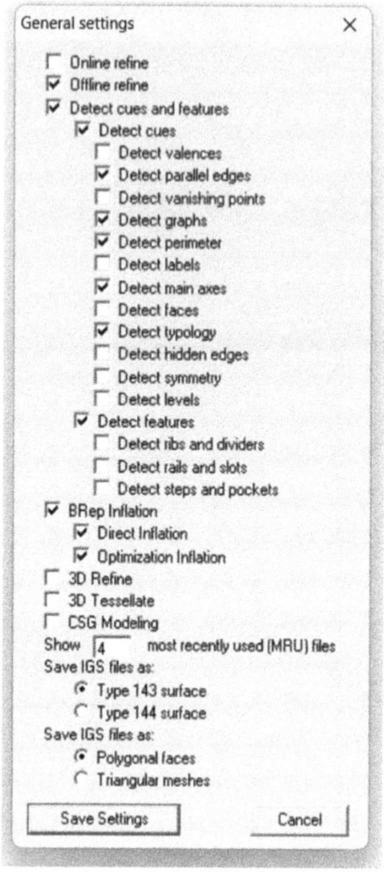

Figure 1.16. *General Settings* dialog box.

saves their settings. The file can also be edited directly with any text editor, as shown in Figure 1.17.

1.5.4.3 *New file*

In the Main toolbar, click *New* to start the creation of a new hand-drawn sketch (see Figure 1.18). Although different input types are supported by REFER, only sketches are accepted as new files. Line drawings cannot be created directly from scratch, but they can be opened as previously created files. Simple line-drawings can be created via the new sketch input module by using the Rule Line Mode.

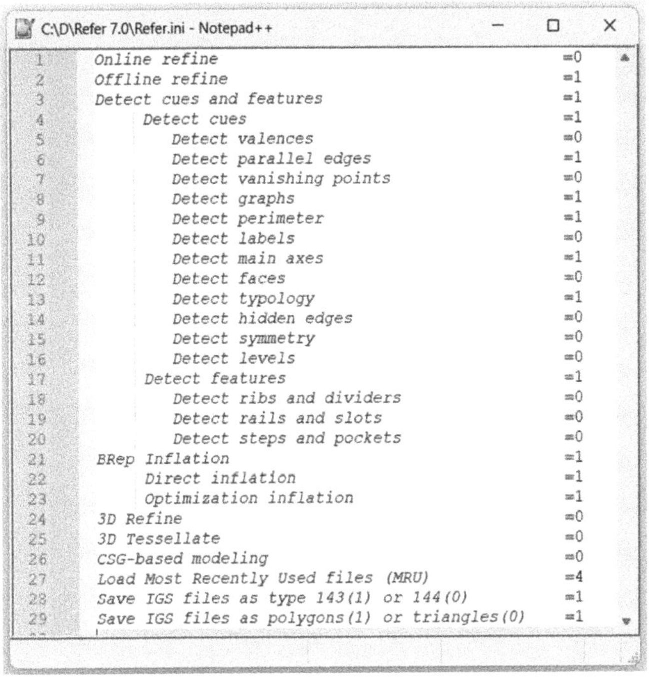

C:\D\Refer 7.0\Refer.ini - Notepad++

```
Online refine                                          =0
Offline refine                                         =1
Detect cues and features                               =1
     Detect cues                                       =1
        Detect valences                                =0
        Detect parallel edges                          =1
        Detect vanishing points                        =0
        Detect graphs                                  =1
        Detect perimeter                               =1
        Detect labels                                  =0
        Detect main axes                               =1
        Detect faces                                   =0
        Detect typology                                =1
        Detect hidden edges                            =0
        Detect symmetry                                =0
        Detect levels                                  =0
     Detect features                                   =1
        Detect ribs and dividers                       =0
        Detect rails and slots                         =0
        Detect steps and pockets                       =0
BRep Inflation                                         =1
     Direct inflation                                  =1
     Optimization inflation                            =1
3D Refine                                              =0
3D Tessellate                                          =0
CSG-based modeling                                     =0
Load Most Recently Used files (MRU)                    =4
Save IGS files as type 143(1) or 144(0)                =1
Save IGS files as polygons(1) or triangles(0)          =1
```

Figure 1.17. *REFER.ini* file.

Figure 1.18. New file in REFER's main toolbar.

1.5.4.4 *Open file*

In the main toolbar, click *Open* to use a previously created file (see Figure 1.19). Select the desired file from the windows explorer (only sketch files will be listed by default. Select the *All Files* filter option to select other file types).

Figure 1.19. Open File in REFER's main toolbar.

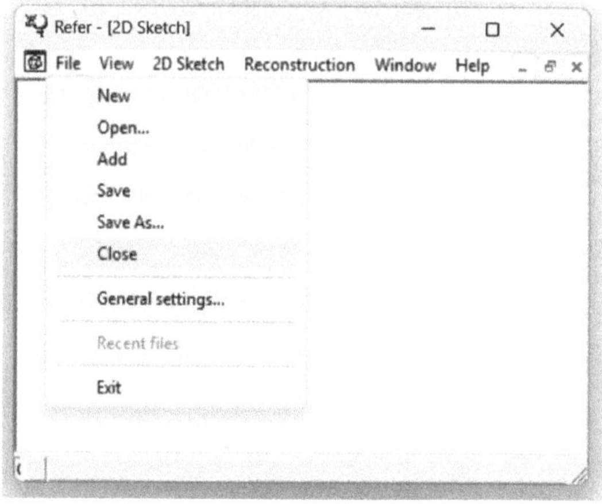

Figure 1.20. Close File in REFER's menu bar.

1.5.4.5 *Close file*

Opened files or "documents," as referred to internally by REFER, can be closed via the *Close* command in the menu bar, as shown in Figure 1.20. If a file is created or opened while another file is already open, both files will share the current application session. However, only the file with the active window will have the focus. The *Close* command closes only the currently active file.

Chapter 2

Input

2.1 Introduction

Sketch-Based Modeling (SBM) studies the automatic creation of 3D models from freehand line-drawings. The input is thus expected to be a sketched drawing, replicating the experience of working in a "virtual paper" environment.

In the words of Liu and Bessmeltsev (2025), "the distinction between clean and rough sketches is approximate but crucial for certain methods, as some algorithms may assume clean continuous strokes drawn on an empty background and fail when this assumption is not met." This distinction is particularly relevant in the context of preprocessing and interpretation. In this regard, REFER does not process raster-based 2D sketches (composed of pixels), but rather vector sketches, which are composed of geometric strokes defined by sequences of points. This vector format assumes a more structured and semantically meaningful input, where the geometry of the strokes is explicitly available.

For readers interested in how raster sketches—such as hand-drawn images on paper or digital canvases—are converted into vector form, we recommend the classical approach introduced by Bartolo *et al.* (2007). Their work lays the groundwork by applying heuristic and geometric rules to trace pixel-based inputs into continuous vector representations. Readers can also explore more recent AI-driven techniques, such as the deep learning-based approach proposed by Simo-Serra *et al.* (2016). Together, these references offer a useful overview of the evolution of sketch vectorization techniques, from rule-based systems to data-driven methods that more effectively mimic human interpretation.

To support SBM processes, REFER provides a default sketching environment, where users can interactively create hand-drawn sketches directly within the sketching area, using a mouse, stylus, or any other device capable of controlling the cursor. However, REFER goes beyond simply replicating traditional paper: it advocates for an enriched virtual paper environment with different 2D operating modes. A subset of 2D Input Modes is particularly relevant here. These modes are managed through the 2D Modes toolbar located inside the 2D Sketch window.

Since correct input is not always achieved on the first attempt, REFER also incorporates basic editing capabilities to refine sketches as needed. In addition, a single editing command (delete) has been implemented as a gesture. This feature was developed as a proof-of-concept intended to explore interaction paradigms based on pen-driven inputs rather than traditional mode-based control. This foundation prepares REFER for future extensions, aimed at enhancing Pen-based input functionalities, as outlined in the Sketch-Based Modeling section.

Input in REFER is a multifaceted process, involving both the management of different types of content ("what") and procedures ("how"). These two aspects are briefly introduced below and further detailed in their respective sections:

- Regarding the content, SBM can address a variety of views. REFER is designed for single, pictorial, parallel projections, where such constraints guide the reconstruction process.
- Regarding the procedure, REFER can process a variety of input types beyond freehand sketches. Alternative input formats are supported to accommodate different user needs and workflows, and to allow users to explore different variations of the reconstruction flow for generating 3D models. For example, controlled line-drawings can be used as input to isolate reconstruction errors from vectorization errors, thus enabling more targeted analysis and testing.

2.2 Operating Modes

It is commonly agreed that the main goal of Sketch-Based Interfaces is to achieve modeless interaction. However, some contributions have also highlighted the value of modes (Company *et al.*, 2009). We argue that full

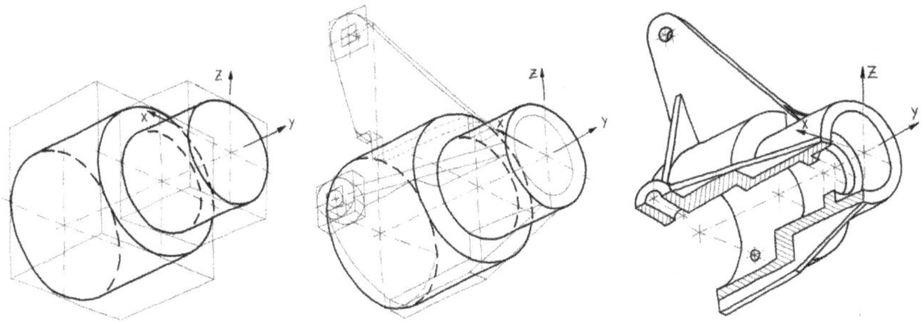

Figure 2.1. Sketch modes.

modelessness should not be considered the final objective. Instead, if the aim is to replicate real paper-and-pencil scenarios in virtual environments, it must be recognized that the use of actual paper-and-pencil naturally involves a rich variety of modes, as shown in Figure 2.1. Indeed, in the hands of a skilled user, a simple pencil can become a complex tool capable of operating in several distinct modes.

Computer-Aided Sketching (CAS) tools must support the various operating modes enabled by pencil and paper which skilled design engineers routinely employ (Company and Varley, 2009). Ideally, if current non-invasive CAS tools were truly comparable to traditional paper and pencil, users would naturally adopt them, given the additional advantages offered by computers. There is general consensus regarding some of the key benefits of computer-based systems:

- Editing content becomes easier.
- Storage and organization are simplified.
- Interfacing with other applications is simpler.

However, replicating the full capabilities of traditional paper and pencil is not straightforward. A pencil is a remarkably versatile tool. It can be used intuitively by novices while offering powerful possibilities to skilled users. In contrast, Computer-Aided Sketching (CAS) tools often overwhelm new users, and several authors have suggested that this complexity has limited the broader adoption of CAS systems (Lim *et al.*, 2004). At the other end of the spectrum, it is sometimes assumed—often without explicit justification—that CAS tools already provide all the operating modes an expert user may need. In our view, this assumption is questionable.

REFER was designed to replicate the modes from the traditional paper and pencil paradigm, while incorporating a nonintrusive strategy for switching between modes. However, determining the precise requirements for these modes requires further studies with actual designers. For example, some actions that interrupt the ideation flow are perceived as intrusive, while others are not. A rigorous investigation is needed to identify which actions are intrusive, and to understand how and why they interfere with the creative process.

In REFER, operating modes are organized as 2D input modes and managed via the 2D Modes toolbar located in the 2D Sketch window.

An alternative approach worth exploring is the Sketch Input Tracer (SKIT) (Ito *et al.*, 2022), a sketch-based CAD interface built on the Fuzzy Spline Curve Identifier (FSCI) (Saga and Makino, 1993) and Infinite-resolution Fuzzy Grid Snapping (IFGS) (Watanabe *et al.*, 2020). The FSCI algorithm classifies input strokes into seven types of geometric primitives (i.e., lines, circles, circular arcs, ellipses, elliptic arcs, closed free curves, and open free curves), while the IFGS algorithm aligns these primitives by snapping their feature points to a square grid.

2.3 2D Input Modes

2.3.1 *Introduction to 2D input modes*

To remain competitive, computer-aided sketching tools must support the various 2D operating modes that pencil and paper offer and that skilled design engineers routinely use (Company and Varley, 2009).

We define an activity as a set of actions that accomplish a particular job, while a task is an individual action within that set. In the context of conceptual design of engineering products, two main activities are typically recognized by designers during the sketching process:

- *Inking*: the process of making permanent, emphasized, or clearer lines on a sketch or drawing to highlight certain features or elements. This process is commonly used to distinguish primary features from background or secondary elements, such as the main outline of a shape, edges, or components that are critical to the design.
- *Erasing*: the action of removing or modifying parts of a sketch or drawing. Erasing is essential for refining designs, correcting mistakes, or clearing sections that are no longer needed in the sketch.

Two additional activities emerge when interactive tools are available, allowing designers to select different parts of the sketch for modification:

- *Selecting*: choosing one or more strokes, marking them for subsequent editing.
- *Editing*: correcting or modifying the characteristics of the strokes that are selected at a given time.

These four activities are managed by REFER as Modes, and users can switch between them using the *2D Modes* Toolbar, located in the 2D Sketch window. Additional activities and sub activities (Company *et al.*, 2011) to be considered in future SBM applications include:

- *Inking construction lines ("scaffolding")*: scaffolding refers to temporary, supportive structures or guides used during the sketching process to assist the designer in creating accurate and well-aligned sketches. The concept of scaffolding in design typically involves providing intermediate steps or visual aids to help achieve a final result, much like physical scaffolding supports construction work. Scaffolding may include grid systems or guidelines that assist the designer in maintaining proportion, alignment, and perspective while sketching. These grids provide structure to the design, ensuring that objects are placed correctly within the sketch.
- *Inking main lines ("highlighting")*: highlighting involves emphasizing or outlining the primary components of a sketch. This action is typically used to distinguish key features or important boundaries within a design. The inking process highlights these main lines to provide clarity and help identify the critical elements of the design.
- *Inking hidden edges ("highlighting the back of the object")*: technique used to represent parts of an object that are not directly visible in the sketch but are crucial for understanding the complete 3D structure. These edges typically represent the backside or internal contours of an object that are obscured by other parts of the sketch, but they help convey depth, structure, and full geometry of the model.
- *Inking reference axes*: the process of manually drawing axes (usually X, Y, and Z in 3D space) on a 2D sketch to indicate the orientation and directionality of a design. These reference axes help establish the coordinate system and provide a clear frame of reference for the designer, aiding in the interpretation and conversion of the 2D sketch into a 3D model.

- *Hatching*: drawing technique used to create tonal or shading effects by drawing closely spaced parallel lines or other form of pattern. Hatching is used to suggest depth, material properties, or other characteristics of the surface being represented.
- *Overtracing for decoration*: the act of tracing over an existing sketch or design (in some cases, multiple times) to enhance its visual appearance, often with the intent of making it more aesthetically pleasing rather than for functional purposes. This technique is commonly used in art and design to emphasize certain features or add detail without altering the underlying structure or function of the original drawing.
- *Annotating*: the process of adding supplementary information to a sketch or model to clarify, explain, or provide additional details about specific parts of the design. Annotations are typically textual or symbolic and serve as a communication tool between the designer and other stakeholders (e.g., collaborators, engineers, clients, etc.) to ensure that the intent behind the design is clear.

2.3.2 *2D modes toolbar*

The *2D Modes* toolbar (shown in Figure 2.2) is located in the *2D Sketch* window, as its controls affect only the information displayed within that window. This toolbar allows users to switch between different modes for interacting with and modifying a 2D sketch, ensuring that the tools provided are relevant to the task in the sketching process.

The first two buttons in the 2D Modes toolbar allow switching between two alternative modes. *Sketch* mode is the default, which enables freehand drawing of strokes. Alternatively, *Select* mode allows users to erase or edit the current sketch to modify elements in it.

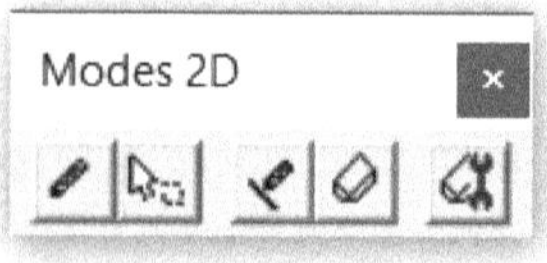

Figure 2.2. *2D Modes* toolbar.

Additionally, the *2D Sketch* window supports two complementary modes, which provide greater flexibility for customizing sketches and drawings:

- Straight line allows users to draw straight segments by specifying their initial and final points.
- Erase stroke nodes allows users to partially erase sections of sketched strokes.

A configuration button is also available to adjust the settings of the *Erase* tool. These complementary modes were originally introduced to equip REFER with more advanced tools for creating and customizing examples to test the behavior of the reconstruction flow. Although they can also be useful to regular SBM users, further experimentation is needed to assess their true value and identify potential improvements and additions.

2.3.3 *Sketch mode*

Freehand sketching can be performed in the *2D Sketch* window. To start sketching, *Sketch Mode* must be active in the *2D Modes* toolbar, as shown in Figure 2.3.

The process for inking a new stroke involves the following steps:

(1) Positioning the cursor at the desired starting point of the stroke.
(2) Moving the cursor along the intended path of the stroke.
(3) Releasing the cursor at the endpoint of the stroke.

The results will be interactively displayed on the screen, as shown in Figure 2.4.

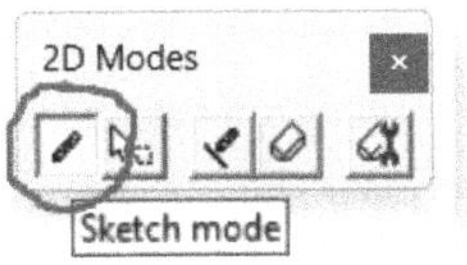

Figure 2.3. *Sketch* Mode in the 2D Modes toolbar.

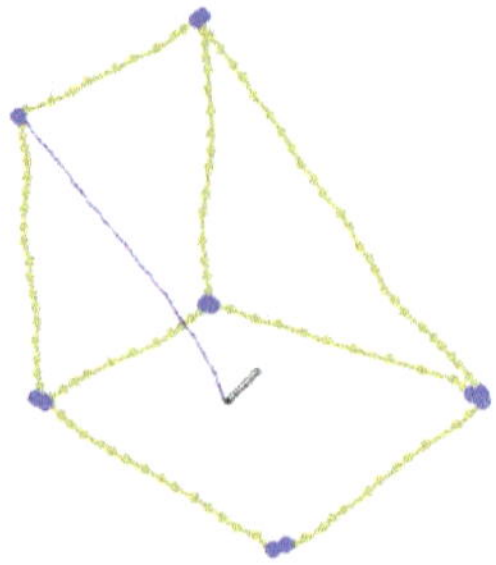

Figure 2.4. Creating a new stroke in Sketch mode.

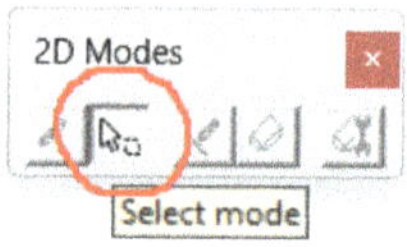

Figure 2.5. *Select* Mode in the *2D Modes* toolbar.

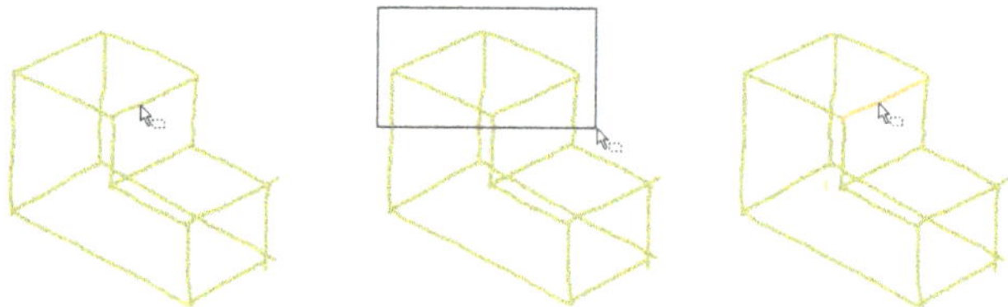

Figure 2.6. Selection tool in REFER.

2.3.4 *Select mode*

The *Select* mode enables the selection of entities in the current sketch for subsequent erasing or editing (see Figure 2.5). Once this mode is active, the *Select* Mode button must be switched off to return to *Sketch* Mode, as these two modes are mutually exclusive (i.e., only one can be active at a time).

The selection tool in REFER works similarly to other drawing applications (see Figure 2.6). A single stroke can be selected by left-clicking on it. Multiple strokes can be selected at once by drawing a selection rectangle. To do this, click the left mouse button at one corner of the desired area, drag the mouse while holding the button, and release it at the opposite corner. To unselect strokes, repeat the same steps. Selected strokes are highlighted in a different color (orange, by default). This visual cue helps distinguish the selected elements from the rest of the sketch.

2.3.5 *Straight line mode*

The *Straight Line* mode can be activated in the *2D Modes* toolbar by pressing the corresponding button. This mode allows drawing straight segments by holding down the left mouse button (or the equivalent action on another device) and moving the cursor from the initial tip to the final tip of the segment. After the initial click to set the first endpoint of the line, an “elastic” or “rubber-band” type of line is displayed, connecting the initial tip to the current cursor location (see Figure 2.7). This dynamic line updates interactively and in real time as the cursor moves. The final line is created once the user clicks to set the location of the final tip.

The *Straight Line* mode functions similarly to the line-drawing features found in commercial 2D CAD software. REFER’s mode, however, is a more basic implementation. For example, the tool does not provide any snapping capabilities commonly found in more advanced CAD systems. As a result, line drawings may have close junctions that are not automatically merged into common vertices (see Figure 2.8). This behavior aligns with the philosophy of SBM, where vertices are merged while refining the close junctions of strokes.

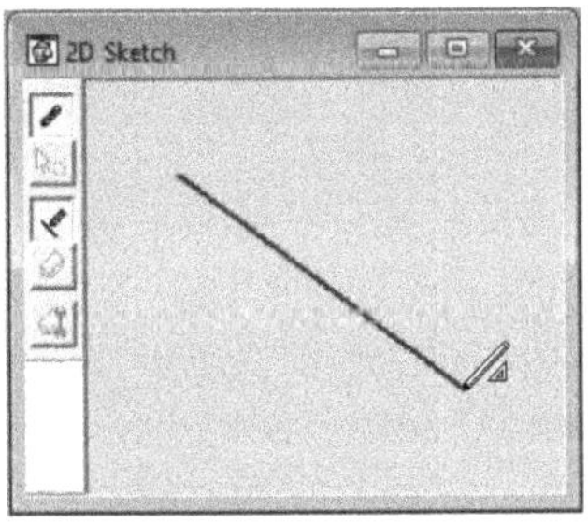

Figure 2.7. Creation of a straight line in *Straight Line* mode.

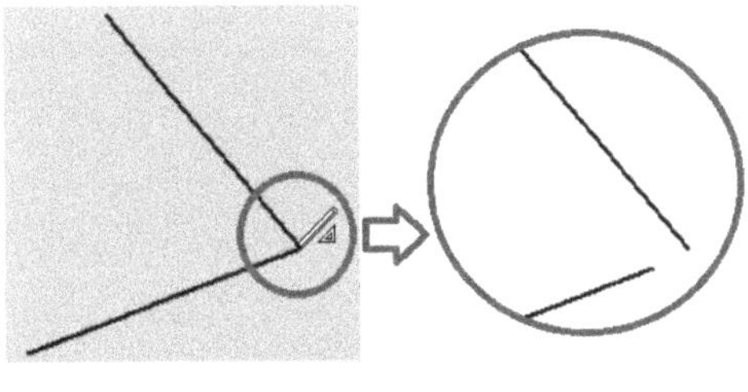

Figure 2.8. Example of a close junction not automatically merged into a common vertex.

2.3.6 *Erase mode*

Strokes can be removed from a sketch by selecting them and pressing the *Delete* key. Alternatively, the *Erase stroke nodes* tool, available in the *2D Modes* toolbar, enables the partial deletion of sketched strokes. Activating the *Display of nodes and corners of strokes* option provides visual feedback on the nodes of the stroke that are being deleted by the tool.

A configuration button is also provided to customize the *Erase* tool. Through the corresponding dialog (shown in Figure 2.9, the user can select the radius of the tool and choose between two methods for handling partially deleted strokes: splitting strokes when part of their intermediate nodes are deleted and turning them into separate strokes or removing the selected intermediate nodes while keeping the strokes as single units.

2.3.7 *Edit mode*

2.3.7.1 *Introduction to edit mode*

Editing tools are located in the *2D Sketch* window and are organized in the *2D Edit* toolbar. Two different types of edit tools are available:

(1) Tools for editing 2D Sketches.
(2) Tools for manipulating Line Drawings.

The available editing capabilities depend on the input format. Entering a sketch enables the corresponding editing tools for sketches. Likewise, the editing of line-drawings is possible only for line-drawings entered as direct input, not for those generated after processing sketches.

When selection mode is active, the selected strokes can be transformed through various sketch operations, including moving, copying,

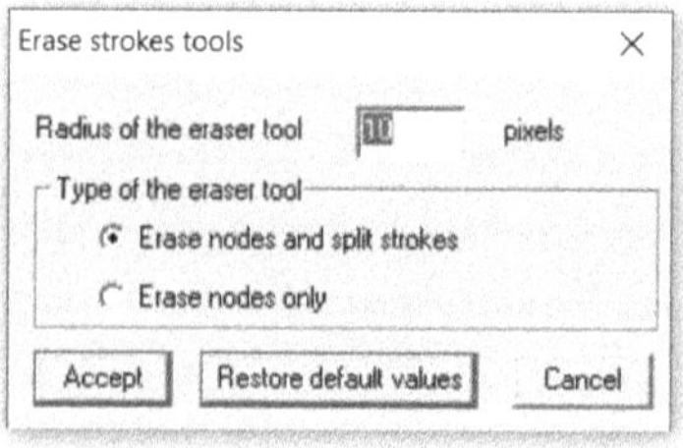

Figure 2.9. Erase tool customization options.

rotating, scaling, and filtering. Additionally, the *Add file* tool located in the *Main* Toolbar is useful for merging a previously saved sketch with the current sketch.

Sketch editing tools were added to REFER to provide a more advanced sketching environment. These tools have proven valuable in helping users create more sophisticated sketches more quickly. Nevertheless, further studies are needed to determine the least invasive method for integrating them into future SBM applications. Additional details are discussed in the *2D Operating Modes* section.

Two methods are available for editing line-drawings, but only as complete entities (individual parts cannot be edited separately). The first method allows rotating the entire line-drawing. The second method, called "distortion," involves randomly modifying the coordinates of the line-drawing's vertices. This technique is useful for testing the robustness of reconstruction approaches, as it introduces artificial geometric imperfections into the line-drawing, mimicking the imperfections that might occur when vectorizing imperfect sketches.

The tools for editing line-drawings were primarily added as a utility for developers working on SBM code. They facilitate the use of line-drawings originally created in 2D CAD applications and help produce imperfect line-drawings that replicate those resulting from the vectorization of 2D sketches. As such, these tools are not intended for general use in future SBM applications but rather serve a specific purpose for development and testing.

It is important to note that edit operations cannot be undone (no Undo command is provided). Therefore, it is recommended to save the original file before making any edits.

2.3.7.2 *2D edit toolbar*

The *2D Edit* Toolbar (shown in Figure 2.10) can be found in the *2D Sketch* window.

The toolbar includes the commands shown in Figure 2.11.

Figure 2.10. 2D Edit toolbar.

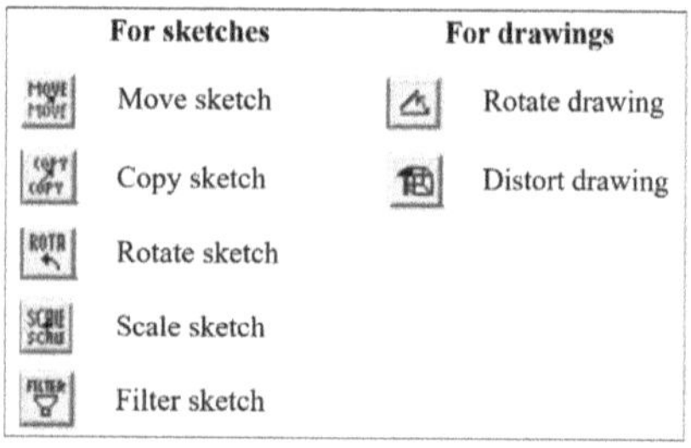

Figure 2.11. Editing commands in the 2D edit toolbar.

2.3.7.3 *Move/Copy sketch*

The *Move* command allows the user to reposition a group of selected strokes in a sketch. Similarly, the *Copy* command creates a copy of the currently selected strokes in the sketch. These strokes must first be selected in *Select* mode. The process is as follows:

(1) Enter *Select* mode to select the strokes in the sketch.
(2) Activate the *Move* or *Copy* command.
(3) Click at an arbitrary point, which will serve as the tail of the displacement vector.
(4) Move the cursor to the desired location. The displacement vector will update accordingly.
(5) Place the strokes in their final position by clicking the left mouse button (or the equivalent action on your input device).
(6) When done, press the *Escape* key to deselect the strokes.

2.3.7.4 *Rotate/Scale sketch*

The *Rotate* command allows the user to rotate the currently selected strokes in a sketch. Likewise, the *Scale* command can be used to resize the currently selected strokes in a sketch. The process is as follows:

(1) Enter *Select* mode to select the strokes in the sketch.
(2) Activate the *Rotate* or *Scale* command.
(3) For rotating, enter the desired rotation angle. For scaling, enter the desired scale factors for the two main directions (X and Y axes).
(4) Click *Accept* to confirm the operation.
(5) When done, press the *Escape* key to deselect the strokes.

2.3.7.5 *Filter sketch*

The *Filter* command can be used to reduce the number of nodes in the currently selected strokes in a sketch. This operation is useful for simplifying strokes with high node density, which typically occur when the user moves the cursor very slowly while sketching. Two methods are available:

- In filter mode, the nodes of the strokes are reduced according to the specified ratio (e.g., a factor of 2 means every other node in the sequence will be removed).
- In resample mode, the stroke nodes are recalculated to produce the same stroke with the number of nodes specified by the user (the default is 20 nodes), approximately equally spaced. Note that chords may have unequal lengths (even if arcs have the same length) when curvature changes abruptly. This procedure is automatically launched when calculating vectorization merits, stroke corners, and other procedures that require resampled strokes. It is available here to allow direct control over the strokes.

The filtering process is as follows:

(1) Enter *Select* mode to select the strokes in the sketch.
(2) Activate the *Filter* command.
(3) Select the desired mode (filter or resample) in the dialog box and enter the corresponding factor.
(4) Click *Accept* to confirm the filter operation.
(5) When done, press the *Escape* key to deselect the strokes.

2.3.7.6 *Rotate/Distort drawing*

The *Rotate Drawing* and *Distort Drawing* commands can be used to rotate or distort a drawing brought into REFER as direct input through one of the suitable input formats. In the case of *Distort Drawing*, the application randomly changes the horizontal and vertical coordinates of each vertex of the drawing. The change is arbitrary, but the range is controlled through a distortion factor, which determines the maximum percentage of the dimensions of the drawing's bounding box that will be applied as an increment or decrement for each coordinate. The default percentage is 0.5%, which is the value assigned by the application when restoring default values. It is important to note that distortions are cumulative, as the drawing coordinates are updated after each distortion. Therefore, each new distortion uses the previously updated coordinates as input. An example of a rotated and a distorted drawing is shown in Figure 2.12.

The process to rotate or distort a drawing is as follows:

(1) Select the *Rotate Drawing* or *Distort Drawing* command.
(2) For rotating, enter the rotation angle in the dialog box. For distorting, enter the distortion factor (as a percentage) in the dialog box.
(3) Click *Accept* to confirm the operation.

2.3.7.7 *Add file*

The Add file command can be used for editing the current sketch by merging it with a previously saved one. The process is as follows:

(1) Open an existing sketch, or, alternatively, create a new one.
(2) Select *Add file* from the *Main* toolbar, as shown in Figure 2.13 (or click *File* in the *main menu*).
(3) Select a previously created sketch.
(4) The selected sketch will be inserted into the current sketch. The combined sketch appears *Selected* and ready to be edited.

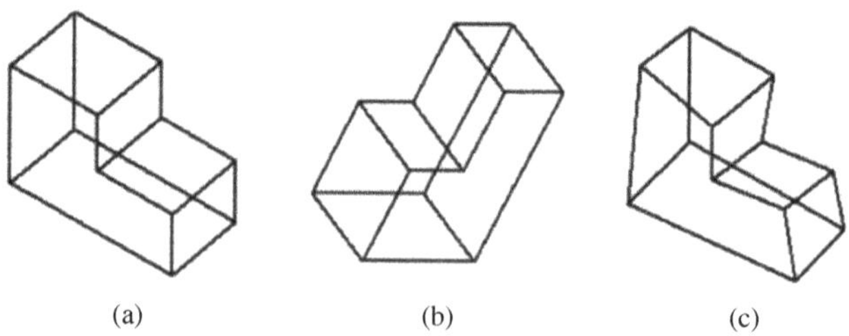

Figure 2.12. Original drawing (a), drawing rotated 90 degrees about its center point (b), and distorted drawing, 0.5% (c).

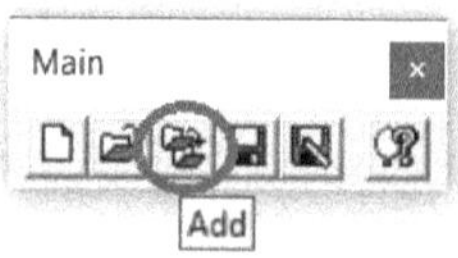

Figure 2.13. *Add File* command in the *Main* toolbar.

2.4 Gestures

2.4.1 *Introduction to gestures*

Two major approaches can be leveraged in SBM. The first, known as geometric reconstruction, is the primary approach used in REFER. It originated in the field of computer vision and employs algorithms to reconstruct geometric objects from sketches of their two-dimensional projections. The second approach is based on the calligraphic interface concept, which interprets gestures and pen-input as commands (Rubine, 1992; Long *et al.*, 2000, Fonseca and Jorge, 2001). In these systems, gestures are used to generate solids directly from 2D sections. Some early gestural modeling systems include:

- *Sketch* (Zeleznik *et al.*, 1996): This system uses a sequence of gestures to input the geometric model, following a set of conventions regarding the order in which points and lines are entered, as well as their spatial relationships.
- *Quick-Sketch* (Eggli *et al.*, 1997): A system focused on mechanical design that features a 2D drawing environment based on constraints. From these constraints, 3D models are generated using modeling gestures.
- *Teddy* (Igarashi *et al.*, 1999): This system is aimed at free-form surface modeling, providing a simple interface of sketched curves, pockets, and extrusions. Users draw silhouettes with pen strokes, and the system automatically proposes a surface using a polygonal mesh that matches the object's contour.
- *GIDeS* (Pereira *et al.*, 2000): This system allows data input from a single-view projection. Additionally, the dynamic recognition of modeling gestures provides users with contextual menus and icons, enabling modeling with a reduced command set.
- *Digital Clay* (Schweikardt and Gross, 2000): Supporting basic polyhedral objects, this system combines a calligraphic interface for data input with Huffman-Clowes algorithms to derive three-dimensional geometry.
- *Stilton* (Turner *et al.*, 2000): Implemented directly in a VRML environment, this system uses a calligraphic interface for data input and employs an optimization approach based on genetic algorithms for the reconstruction process.

A recent review by Bhattacharjee and Chaudhuri (2020) explored the impact of Augmented Reality (AR) and Virtual Reality (VR) in transforming sketching into a more powerful and user-friendly modality for content creation. The approach is also beneficial for CAD models based on free-form shapes (Xu *et al.*, 2014). However, REFER focuses exclusively on analytical shapes used for engineering purposes. The idea of using agent-based architectures to support the definition and real-time recognition of sets of 2D graphical symbols has also been explored (Avola *et al.*, 2011).

The gestures implemented in REFER are based on those used in the Calligraphic Interface for Geometric Reconstruction (CIGRO), a system developed by Naya *et al.* (2003). CIGRO provides an integrated 2D-3D environment that allows users to sketch and immediately switch the viewpoint to visualize the corresponding 3D model. Real-time sketching is supported through a minimal gesture alphabet, automatic line-drawing beautification, and a fast, robust axonometric-inflation engine. The gesture recognition in CIGRO was inspired by Jorge and Fonseca's CALI (1999).

REFER's implementation is currently limited to a single gesture intended to provide a test environment for evaluating whether modes are more or less intrusive than gestures, and, consequently, more or less useful for an SMB system. This investigation aligns with previous studies (Company *et al.*, 2006, 2008; Naya *et al.*, 2007, 2008). The only explicit gesture available in REFER is *Delete*. However, the system also implements the concept of *Zoom* as a gesture, which allows users to convey intent and context by adjusting the zoom levels while drawing. It is important to note that gestures are only interpreted by REFER if both Sketch Mode and Interactive Vectorization are active.

2.4.2 *The delete gesture*

The *Delete* gesture involves making a scratch motion with the mouse to instruct the application to delete all strokes in the sketch that intersect the scribble, as shown in Figure 2.14.

Detecting gestures introduces certain challenges that are simple to describe but require further research to completely resolve. Firstly, continuously parsing all the input strokes is time consuming and computationally intensive. This is the case in REFER, where canceling the gesture

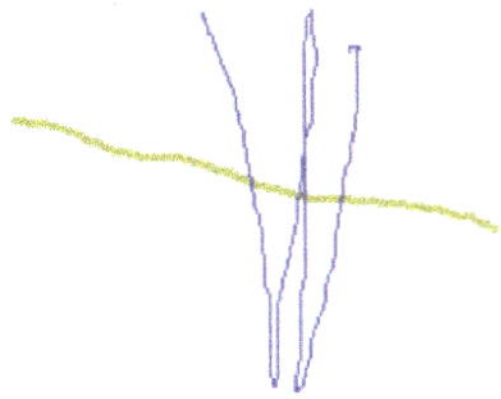

Figure 2.14. The *Delete* gesture creates a scribble (represented in blue) that intersects lines in the sketch (in green).

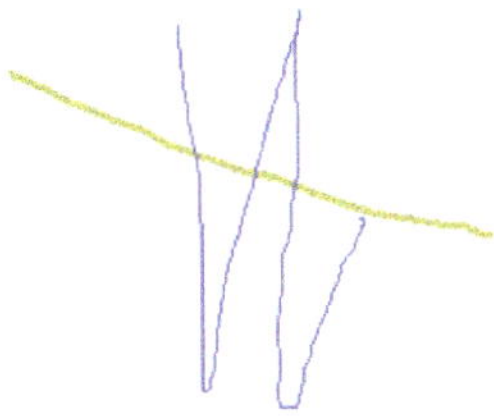

Figure 2.15. Example of a gesture that was not recognized as *Delete*.

parsing task is not allowed. Additionally, the stroke must meet specific criteria to be recognized as a valid gesture. The example shown in Figure 2.15 illustrated a gesture that was not recognized as a *Delete* signal.

In REFER, a stroke is parsed as a delete gesture if the following four conditions are met:

(1) The shape is a polyline.
(2) The polyline consists of more than two segments.
(3) The angles between consecutive segments are all acute (less than 22.5 degrees).
(4) The angles between consecutive segments are opposite, causing the polyline to zigzag.

All strokes of the sketch that intersect the bounding box of the gesture are considered for deletion. However, this seemingly straightforward approach hinders precise control over which strokes to delete, especially in densely populated sketches. This effect may lead to "unnatural" behavior in some situations. For example, Figure 2.16 illustrates a *Delete* gesture that failed to capture the "horizontal" stroke. This failure happens because both the horizontal and vertical segments belong to a single

Figure 2.16. *Delete* gesture that failed to capture the "horizontal" stroke of the sketch.

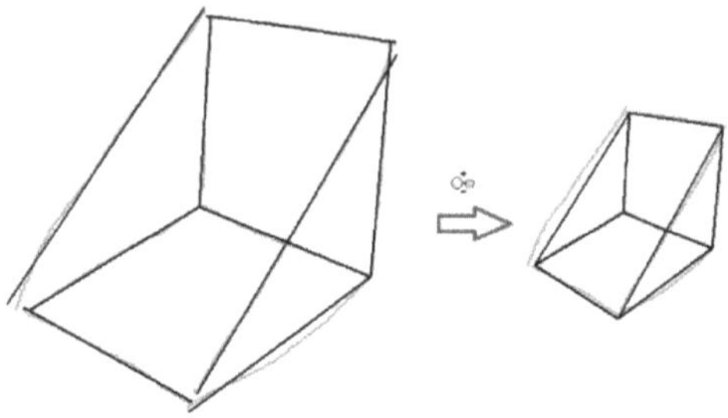

Figure 2.17. Zoom as a gesture to increase the tolerance for merging nearby tips.

"polyline" stroke, and the gesture is too far from the center of this polyline.

A related issue is that strokes that are not recognized as gestures are not automatically deleted or even processed as commands. As a result, these strokes are added to the current sketch and must be manually deleted in a subsequent *Delete* gesture.

2.4.3 *Zoom as a gesture*

Delete is the only explicit gesture currently supported in REFER. However, in specific contexts, REFER interprets zooming as an indirect gesture to dynamically influence algorithmic behavior. One such case occurs in the 2D refinement stage, specifically to control the threshold when merging triplets of neighboring vertices to identify cubic corners.

The example shown in Figure 2.17 illustrates a scenario in which reducing the zoom level effectively increases the tolerance for merging nearby tips. This function allows the system to better handle minor sketching imperfections, such as slightly misaligned or disconnected line endings, by implicitly relaxing the discrimination threshold.

Although this mechanism aligns with the well-known "what you see is what you get" (WYSIWYG) design principle, we have found no references in the literature describing the use of zoom as a means of controlling precision thresholds in SBM algorithms. One technical challenge worth noting is the need to reconcile viewport-based metrics with algorithms that typically operate in world coordinates.

In practice, this approach provides a fast and intuitive interface for users to explore how merging thresholds influence the detection and correction of imperfect junctions. It also invites further exploration of zoom-based interaction as a method for dynamically adjusting other thresholds throughout the SBM process.

2.5 Views

2.5.1 *Introduction to views*

Creating a two-dimensional representation of a 3D object involves two primary stages: modeling and projection. Modeling refers to the process of simplifying the infinite complexity of a real object by retaining only the most relevant features needed for a specific application. A detailed study of this topic is beyond the scope of this book, but readers interested in further exploration can find a useful introduction in Camba and Company (2020). In this section, we briefly review basic concepts related to line drawings used for engineering purposes and their fundamental geometric elements. We also discuss the differences between the most common models used in SBM: Wireframe and Natural models.

Projection is a geometric procedure used to transform a three-dimensional model into a two-dimensional figure. For example, a cylindrical oil barrel can be simplified to a cylinder, which is then projected onto a two-dimensional surface (e.g., a sheet of paper) as a pair of ellipses and two straight segments, as shown in Figure 2.18.

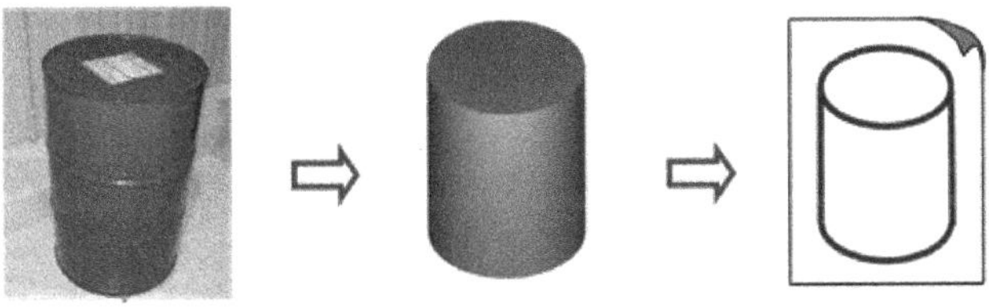

Figure 2.18. Representation of an object through projection.

To a certain extent, SBM can be considered the inverse of projection, as the goal is to generate 3D models from drawings. Consequently, researchers interested in SBM should have a general understanding of projections and their associated invariants.

In essence, engineering drawings are organized into figures, which are composed of geometric elements that represent three-dimensional (3D) shapes through two-dimensional (2D) projections. These figures are the result of—or should correspond to—images derived from projections of 3D models onto a 2D plane. The process of creating these figures from 3D models is referred to as *projection*, and the resulting 2D representations are known as *views* in the context of engineering drawings and Computer-Aided Design (CAD). Views are not interchangeable or equivalent; each view represents a unique interpretation of the 3D model. The images produced by projection are not singular or unique because the transformation from 3D to 2D is inherently non-unique. Various parameters—such as the angle, scale, and position of the projection plane—govern the behavior of the mathematical transformation, leading to different possible views of the same model. As a result, multiple types of views can be generated from a single 3D model, each associated with specific invariants that capture distinct aspects of the model's geometry. Two dominant types of views have been used in the literature to reconstruct models from sketches, as illustrated in Figure 2.19:

- One approach reconstructs 3D shapes depicted by *pictorial* views. This is the only approach currently implemented in REFER.
- The second approach reconstructs 3D shapes depicted by multiple *orthographic* views.

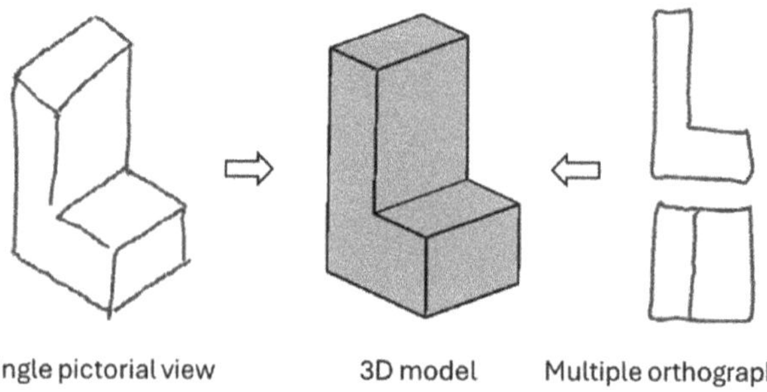

Figure 2.19. Types of views used to reconstruct 3D shapes from sketches.

2.5.2 *Modeling*

2.5.2.1 *Introduction to modeling*

Models, as representations of objects, are central to many disciplines and professions. For instance, designers and inventors create descriptive models in the form of sketches, while artists use materials like clay to craft sculptures. Artisans build objects from a variety of materials, and scientists generate knowledge, typically expressed as laws or theories, that govern how models should be constructed.

Across all these fields, the tools used to create models have evolved significantly. What once began with basic tools like paper and pencil (or hammer and chisel) has progressed to advanced computing technologies and specialized software designed specifically for managing and manipulating virtual models. However, the purpose and intended use of a model heavily influence both the modeling approach and the tools employed in its creation. For readers interested in understanding the different types of geometric models, their evolution, scope, and limitations, a comprehensive discussion can be found in Camba and Company (2020).

Models reconstructed through SBM approaches are primarily derived from line drawings, which serve as the foundational representation. These drawings are constructed using various geometric elements that form the building blocks of the model. Depending on the approach, these elements are typically arranged in one of two primary styles: wireframe or natural views.

Wireframe views emphasize the outline of a shape, focusing on the edges and structure of the object, while natural views aim to provide a more intuitive, realistic depiction by incorporating additional elements such as shading or curvature. These different methods of arrangement are crucial in defining how the 3D model is interpreted and reconstructed from its 2D sketch representation.

2.5.2.2 *Line drawings*

Line art refers to a monochrome image composed of various straight and curved lines on a plain background, devoid of shadows or tonal variations. This simple yet effective visual style focuses on outlines and contours, emphasizing clarity and precision without additional visual complexity. Line drawing is a specific type of line art that is meticulously crafted to adhere to strict geometric principles. Its primary goal is to accurately and unambiguously represent the geometry of an object or component. In technical fields like engineering and design, line drawings are essential for conveying precise information about the shape, dimensions, and structure of products. In the context of SBM, interpreting line drawings is a fundamental task. The aim of SBM is to extract meaningful geometric information from these drawings, enabling the reconstruction of 3D models or further analysis of the design. Line drawings are therefore not only a visual medium but also a critical communication tool in engineering and product design.

However, the practical needs of engineers demand an even more sophisticated form of communication. Technical drawings represent a specialized category of line drawings that rely on a highly standardized graphical language. These conventions are critical for ensuring that technical information is conveyed clearly, consistently, and unambiguously across different industries and teams. In technical drawings, the shape of a product is represented using the simplest geometric elements: vertices, edges, and faces. According to the standard conventions, visible edges are illustrated as continuous thick lines, while hidden edges are depicted with dashed lines. Vertices are represented implicitly, as the points where two or more edges converge.

Surface details, such as textures or shading, are generally omitted; only the essential borders and contours needed to define the shape are included. Based on the level of detail and style of representation, line drawings are classified as wireframes or natural. Wireframe drawings display the complete edge structure, including hidden edges, whereas natural views portray objects with opaque surfaces, hiding parts not visible from the chosen viewpoint.

Finally, engineering drawings are annotated technical drawings used to define and communicate the full set of technical requirements for a product. Beyond depicting the geometry of an object, these drawings incorporate a wide range of annotations, such as dimensions, tolerances, material specifications, surface finishes, assembly instructions, and other critical notes.

The process of annotating drawings introduces an additional layer of complexity to the communication of information, requiring not only graphical precision but also textual and symbolic clarity. Interpreting and managing these annotations demands an understanding of both geometric conventions and specialized domain knowledge. At present, most SBM systems are focused on geometric reconstruction and do not possess the capability to interpret or generate annotated engineering drawings. Handling this level of detailed technical communication remains an open and challenging area for future research and development.

2.5.2.3 *Geometric elements*

Geometry provides the fundamental framework for representing and studying engineering shapes. At its core, geometry is built upon a set of geometric elements—primary concepts that are considered intuitive and cannot be formally defined by reference to simpler notions. These elements (shown in Figure 2.20) form the foundation upon which all further geometric reasoning is constructed:

- We postulate the existence of infinite entities known as points, and the complete collection of all points constitutes what we refer to as space.
- From points, we recognize two critical aggregates: straight lines and planes. These are fundamental groups of points that also remain undefined but are accepted as basic constructs essential to describing spatial relationships.

These primary elements serve as the foundation for developing more complex geometric concepts, which, in turn, enable the precise modeling and interpretation of engineering designs. By building upon these elements, we can construct geometric figures, which are nonempty sets of geometric elements connected through specific relationships. The structure and properties of these figures are determined by the relationships among their constituent elements.

To formalize these connections, we accept certain fundamental associations—called axioms—which are intuitive truths used to establish the basic rules governing how geometric elements interact. These axioms,

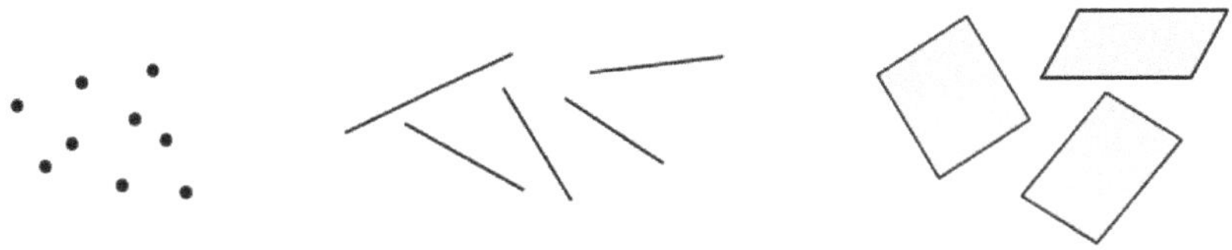

Figure 2.20. Geometric elements: points (left), straight lines (middle), and planes (right).

illustrated in Figure 2.21, allow the systematic definition and study of more advanced figures and structures within geometry:

(1) One, and only one, straight line passes through two distinct points. This axiom establishes the fundamental relationship between points and lines, ensuring that any two distinct points uniquely determine a straight line.
(2) Three non-collinear (non-aligned) points define a plane. This principle asserts that exactly one plane can be determined by three points that do not all lie on the same straight line, providing the foundation for understanding planar surfaces in three-dimensional space.

A solid is typically represented by a combination of its boundary entities—namely faces, edges, and vertices—along with their topological relationships (how these entities are connected) and the geometric definitions (the mathematical descriptions) that specify the precise shape and position of the boundaries.

In particular, polyhedral shapes—solids composed entirely of flat polygonal faces—are represented in line-drawings through a systematic depiction of their structural features:

- Edges are depicted by lines in the drawing. In many cases, connected loops or circuits of lines suggest the presence and boundaries of individual faces.
- Vertices, the points where edges converge in the 3D shape, are represented as junctions in the drawing, where line endpoints meet.

The correct interpretation of these lines and junctions is crucial for reconstructing the original 3D geometry from the 2D drawing.

REFER adopts this representation because it is the most intuitive and best suited for the visual management of geometric shapes. It is also particularly well adapted for operations such as inflation, where the boundary

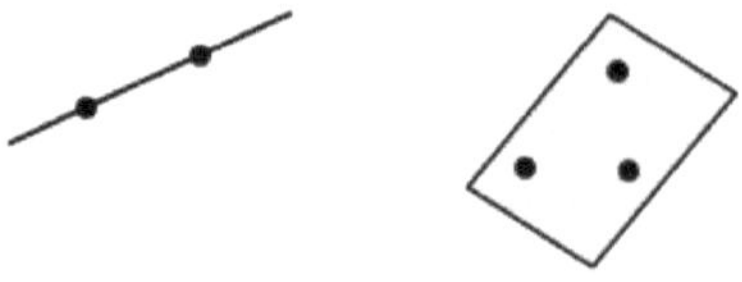

Figure 2.21. Axiom 1: one, and only one, straight line passes through two distinct points (left) and axiom 2: three non-collinear (non-aligned) points define a plane (right).

representation is expanded into a 3D solid. However, while this approach is ideal for human interaction and visual tasks, alternative internal representations could be explored for computational purposes. Applications that require efficient "geometry crunching"—such as performing complex queries, advanced analyses, or real-time updates—may benefit from different modeling strategies that are better optimized for processing speed and storage efficiency.

Constructive Solid Geometry (CSG) is an alternative approach that describes a physical object as a combination of solid primitives using Boolean operations such as union, intersection, and difference. CSG models offer advantages for certain types of analyses and procedural generation tasks, where the structure of the model can be manipulated at a higher level of abstraction. Readers interested in exploring this representation in more depth are encouraged to consult the classical works by Falcidieno and Giannini (1989), Salomons *et al.* (1993), and Heath *et al.* (1993).

2.5.2.4 *Wireframe vs. natural*

Line-drawings that do not include surface information are referred to as *wireframes*. Wireframe drawings represent only the edges of an object, without distinguishing between visible and hidden surfaces. In contrast, line-drawings that depict opaque surfaces—by omitting the back parts that would otherwise be visible through transparency—are called *natural* line-drawings. A hybrid approach is often used, where *hidden* edges are represented with dashed lines, providing a compromise between the completeness of wireframes and the clarity of natural views. Examples of these three types of line-drawings are shown in Figure 2.22.

Wireframes represent the projections of all the edges of a polyhedron, including both visible and hidden, without distinction. This simplicity makes wireframes particularly useful for initial design phases and for visualizing the skeleton structure of 3D objects.

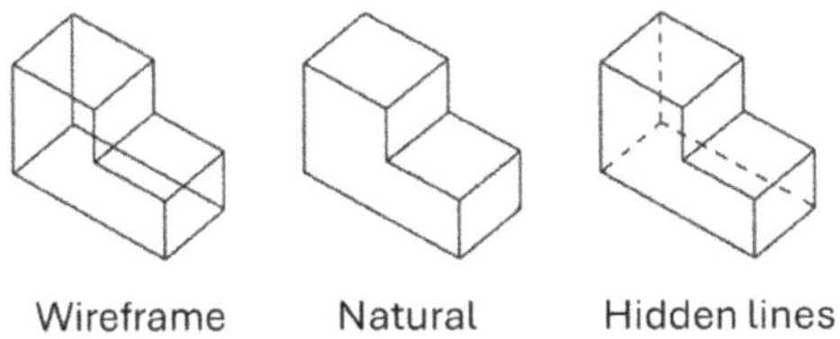

Figure 2.22. Types of line drawings.

Due to their minimalistic nature, wireframe models were among the first types of models used in computer graphics. Their simplicity allows for fast generation and manipulation within a computer system, making them efficient for early-stage design work. However, wireframes are limited in their ability to represent complex geometries accurately. While they are perfectly suited for representing polyhedral shapes, curved surfaces need to be approximated. An iconic example of such an approximation is Uccello's famous chalice, where curved shapes were represented using wireframe techniques (shown in Figure 2.23).

It has been proven that certain realizability problems—specifically those related to trihedral polyhedra without holes—can be solved efficiently in polynomial time (Alevizos, 2013). These polyhedra can be modeled as connected graphs, denoted as P = (V, E), where |V| represents the number of vertices (nodes) and |E| represents the number of edges, with each vertex having a degree of 3. This insight highlights the feasibility of solving complex geometric problems in computational geometry for well-structured polyhedral objects.

The primary limitation of wireframe models is their inability to convey essential information about surfaces and volumes. This lack of surface representation can lead to ambiguity, as the same wireframe may

Figure 2.23. Uccello's chalice.

correspond to multiple different 3D shapes, depending on how the surfaces and volumes are interpreted. For instance, the wireframe model shown in Figure 2.24 (left) could represent several different solid shapes, each with a distinct geometry and structure, as illustrated in the figures to the right. Without explicit surface data, it is difficult to accurately infer the complete geometry of the object, which can be problematic in practical applications where precise visualization and analysis of 3D shapes are required.

Hidden edges are edges that are not visible when observing an object from a specific viewpoint. These edges are typically removed in a drawing to present a cleaner, more intuitive representation of the object, one that appears "solid" and "opaque." As a result, the drawing becomes more "natural," reflecting how we usually perceive objects in the real world, where most things are not transparent. This approach makes it easier for humans to interpret the shape, as we are accustomed to viewing objects from one side and naturally filling in the missing details. However, from the perspective of conveying precise geometric information, natural line-drawings are less informative than wireframes. Since the rear part of the object is not depicted, viewers must infer the hidden geometry, leading to potential ambiguities in understanding the full structure of the object.

An effective artifact used to convey information about the hidden parts of an object, while enhancing the perceptual sense of its volume, involves the use of dashed lines. Dashed lines serve to differentiate between surfaces, edges, or corners of the object that are not visible from the current viewpoint, helping to maintain the illusion of three-dimensionality while communicating the hidden aspects of the shape. This method is particularly useful for providing a clearer understanding of an object's full structure without overwhelming the viewer with unnecessary details.

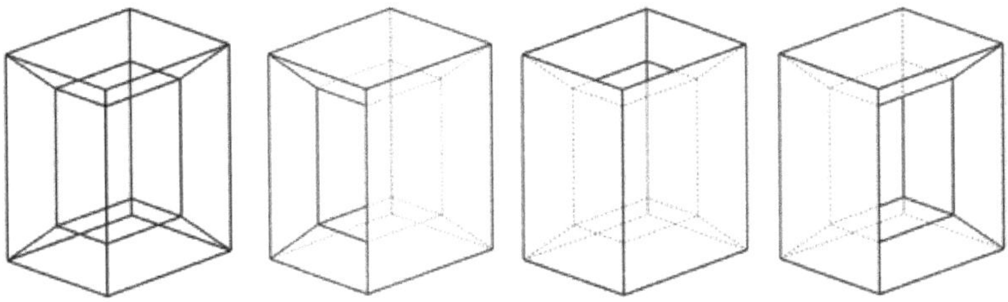

Figure 2.24. The wireframe representation of an object (left) and its possible interpretations.

An interesting approach proposed by Kyratzi and Azariadis focuses on extracting a polyhedral shape from a single view natural sketch. Their method aims to provide "the 2D geometric definition of the hidden part in a topologically reconstructed line drawing (intermediate wireframe sketch) as this is generated from the given natural sketch" (2015). This approach highlights the potential of using advanced geometric reconstruction techniques to infer the missing parts of a shape based on partial, 2D sketch data, thereby bridging the gap between visual perception and accurate model creation.

2.5.3 *Projection*

2.5.3.1 *Introduction to projection*

Projection is a highly effective method for representing models, particularly those with edges and contours. As a homographic transformation, it converts points into points and straight lines into straight lines. This mathematical property aligns well with how humans perceive the world, as projection generates two-dimensional images that capture the essential three-dimensional characteristics of the original model. Humans, through experience and intuition, can typically deduce much of this information without specialized training in geometry or topology (Camba and Company, 2020).

Projection plays a foundational role in the realm of computer graphics. Indeed, it is the backbone of 3D graphics, enabling the display of three-dimensional models on a two-dimensional surface such as a computer screen (Salomon, 2011). Projective transformations are a prerequisite for many tasks such as rendering, where a 3D scene is converted into a 2D image; and clipping, which involves removing elements that fall outside the camera's field of view. In addition to these computational applications, projections have long been used in design and engineering fields. In particular, technical drawings rely heavily on projection methods to communicate and document designs. Modern Computer-Aided Design (CAD) systems automate the generation of such drawings, transforming complex 3D models into detailed 2D representations. For readers interested in a deeper exploration of this topic, Peter Booker's book offers a thorough study of the evolution and significance of technical drawings in engineering (Booker, 1979).

Geometric projection is a process defined by two primary elements: the point of view (V) and the picture plane (π). By adjusting either of these elements, it is possible to generate drastically different images of the same three-dimensional object.

One common type of projection is the central or perspective projection, which occurs when the point of view is positioned at a finite distance from the picture plane, shown in Figure 2.25. In this setup, all the projection lines converge at a specific point known as the center of projection, commonly referred to as the "vanishing point" in the case of perspective projections. This convergence produces a sense of depth and realism, where objects appear smaller as they recede into distance. The central projection is a key tool in fields like architecture, engineering, and computer graphics, as it closely mimics human visual perception and allows for a more lifelike representation of three-dimensional objects on two-dimensional surfaces.

Alternatively, when the point of view is positioned at infinity (often referred to as the "ideal point" in geometric terms), all the projection lines become parallel to one another, resulting in a parallel projection. In this case, the object's size and shape do not change based on distance from the viewer, as there is no vanishing point to create this effect. Parallel projection is often used in technical and engineering drawings because it preserves the true dimensions of objects, regardless of their position in space.

Two main types of parallel projections can be distinguished, each defined by the orientation of the projection lines relative to the picture plane (see Figure 2.26). In an oblique projection, the projection lines are angled relative to the picture plane, resulting in a distorted view of the object. This type of projection is useful for representing objects in a more visually accessible manner, though it sacrifices accurate dimensionality. Conversely, an orthogonal projection involves projection lines that are perpendicular to the picture plane. This type of projection maintains more

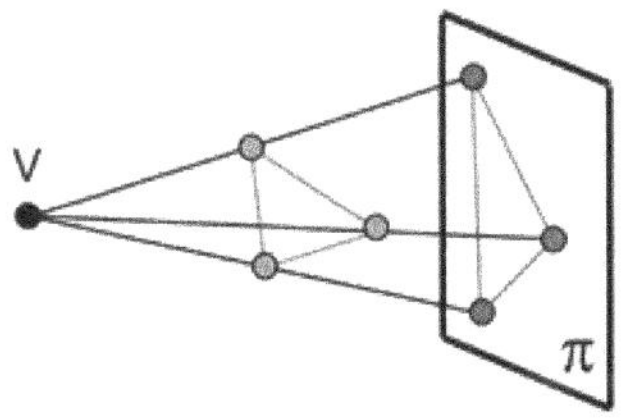

Figure 2.25. Perspective projection.

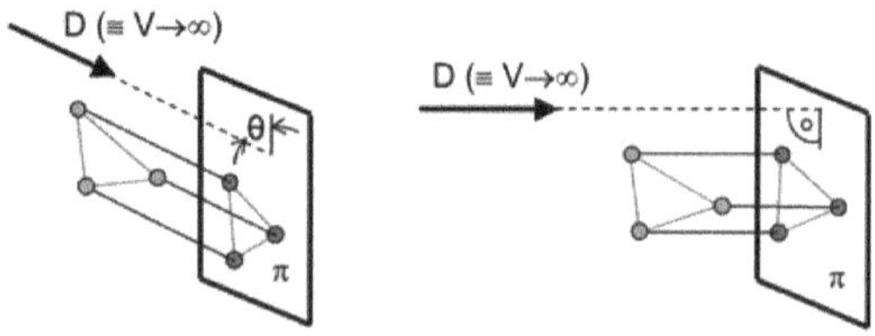

(a) (b)

Figure 2.26. Oblique (a) and Orthographic (b) projections.

consistent geometric relationships and is often used for technical drawings to provide an undistorted view of the object's true proportions in a 2D plane.

Understanding the various types of projections is crucial for researchers and users of SBM, as it aids in anticipating how various views will appear when projected onto a 2D surface. Each projection method can alter the appearance of a 3D model, and being aware of these differences allows for more accurate interpretation and manipulation of models in SBM systems. Additionally, the concept of projection invariants—geometric properties that remain unchanged regardless of the type of projection applied—is also essential. These invariants, such as distances, angles, and relationships between points or lines, provide critical information that can be leveraged to maintain the integrity of the original 3D model during transformation and projection. By recognizing these properties, SBM practitioners can ensure that the essential characteristics of the design are preserved during the modeling process.

2.5.3.2 *Projection invariants*

Invariants are geometric properties of a three-dimensional model that remain unchanged when the model is projected onto a two-dimensional surface. These properties are inherent to the original geometry and persist through the projection process, making them critical for maintaining the integrity of the model in various forms of projection.

All projections share three fundamental invariants (Camba and Company, 2020), which include the geometric properties of *incidence*, *intersection*, and *tangency*, as illustrated in Figure 2.27. These invariants ensure that the relationships between points, lines, and surfaces are preserved, regardless of the type of projection being used. For example, the

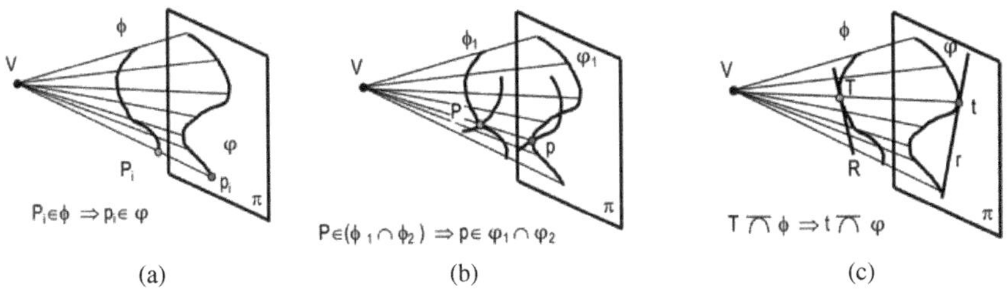

Figure 2.27. Fundamental invariants in projection: incidence (a), intersection (b), and tangency (c).

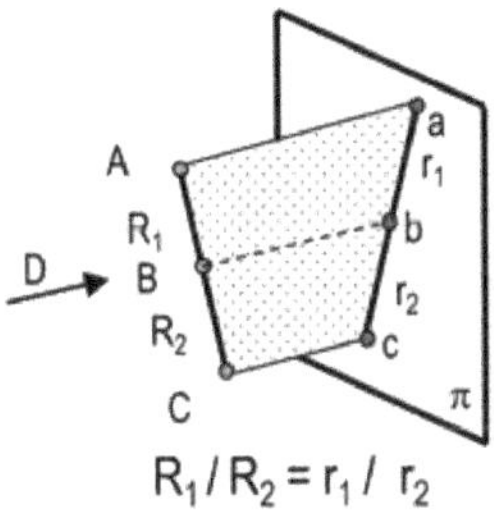

Figure 2.28. Parallelism and proportion are preserved in parallel projections.

incidence of a point on a line, the intersection of two lines, or the tangency between curves remains consistent in both the 3D model and its 2D projection.

However, some properties are specific to particular types of projections. For instance, in parallel projections, both parallelism and proportion are preserved, i.e., parallel lines in 3D space will remain parallel in the projection, and the relative proportions between objects are maintained. In contrast, perspective projections do not preserve these properties. Parallel lines may converge, and the proportional relationships between objects can appear distorted, as shown in Figure 2.28.

The differences in invariants between projection types are of fundamental importance when searching for cues—implicit pieces of information about the model embedded within its views. Projections that preserve a greater number of invariants tend to convey more reliable geometric information, aiding both interpretation and reconstruction tasks. However, invariant properties are not necessarily biunivocal: for instance, two parallel lines in the projection may have originated from lines that were not parallel in the three-dimensional model, as shown in Figure 2.29.

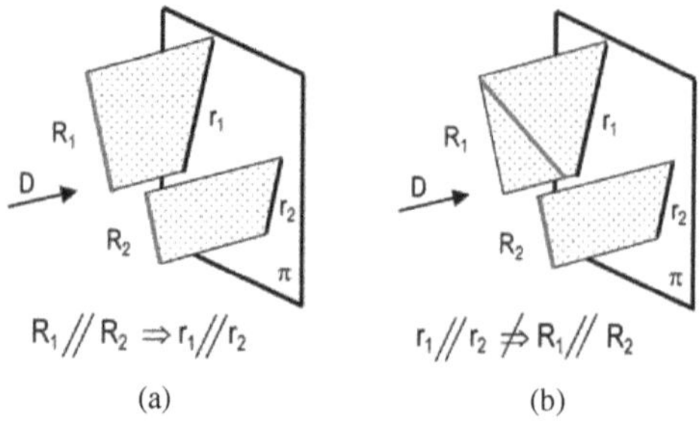

Figure 2.29. Parallel lines in projection originating from lines in the model that are parallel (a) and not parallel (b).

The implication is that two parallel lines in a view may or may not represent two spatially parallel edges in the 3D object. Conversely, two lines that are not parallel in the projection almost certainly do not represent parallel lines in space—assuming a parallel projection was used. Therefore, while invariants provide a strong foundation for identifying cues, cue-finding cannot rely exclusively on them and must consider additional context and information.

In any case, invariants are critical for several reasons: they guide the interpretation of projected images (e.g., determining whether parallel lines in an image represent parallel edges in space), inform the selection of appropriate projection methods based on the intended use (e.g., whether to use perspective or parallel projection to detail mechanical components), and support the extraction of structural information necessary to reconstruct the original 3D model from its views.

2.5.3.3 *Types of views*

Two principal types of views correspond to the two main categories of practical projections: orthographic and pictorial views (which should not be confused with parallel and perspective projections). Although all orthographic views arise from parallel projections, not every parallel projection is considered orthographic. Strictly speaking, the term "orthographic" applies only when the main axes of the geometry are either parallel or perpendicular to the picture plane. Otherwise, if the orientation is more arbitrary, the resulting projection is classified as pictorial. The two types of views are shown in Figure 2.30.

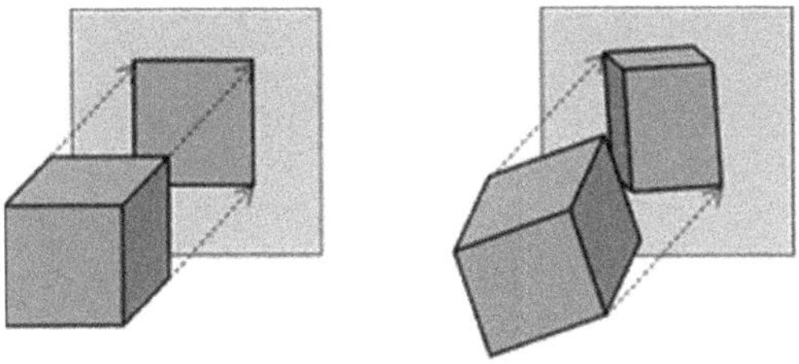

(a) (b)

Figure 2.30. Orthographic (a) vs. Pictorial (b) views.

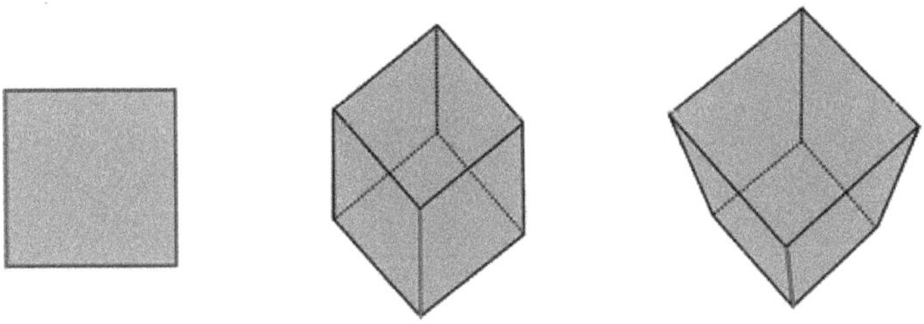

(a) (b) (c)

Figure 2.31. Orthographic view (a), parallel pictorial view (b), and perspective view (c).

Each type of view has distinct advantages and disadvantages, and it is the responsibility of the user to select the most appropriate one for any given task. In an orthographic view, only two dimensions of the three-dimensional object are represented per view. The main advantage is that measurements remain true to scale, since the preserved invariants ensure that the image is a faithful geometric representation of the original object. In contrast, a pictorial view presents all three dimensions in a single image, offering a more intuitive and holistic visualization, albeit with distortions in scale and angles. These distortions are typically more pronounced in perspective pictorials than in parallel pictorials. An example is shown in Figure 2.31.

Thus, while orthographic views are appropriate for accurate measurement and technical specification, pictorial views excel at providing an overall understanding of the object's shape and structure. In practice, both types are often used together to leverage the strengths of each.

2.5.3.4 *Orthographic views*

Orthographic projection forms the basis for certain representation systems—sets of views that collectively convey the complete geometric information of an object—such as the dihedral system.

The dihedral system is one of the most widely used tools in engineering and architecture for communicating, analyzing, and documenting designs prior to manufacturing or construction. It presents multiple orthographic views of an object in a systematic manner to provide a full description of its shape and dimensions.

Orthographic views are particularly effective for measuring tasks, since dimensions along the major axes are preserved without any distortion or foreshortening. Foreshortening refers to the visual compression that occurs when features extend away from the viewer, causing them to appear shorter than they are. Since orthographic projections eliminate this distortion, measurements can be taken directly from the drawings without additional calculations. Furthermore, creating orthographic views is typically simpler and more precise than generating pictorial views, especially for technical documentation.

The primary advantage of orthographic projections lies in the strong preservation of invariants: parallelism and proportionality not only remain intact but, in certain cases, become invariants of identity. Specifically, any figure lying in a plane parallel to the picture plane will be projected onto it without any alteration in shape or size. Thus, a planar figure parallel to the projection plane produces a projected image identical to the original figure, as shown Figure 2.32.

The main limitation of orthographic projection is the inability to represent depth within a single view. The third dimension is entirely lost,

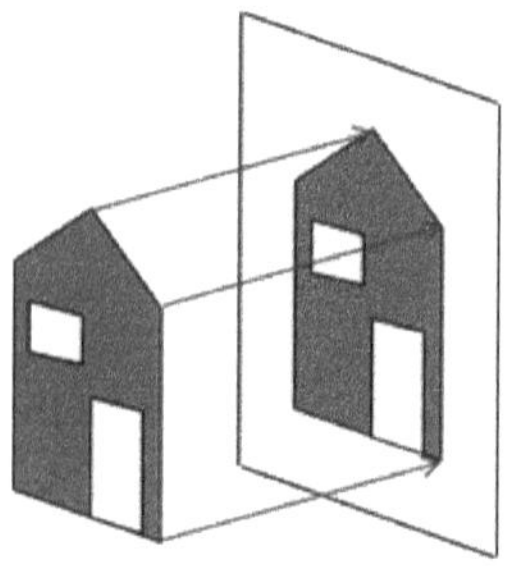

Figure 2.32. Parallel projection of a planar figure.

requiring multiple carefully coordinated views (e.g., front, top, and side) to fully convey the three-dimensional structure of the object. Without familiarity with interpreting such drawings, understanding the overall shape of the model can be challenging.

An established approach for reconstructing 3D shapes leverages the interpretation of multiple orthographic views. Early efforts in this direction were pioneered by Nagendra and Gujar (1988), who explored methods to infer three-dimensional geometry from a set of standard orthographic projections. More recent contributions were made by Governi *et al.* (2013), as well as Han *et al.* (2020), who extended the methodology by incorporating multiple views that are not necessarily orthographic, broadening the flexibility and applicability of the reconstruction process. Unfortunately, REFER does not currently implement any procedure for reconstructing 3D models from orthographic views.

2.5.3.5 *Pictorial views*

Since the pioneering work by Lawrence Roberts (1963), pictorial views have been the most common input for SBM applications. Pictorial views are usually classified in two different types: (1) parallel or axonometric, and (2) perspective.

Axonometric pictorials are less realistic than perspective pictorials but convey information in a way that enables direct measurements and comparisons between features. In contrast, perspective pictorial views are more realistic and are often considered more "artistic," as opposed to the more "technical" nature of parallel pictorial views. Perspective views present additional challenges for SBM, as they convey geometric information in a manner that is more difficult to interpret: detecting cues or features in perspective projections is notably more demanding.

It is important to note that axonometric views can be extended into a complete representational system comprised of four complementary views, which together fully define a model's geometry and spatial orientation. However, in most practical contexts—including SBM—only the principal view is typically depicted, which is why the term "single view" is commonly used.

Two main approaches can be used to represent models in pictorial views: wireframe and natural line drawings. Wireframes display all the edges of the model, regardless of whether they are occluded from view or

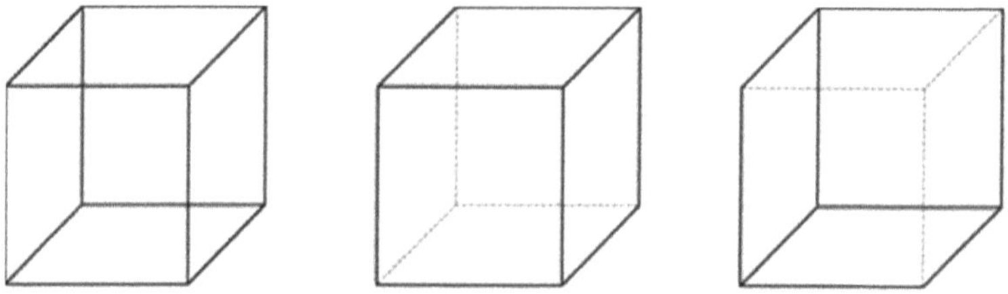

(a) (b) (c)

Figure 2.33. Necker reversals: wireframe model (a), perceived model (b), and alternative model (c).

not. The term "wireframe" evokes the idea of constructing objects with metal wires representing their three-dimensional edges. In contrast, natural line drawings assume that solids occlude their own back features, thus depicting only the edges that are visible from the current viewpoint.

Orthographic views can also be represented either with or without hidden lines. Since multiple orthographic views are typically employed to fully describe the object, the choice between including or omitting hidden lines is not as critical as it is in pictorial single views. In single-view pictorial representations, both wireframes and natural line drawings are used, each offering distinct advantages and disadvantages.

As the name suggests, natural line drawings are highly intuitive for human perception; viewers typically have no difficulty interpreting them. However, they provide incomplete information, as the occluded (back) parts of the model are not represented.

Wireframe views provide complete information about all edges of the model. Nevertheless, this information is more ambiguous, as it lacks visual cues to distinguish depth and occlusion. A classic example is Necker's cube, which illustrates the phenomenon of multi-stability—the perceptual behavior where a single image supports two alternative but equally valid interpretations. These interpretations typically alternate in the viewer's perception every few seconds, producing two distinct interpretations known as Necker reversals, illustrated in Figure 2.33.

2.5.3.6 *Axonometric pictorial views*

Axonometric pictorial views can be classified based on the angle at which the view deviates from the orthogonal. Since there are many useful orientations, only the most common ones have been given special names, such as isometric, dimetric, and trimetric (see Figure 2.34).

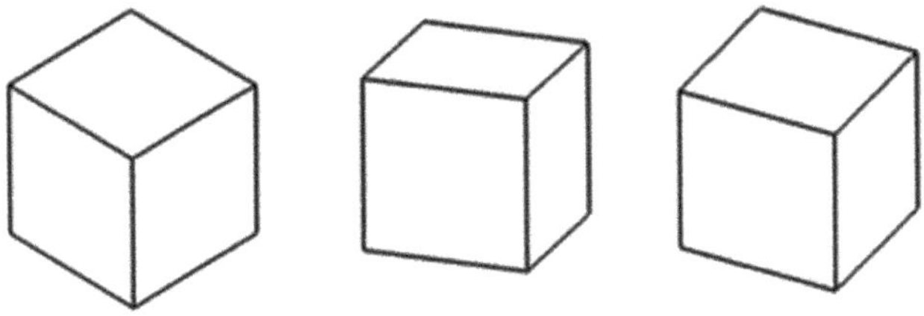

(a) (b) (c)

Figure 2.34. Isometric (a), dimetric (b), and trimetric (c) views.

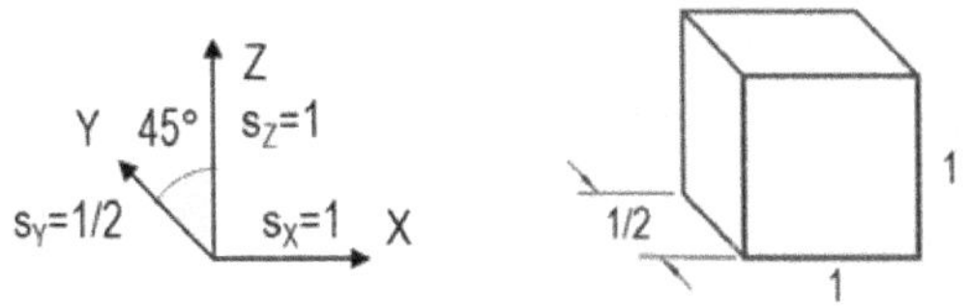

Figure 2.35. Cabinet projection.

Orthogonal axonometry results when the projection plane is perpendicular to the line of sight. This is the case with isometric views. When the projection plane is not perpendicular to the line of sight, the axonometry is referred to as oblique. In oblique projections, dimensions along the receding direction may be intentionally reduced (foreshortened) to create a more visually balanced image. A common foreshortening ratio is 1/2, meaning it is drawn at half its actual length. This specific variation is known as a cabinet projection and illustrated in Figure 2.35.

If no foreshortening is applied to the receding direction (ratio=1), the oblique projection is known as a cavalier projection. For other ratios, such as 3/4, the oblique projection is called a general oblique projection. A notable variant, the military projection, represents the orthogonal axes according to the top view of the object rather than the front (see Figure 2.36).

In principle, REFER accepts only single pictorial, parallel, orthogonal projections. However, some of the algorithms implemented in the system have extended capabilities that allow limited handling of (1) oblique axonometric (see axonometric inflation), and (2) perspective projections.

Requiring parallel projections should not be seen as a significant limitation in REFER. In fact, many reconstruction approaches intentionally impose this constraint to simplify the interpretation and processing of

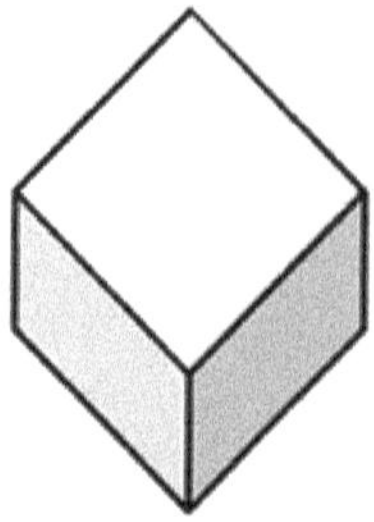

Figure 2.36. Military projection.

projected views. Likewise, the requirement for orthogonal projections is a common restriction adopted across various methods, as it aligns with standard practices in technical drawing and facilitates geometric consistency. However, this restriction is not always obvious to many users or even some developers. Isometric views can often be interpreted reasonably well by certain methods such as the Direct axonometric inflation approach provided by REFER. Nevertheless, many users continue to apply this method to inflate cavalier views, despite not being inherently suited for such projections. The method assumes uniform foreshortening along the receding direction, which does not hold true for cavalier views, where dimensions are typically preserved without reduction.

2.5.3.7 *Perspective pictorial views*

The foundational principles of perspective were first systematically articulated in Albrecht Durer's Four Books on Measurement (1522), laying the groundwork for how three-dimensional forms are represented on two-dimensional surfaces (Wright, 1983; Palmer, 1999). Today, the influence of perspective is widely recognized and thoroughly embedded in both artistic practices and technical fields.

Perspective views are considered the most "natural" because they closely replicate the way humans perceive the world. However, they inherently "distort" geometry. Parallel and proportional invariants, which are preserved in parallel projections, no longer apply under perspective projection. As a consequence, lines that are parallel in the three-dimensional model appear to converge towards a common point in the two-dimensional image. This point is known as a vanishing point. Moreover, objects appear larger the closer they are to the observer, and

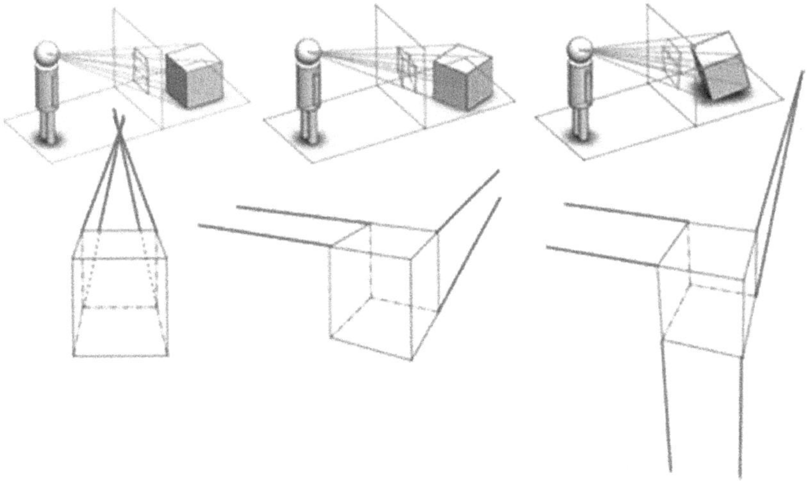

Figure 2.37. Vanishing points in perspective views.

progressively diminish in size with increasing distance, as a result of foreshortening.

There are theoretically as many vanishing points as there are distinct sets of parallel lines in the original model. In practice, we typically focus on three principal directions within a model, which correspond to three main vanishing points in the image. The concept of the vanishing point is intrinsically linked to the horizon line, as parallel lines in space seem to converge toward the horizon when observed in the distance. In a typical setup, the two primary vanishing points (for the horizontal directions) lie on the horizon line, whereas the third vanishing point (for vertical lines) may be located above or below it, depending on the observer's viewpoint. Depending on the orientation of the line of sight relative to the object, three types of perspective views are distinguished, as illustrated in Figure 2.37.

Understanding and detecting vanishing points is a key aspect in the field of geometric reconstruction of three-dimensional shapes from two-dimensional images. Research suggests that designers and engineers often favor axonometric projections over perspective ones, particularly for moderately sized objects. Nonetheless, REFER includes algorithms with limited support for handling perspective projections (Plumed *et al.*, 2013; Company *et al.*, 2014). Additionally, other scholars have developed noteworthy techniques for reconstructing models from perspective views (Yang and Zhang, 2017).

2.6 Input Formats

2.6.1 *Introduction to input formats*

REFER supports three main types of input, which are stored in different formats:

(1) *Sketches*:
Sketches are meant to be drawn directly within REFER. They must go through a vectorization stage during the reconstruction process. Although REFER includes a basic drawing feature (*straight line tool*) its focus is on internally created sketches. REFER is also capable of handling certain types of externally generated sketches.
Sketch files have the extension .stk, which are files created and saved directly within REFER.

(2) *Line drawings*:
Line drawings are typically created using external tools and then imported into REFER. Line drawings can bypass the vectorization stage and proceed directly to the reconstruction stage.
Externally produced line drawings come in two formats:

(a) DXF files: A CAD file format designed to enable interoperability between AutoCAD(R) DWG files, which has become de facto standard for 2D CAD exchange.
(b) Ribald files (with extension *.sketch*), the native file format used in the RIBALD software (Varley *et al.*, 2004).

(3) *Background images*:
Bitmap files (in *bmp* format) can be imported into REFER as background images.

2.6.2 *Sketches*

Sketches are hand-drawn images, typically created without the aid of instruments or tools, that represent drawings made directly by hand. These sketches are distinct in their simplicity, capturing the essence of the model without the precision of tools or digital means. An example of a sketch is shown in Figure 2.38.

Sketches are composed of individual strokes, which are the basic building blocks of the drawing. These strokes usually represent the edges

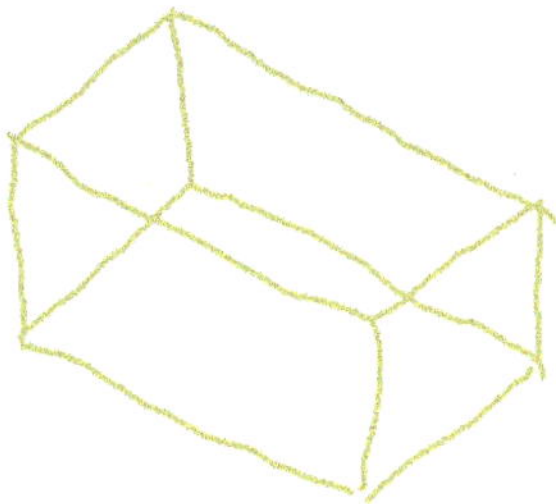

Figure 2.38. Example of a hand-drawn sketch.

Figure 2.39. Representation of a stroke in REFER.

of the object that is being depicted. They can also include gestures or other elements like shadows and textures (some of these features are not currently supported by REFER). Chansri and Koomsap (2012) classified this type of input as "Non-overtraced online freehand sketches," to refer to sketches that are drawn without any overlapping or automated tracing, focusing on the raw, natural representation of the subject.

2.6.3 *Strokes*

Strokes are hand-drawn lines formed by sampling a scribbled path as a pointing device, such as a mouse, pen, or stylus, moves across the screen. The system automatically records a sequence of consecutive nodes between the pen-down and pen-up actions, capturing the stroke's trajectory.

The result is a series of ordered points, which are then connected by straight line segments to approximate the original, freehand scribbled motion. This process effectively converts the continuous and fluid nature of freehand sketches into a discrete set of data points that represent the sketch and enable further processing and analysis. A representation of a stroke in REFER is shown in Figure 2.39.

When the working sketch is being displayed, and the option to display Nodes and Corners of strokes is activated, the endpoints of each stroke are highlighted. The first endpoint of the stroke is shown in green, and the last endpoint, in red. This scheme helps users easily identify the starting and

Figure 2.40. From scribbled line to vectorized line.

ending points of each stroke in the sketch, facilitating more intuitive interaction with the drawing.

These strokes correspond to what Onkar and Sen (2010) describe as "defining strokes." They carry minimal data-primarily the coordinates of sampled points and the time intervals between them. Serving as the basic building blocks of a sketch, these defining strokes represent the unprocessed, raw input generated by the motion of the sketching tool.

Subsequently, a vectorization process is applied to these defining strokes. This process extracts and refines the implicit geometric information conveyed by the scribbled lines, transforming them into higher-level semantic lines that better represent the intended meaning of the sketch. The process is illustrated in Figure 2.40.

2.6.4 *Sketch files*

The preferred method to create sketch files is through the *Save as* feature in the *2D Sketch* window, which saves the sketches drawn within REFER. However, generating new sketch files or examining existing examples can also be accomplished without using the graphical interface. Sketch files are plain text documents that require only the definition of strokes and their associated points.

The information that is saved for every stroke point includes (x-coordinate, y-coordinate, time), where "time" is the time elapsed (in milliseconds) since the previous point was drawn. An example of the structure of a sketch file in REFER is illustrated below:

```
REFER SKETCH
number_of_strokes 1
STROKE 0
number_of_points 8
104.250000 404.750000 47.000000
150.750000 374.750000 63.000000
204.750000 340.250000 78.000000
```

278.250000 302.750000 94.000000
353.250000 265.250000 109.000000
389.250000 248.750000 125.000000
416.250000 236.750000 125.000000
432.750000 229.250000 141.000000

2.6.5 *DXF line drawings*

The standard reconstruction process typically includes vectorizing a 2D sketch to produce a line drawing, which then serves as an intermediate step toward building the 3D model. However, REFER also supports the direct import of line drawings, allowing users to bypass the vectorization phase altogether. Compatible line drawings can be created using any 2D CAD software and are usually imported in the DXF file format. It is important to note that REFER processes only the geometric data contained within a DXF file. Non-geometric information such as attributes, layers, or visibility settings is ignored. As a result, all lines, including those intended to be hidden, are interpreted as visible, effectively converting hybrid drawings into wireframe representations. Additionally, any face information present in the original CAD file will not be imported.

Additionally, DXF files can also be generated as an output from REFER using the *Save As* command. An example of a DFX file created with REFER is shown in Figure 2.41.

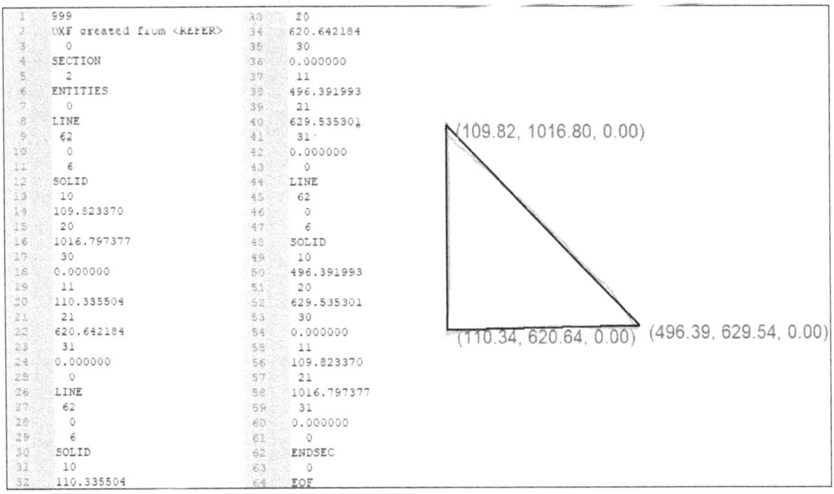

```
999
DXF created from <REFER>
0
SECTION
2
ENTITIES
0
LINE
62
0
6
SOLID
10
109.823370
20
1016.797377
30
0.000000
11
110.335504
21
620.642184
31
0.000000
0
LINE
62
0
6
SOLID
10
110.335504
20
620.642184
30
0.000000
11
496.391993
21
629.535301
31
0.000000
0
LINE
62
0
6
SOLID
10
496.391993
20
629.535301
30
0.000000
11
109.823370
21
1016.797377
31
0.000000
0
ENDSEC
0
EOF
```


Figure 2.41. DXF file generated with REFER.

2.6.6 *RIBALD files*

Ribald files were originally developed to provide a quick and efficient method for generating example models to test the RIBALD system. More information about RIBALD can be found in Peter Varley's Test Sets of Sketches of Polyhedral Solids. This simple, text-based format enabled users to define basic models without the complexity of traditional CAD software. Within REFER, Ribald files continue to serve this purpose, providing an accessible mechanism to create or import line drawing examples for testing and experimentation.

RIBALD files only require the specification of vertices and lines. This streamlined format simplifies the setup of custom models and the adaptation of existing examples for targeted testing.

First, each vertex should be listed with its coordinates, following the expected format. Next, lines are defined by referencing the indices of the previously defined vertices. After all vertices and lines have been specified, the keyword "End" must be included to indicate the end of the file content. Ribald files are saved with a *.sketch* extension.

Vertices are defined using the keyword "Vertex" followed by the vertex number. After a colon (:), the two coordinates (x and y) are provided, separated by a comma. Coordinates are given with respect to a 2D plane, where (0,0) corresponds to the top-left corner.

After specifying the coordinates, the letter "D" is used to indicate the line number associated with the vertex. While this line is often the first (i.e., lowest-numbered) line connected to the vertex, it is not required. It simply designates one of the incident lines.

Lines are defined using the keyword "Line" followed by the line number. After a colon (:), the two endpoints of the line are specified. Each endpoint is indicated by its associated vertex number, preceded by the letter "V." The two vertices are connected by the keyword "to."

For convenience and readability, we recommend starting each new entry (whether a vertex or a line) in a new line. Line breaks are ignored by the parser, and continuous sequences without breaks are also valid. As a simple example, the input needed to define a cube is shown in Figure 2.42.

Additional information must be provided for T-junctions. For example, in the file shown in Figure 2.43, the line "Vertex 12: 252.45068,369.294278 D:14 Intercept 8" means that vertex 12 is a T-junction where line 14 is occluded by line 8.

```
Vertex 0:  0.000000,727.659575 D:0
Vertex 1:  12.765958,224.680852 D:1
Vertex 2:  423.829787,0.000000 D:2
Vertex 3:  834.893617,240.000000 D:3
Vertex 4:  822.127660,737.872340 D:4
Vertex 5:  413.617021,960.000000 D:5
Vertex 6:  418.723404,480.000000 D:6
Line 0:  V0 to V1
Line 1:  V1 to V2
Line 2:  V2 to V3
Line 3:  V3 to V4
Line 4:  V4 to V5
Line 5:  V5 to V0
Line 6:  V1 to V6
Line 7:  V6 to V5
Line 8:  V6 to V3
End
```

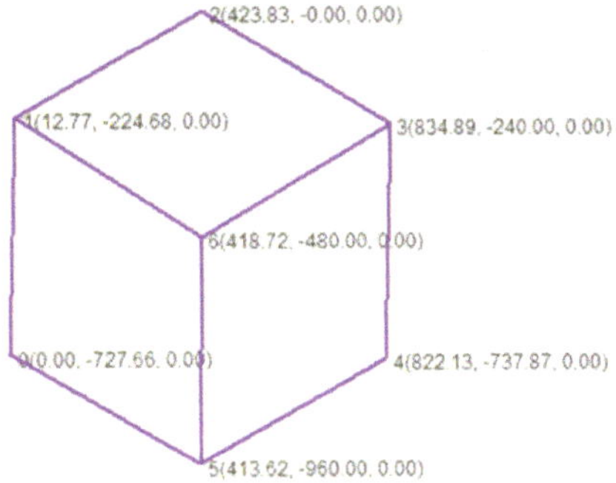

Figure 2.42. Ribald representation (left) of a cube (right).

```
...
Vertex 8: 252.425068,630.408719 D:7
Vertex 9: 252.425068,368.828338 D:8
Vertex 10: 504.850136,436.839237 D:10
Vertex 11: 184.414169,377.983651 D:13
Vertex 12: 252.425068,396.294278 D:14   Intercept 8
...
Line 8: V8 to V9
...
```

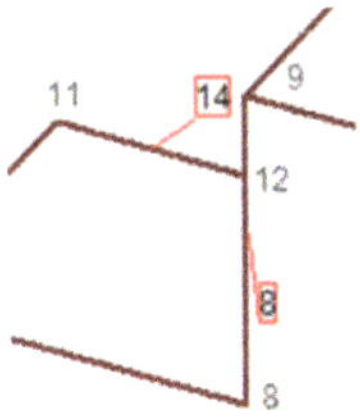

Figure 2.43. Representation of T-junctions in a Ribald file.

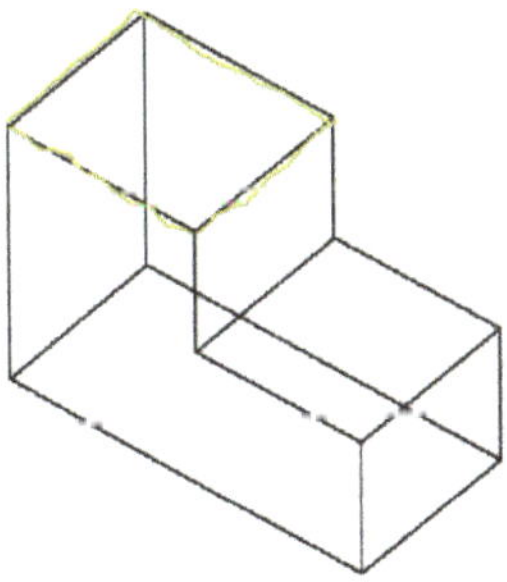

Figure 2.44. Sketch with black lines loaded into REFER as a background image and traced over by the user (in green lines).

2.6.7 *Background images*

Bitmap images can be added as backgrounds in the 2D Sketch window. This feature is particularly useful for manually tracing over the original image to create sketch-based replicas, as shown in Figure 2.44.

This method enables users to generate sketches with controlled levels of deviation. Such control is valuable for systematically testing the performance and robustness of various reconstruction strategies under varying conditions. The general procedure to work with background images is as follows:

(1) Click *Open* to load the desired bitmap (in *bmp* format) as a background image.
(2) Draw the sketch by tracing over the background image, using it as a reference template.
(3) Save the resulting sketch to a new file.
(4) Close the current file to remove the background bitmap when it is no longer required.

Chapter 3

Reconstruction

3.1 Introduction

Two main strategies are employed in Sketch-Based Modeling (SBM) to produce three-dimensional engineering models from two-dimensional sketches: reconstruction and recognition. "Reconstruction is the task of creating a complete description of the 3D geometry of an object based on a 2D representation. Recognition is identifying which class of object an image represents based on shape memory" (Olsen *et al.*, 2009). While recognition is faster and generally more reliable, it requires prior knowledge of object typologies.

REFER does not address recognition approaches. Instead, it focuses on reconstruction, tackling the more challenging task of inferring unknown shapes by replicating human perceptual processes. This line of research can be traced back to the foundational studies by Hochberg and Brooks (1960) and Biederman (1987).

The reconstruction flow implemented in REFER involves multiple stages. Initially, the input sketches must be vectorized and refined. The objective is to fit the input strokes into the most likely types of lines and then to beautify these lines to achieve the most consistent and plausible geometric arrangement. Vectorization and refinement are not isolated tasks; they involve many interdependent processes that must be resolved through concurrent reasoning to achieve satisfactory results.

Once the cleanest possible line drawing is obtained, the focus shifts to perceptual reconstruction. This stage relies heavily on the detection of suitable regularities, cues, and features that guide the inference of the most likely 3D model. The core transformation from 2D to 3D can proceed through either CSG modeling or inflation techniques. In both cases, the resulting 3D model may undergo further refinement to improve accuracy. For models generated via inflation, an additional tessellation stage is required to prepare the geometry for output as a B-Rep model.

A recent state-of-the-art review by Bonnici *et al.* (2019) organizes the machine interpretation of paper-based drawings into three stages: binarization, vectorization, and interpretation (which corresponds to what we call "reconstruction"). The binarization step, however, is not considered in REFER because it is intended for processing scanned hand-drawn images to separate foreground (lines) from background (paper). Since sketches in REFER are interactively drawn in a digital environment, binarization is not necessary. Nevertheless, Bonnici and Camilleri (2015) proposed a vectorization approach capable of operating directly on grey-level images, thus bypassing the need for binarization altogether.

The reconstruction flow typically demands careful fine-tuning of various parameters, achievable using multiple configuration toolbars. Moreover, it may be necessary to execute the reconstruction process multiple times, adjusting settings between successive runs. To facilitate this iterative workflow, REFER provides a *Reconstruction* toolbar, which allows repeated launches of the reconstruction process with updated configurations.

3.2 Reconstruction Flow

Reconstruction is performed automatically in REFER. The reconstruction flow can be triggered using the *Reconstruction* toolbar. Each stage within the flow must be carefully configured. Without appropriate tuning, the reconstructed 3D model may differ significantly from the expected result.

The default configuration is designed to successfully reconstruct the most common geometric shapes. However, it is not optimized to always provide the best solution with minimal effort or execution time. Instead, users are encouraged to explore the various configuration dialogs to adapt the application to the specific characteristics of each case, thus achieving a suitable solution.

The ultimate goal in REFER is to extract from the 2D sketch a feature-rich Constructive Solid Geometry (CSG) tree, which captures and organizes the design and manufacturing features implicit in the original drawing. This process begins with the vectorization of the sketch, continues with refinement and detection of significant geometric features, and culminates in the generation of a coherent procedural 3D model. The reconstruction flow proceeds sequentially—or concurrently, if complexity allows—through the following stages:

(1) *Vectorization*: converting the sketch into a structured line-drawing.
(2) *2D Refinement*: eliminating imperfections in the line-drawing caused by the natural inaccuracies of hand-drawn sketches.
(3) *Detection*: making explicit as much relevant information from the drawing as possible, such as symmetries, parallelisms, and feature axes.
(4) *CSG Modeling*: interpreting the detected features to procedurally generate a 3D model.
(5) *3D Refinement*: polishing the resulting model to improve precision, clarity, and conciseness.
This flow results in the generation of procedural 3D models, which represents a novel output in the field of SBM. Traditionally, SBM systems have produced B-Rep models as their standard output. To generate a B-Rep model instead of a procedural one, the process modifies stage 4 and incorporates an additional stage:
(5a) *Inflation*: generating an explicit B-Rep model from the 2D sketch.
(6) *3D Tessellation*: tessellating the B-Rep model to prepare it for output.

Conceptually, information is routed from one stage to the next. However, REFER allows users to configure alternative flows, meaning not all stages must be active at every execution. Users can configure the reconstruction flow based on their objectives. The application typically warns if some required intermediate information is missing. Nonetheless, in certain cases, missing information may cause the application to silently skip a stage, potentially leading to unexpected errors.

Beyond the traditional B-Rep and procedural CSG workflows, hybrid approaches combining stages 4 and 4a offer promising directions. In practice, the two prevailing SBM strategies—inflation (relying on implicit volumetric cues) and perception/interpretation (based on explicit feature detection)—have shown limitations when used independently. A robust reconstruction method must synthesize both approaches: uncovering hidden constraints to inform the inflation process, while concurrently identifying geometric invariants from the drawing and the model itself. Importantly, the balance between these strategies should not be fixed in advance but should instead dynamically adapt to the specific characteristics of each input sketch.

The purely sequential reconstruction flow has limitations, which remain open problems as no optimal, general-purpose solution is currently available. Thus, further variations of the flow should be explored. However, the dependency between stages is high, as later stages often require information generated earlier. For example, cue detection assumes that the sketch has not only been vectorized into lines, but also that vertex merging has already been performed. This dependency hinders the exploration of certain alternative flows—for instance, using the detection of main axes to improve vertex merging strategies for normalon shapes.

Currently, REFER relies on a sequential reconstruction flow. Previous attempts to develop a successful concurrent architecture have proven extremely challenging, mainly due to the high degree of interdependence among tasks. Nevertheless, addressing this complexity remains an important research goal. Historically, some attempts to tackle similar problems—such as sketch recognition—have employed multi-agent architectures (Avola *et al.*, 2011; Fernandez-Pacheco *et al.*, 2012; Conesa and Mula, 2016), which suggest a promising avenue for future development.

3.3 Reconstruction Flow Configuration

The reconstruction flow in REFER can be customized through the toolbars linked to each individual reconstruction stage. These toolbars offer intuitive controls for managing specific steps in the process:

- *2D Vectorize toolbar*: for managing the vectorization of sketches.
- *2D Refine toolbar*: to refine and improve the quality of the 2D line drawing.
- *Detection toolbar*: for detecting relevant features and geometries within the sketch.
- *CSG Modeling toolbar*: for performing Constructive Solid Geometry (CSG) modeling based on detected features.

- *B-Rep Inflation toolbar*: to switch to a boundary representation model generation via inflation.
- *3D Refine toolbar*: for refining the 3D model output to ensure accuracy and precision.
- *3D Tessellate toolbar*: to prepare the 3D model for tessellation and final output.

Alternatively, the reconstruction flow can be configured through the menu bar under the *Reconstruction* option (see Figure 3.1), which provides a more granular approach to configuring the reconstruction stages.

Once the flow configuration is customized, it can be saved and automatically used in subsequent executions of REFER. Settings can be saved through the *General settings* command under *File* in the menu bar.

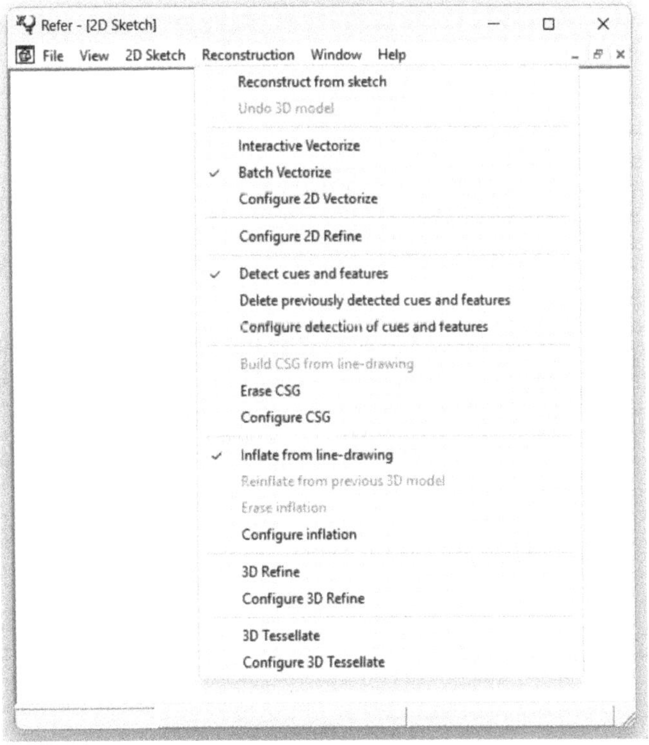

Figure 3.1. Reconstruction option in the main menu bar.

Figure 3.2. *Reconstruction* toolbar.

3.4 Reconstruction Toolbar

The *Reconstruction* toolbar is located over the parent window and enables the configuration and management of the various stages of the reconstruction process across the entire flow. The toolbar includes two commands: one to launch the reconstruction process, and another to reset the generated model, as shown in Figure 3.2. The reset function allows the user to clear the current model, enabling a fresh start for the reconstruction if needed. This option is only available after a model has been successfully reconstructed, ensuring that users are working with a complete model before proceeding.

The current model serves as an input for initiating a new reconstruction process, allowing the user to experiment with different combinations of reconstruction stages. This flexibility enables more in-depth exploration and customization of the reconstruction flow.

The reconstruction flow can be configured to suit specific needs, as described in the previous section. This feature provides the ability to tailor the process to specific project requirements.

After the reconstruction is completed, the Status Bar is updated with relevant information, including details about the file and the reconstruction process itself. This information helps to keep track of the ongoing work and monitor the progress of the models.

3.5 Vectorization

3.5.1 *Introduction*

Representing information on a computer involves optimizing memory usage and simplifying data processing. Since images often require complex calculations, two primary methods can be used to encode and represent images: raster and vector. While both methods are ultimately based on binary code, the manner in which the data is represented and processed in each case differs significantly.

Rasterization is a widely used operation in computer graphics, particularly in 3D rendering techniques. It is a fast and relatively accurate process. Since most graphics hardware relies on raster technology, rasterization is a crucial step for displaying vector images on a computer screen. However, rasterizing a vector image results in a loss of geometric information, such as the shapes' precise definitions, which in turn limits the ability to manipulate these shapes as independent geometric entities.

Vectorization also results in some loss of information. Specifically, the attributes of individual pixels are discarded when they are clustered into a single graphic primitive. The main challenge with vectorization, however, is the loss of high-level geometric information in a raster image, which makes vectorizing a raster image significantly more complex. Although it is feasible to vectorize images with simple geometric elements, producing high-quality vectorized images from raster sources can be problematic, especially for images containing gradients and continuous tones.

The typical reconstruction process generates high-semantic vectorized line-drawings as an intermediate stage between the 2D sketch and the final 3D model. These line drawings can be displayed either online or offline, depending on the computational method employed.

Producing vectorized line drawings with high-semantic geometric elements presents a considerable challenge. The process is not purely geometric but also perceptual. To better simulate the perceptual nature of the vectorization stage, we advocate the use of merits. Moreover, vectorization is inherently complex and often requires multiple stages to be successful. These stages are designed to progressively refine the 2D line drawing, which serves as an intermediary result before outputting a 3D CAD model.

It is commonly accepted that the most effective perceptual fittings depend on context. As such, we can only extract the most accurate line-drawing from a sketch by employing a complex recognition flow. In this process, lines are iteratively fitted according to varying tentative relationships until the most plausible line drawing is identified.

3.5.2 *Raster vs. vector*

A raster image, or bitmap, is represented as a two-dimensional grid composed of individual elements, arranged in rows and columns. Each element in this grid can be independently controlled and assigned a specific value, such as color or brightness. When viewed from a distance, the grid of elements creates the illusion of a continuous image, where the individual elements (or pixels) are not perceived as separate or disconnected. This perception occurs due to the small size of the pixels and the way they are displayed in a dense arrangement. The higher the number of pixels in the grid, the finer the image detail, as more pixels are used to represent the same area, resulting in a denser and more detailed representation of the image.

Each individual element in a raster image is called a pixel, which is short for "picture element." The value of each pixel is typically a color or brightness level. In the simplest case, pixels are binary, i.e., they are either "on" or "off," creating black and white images. Modern raster images often use a variety of colors for each pixel, allowing for a much broader color range. Despite this flexibility, the color of each pixel is uniform; a single pixel can only display one color at a time. Gradients or multiple colors cannot be represented within a single pixel.

Raster images are generated through a process called spatial sampling. In this process, a continuous image is captured by selecting representative points from the scene at different spatial locations. Imagine a 3D object in space, with a picture plane positioned between the object and the observer. The picture plane is structured as a grid of pixels, and as the observer's line of sight moves across the pixels, the light intensity (or color) at each point where it intersects the object is recorded as a sample. This process results in a discrete set of color and intensity values, which collectively form the raster image, as illustrated in Figure 3.3. This method is commonly employed by devices such as scanners and digital cameras to convert real-world scenes into digital raster images.

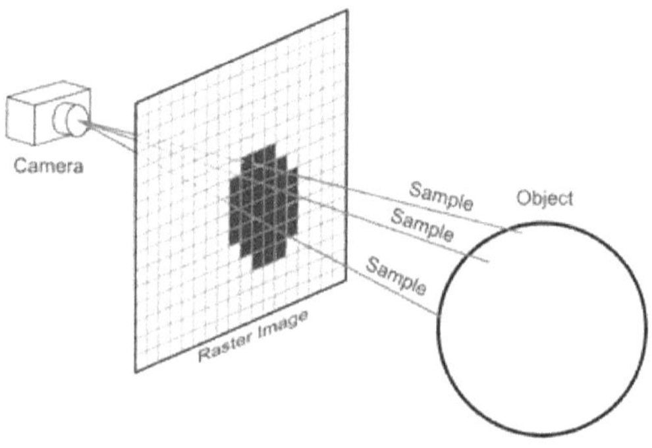

Figure 3.3. Spatial sampling.

A continuous image like a scanned photograph or physical object can also be sampled using this method, converting it into a digital format through discrete sampling points.

The quality of a raster image is determined by its resolution, which refers to the number of pixels per unit of surface area. A higher resolution means more pixels are used to represent the image, resulting in finer details and sharper image quality. Smaller pixels result in greater resolution, allowing for more precise representation of complex visual information.

In contrast, a vector image is represented in terms of geometric shapes, breaking the image down into a series of mathematical primitives, such as points, lines, curves, and polygons. These primitives are described using mathematical equations or formulas that define their positions, sizes, and relationships. Unlike raster images, vector images do not rely on a grid of pixels. Instead, they are stored as a sequence of instructions that describe how to construct the image, as illustrated in Figure 3.4. This method of representation means that vector images do not have an intrinsic resolution, which allows them to be scaled infinitely without loss of quality. Whether enlarged or reduced, vector images maintain their sharpness and precision.

The main advantage of vector graphics is that they allow for exact analytical operations on the image's elements. For example, it is possible to calculate the lengths, areas, angles, and intersections of the shapes within a vector image with mathematical precision. This characteristic makes vector images particularly well-suited for applications that require high accuracy, such as Computer-Aided Design (CAD).

Various algorithms have been developed to convert raster images into vector images and vice versa. The process of converting a raster image into a vector image is called vectorization, while converting a

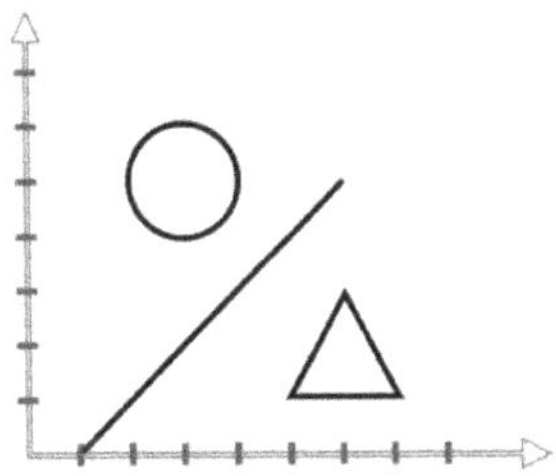

(a)

LINE	1.0	0.0	6.0	5.0		
CIRCLE	3.0	5.0	1.0			
POLYGON	5.0	1.0	7.0	1.0	6.0	3.0

(b)

Figure 3.4. Representation of a vector image.

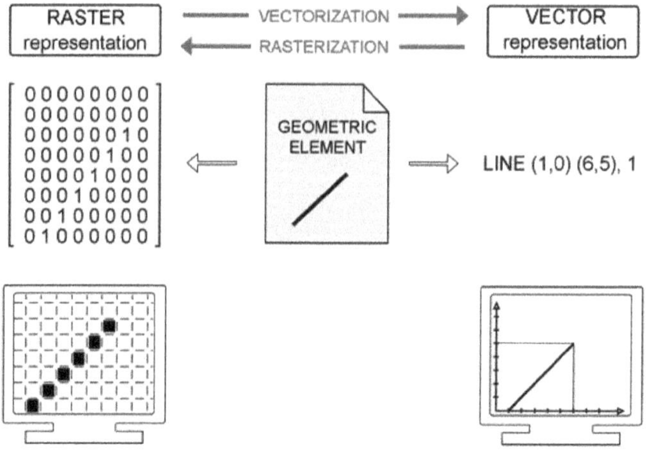

Figure 3.5. Rasterization vs. Vectorization.

vector image into a raster image is referred to as rasterization, as illustrated in Figure 3.5.

3.5.3 *Vectorization merits*

REFER's SBM approach is grounded in perceptual principles, specifically following the "What You Perceive Is What You Get" (WYPIWYG) concept introduced by Saund and Moran (1994). In this paradigm, the goal is to align the system's interpretation of sketches with human perceptual expectations, rather than strictly enforcing geometric accuracy. A related direction based on fuzzy logic was explored by Chen and Xie (1996) and Qin *et al.* (2000), among others, which emphasize the handling of ambiguity and uncertainty inherent in hand-drawn sketches.

We endorse the perspective first proposed by Varley (2002), and later articulated by Plumed *et al.* (2015), who argued that "humans produce sketches which are inherently imperfect from the point of view of geometry, but human perception includes the ability to filter out imperfections and detect the underlying geometry. Current recognition approaches are biased towards geometrical accuracy or ultra-accuracy, and pay insufficient attention to human perception. We argue that algorithms should accept what humans accept, should reject what humans reject, and should doubt where humans doubt." This principle shapes the core philosophy of

REFER, which aims to develop algorithms that mirror human perceptual judgments rather than purely mathematical or geometric criteria.

Since the goal is to "artificially perceive" the geometric information conveyed in sketches of engineering parts, the perceived goodness of each proposed fitting is systematically quantified. That is, REFER seeks to measure how likely it is that a particular geometric fitting would be perceived by a human observer as suitable or acceptable. To this end, the concept of figures of merit is leveraged, which assigns numerical values to each fitting between the range of 0.0 (clearly unacceptable) to 1.0 (clearly acceptable).

In fact, REFER's approach extends this principle further. During the vectorization stage, figures of merit are not simply used to evaluate the quality of a fitting once it has been made; rather, they guide the decision about which fitting to attempt in the first place. Specifically, the merit value quantifies how likely it is that a particular sketched stroke represents a given high-level geometric primitive (such as a straight line or an ellipse). In practice, REFER first assesses multiple potential fittings by calculating their merits, and then selects the primitive type that has the highest perceptual plausibility. Importantly, the primitive may be fitted using a more context-sensitive technique later on, and not necessarily the one initially used to compute the merit.

Unlike systems that propose multiple alternative interpretations and require user selection, REFER's approach seeks to autonomously determine whether a stroke plausibly depicts any instance of a given primitive type. The precise parameters of the fitted primitive are secondary at this stage; they are refined later during the refinement process, which can take broader contextual information into account. In this way, the system prioritizes perceptual plausibility over immediate geometric precision, with the understanding that beautification and adjustment will occur during subsequent stages.

The detailed procedures for quantifying the merits of different types of fittings are separately described for fitting straight segments and fitting elliptical curves. Each case requires tailored criteria that reflect both geometric properties and perceptual expectations.

Moreover, the principle of quantifying perceptual merit is applied not only during initial vectorization, but also in later stages of the reconstruction process, such as during 2D vertex merging refinements. This consistent application of merit-based evaluation across different stages helps ensure that the resulting models maintain a perceptual coherence aligned with human interpretations of the original sketches.

3.5.4 *Vectorization stages*

Converting raster images into vectors often leads to significant gains in both speed and precision when manipulating images. These improvements can also be framed in terms of increased power. A vector-based application is considered more powerful when it provides users with a rich set of geometric primitives. When the appropriate primitive is available, complex shapes can be created directly without the need to manually trace a series of simpler elements to approximate the same result. In this sense, the value of vectorization generally lies less in the visual fidelity of the resulting image and more in enabling efficient and precise image processing operations. For example, in topography and Geographic Information Systems (GIS) applications, aerial photographs and satellite images are automatically vectorized to produce maps. Similarly, SBM tools incorporate vectorization to interpret freehand sketches as a necessary step toward creating 3D models.

However, there is no universal vectorization technique suitable for all purposes. Different methods cater to different needs, and the choice of technique heavily depends on the intended application. When the input drawing is created on physical paper, producing a digital version usually involves two stages: first, the drawing is sampled by scanning it into a raster image; second, the raster image must be vectorized to extract low semantic-level geometrical elements, such as simple strokes (Wenyin and Dori, 1999). The process is illustrated in Figure 3.6.

The main challenge at this stage is that the output consists of low semantic-level primitives. As Fauvreau *et al.* (2016) observed: "existing algorithms tend to produce overly complex drawings composed of many short curves and control points, especially in the presence of thick or sketchy lines that yield spurious curves at junctions." In response, they

Figure 3.6. Stages in the vectorization process of a low-resolution ICO image.

proposed an algorithm that "explicitly balances fidelity to the input bitmap with simplicity of the output, as measured by the number of curves and their degree."

A more recent contribution by Zhang *et al.* (2022) claimed significant improvements for line drawings composed of thick lines. Their method first classifies each pixel as either a line pixel or a background pixel. Then, they extract the skeleton of the line pixels using a classic thinning algorithm, followed by a refinement step that adjusts the skeleton based on boundary information to better match the human-perceived centerlines and overall topology. This approach notably improves the robustness of vectorization at junctions and in areas with thick lines.

Currently, this initial vectorization stage is almost entirely bypassed in SBM workflows. Its primary role today is in fields such as the archaeological recovery of technical knowledge from historical blueprints. It remains active primarily in architecture (Xuetao *et al.*, 2009; Egiazarian *et al.*, 2020), where extracting full geometric information from legacy drawings is still a major goal. Nevertheless, the sub-stage of converting raw sketched input into digital strokes remains a non-trivial problem (Hilaire and Tombre, 2006). Various studies have addressed different facets of this challenge (Dori and Liu, 1999), and one particularly successful low-semantic-level vectorization algorithm was proposed by Bartolo *et al.* (2007, 2008) and Bonnici and Camilleri (2023). For most SBM applications, the primary challenge is not simply converting raw input into strokes, but rather converting strokes directly into lines, i.e., bypassing low semantic-level geometrical elements to recover high semantic-level primitives, shown in Figure 3.7.

This challenge is typically addressed through different fitting approaches. Several practical solutions are available as public libraries

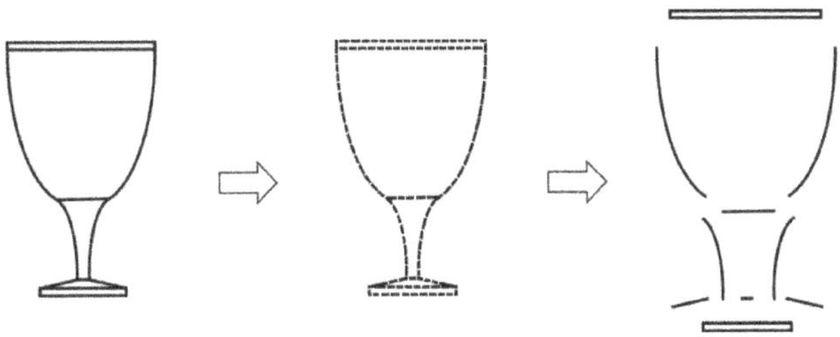

(a) (b) (c)

Figure 3.7. From raw input (a) to strokes (b) to lines (c).

such as CALI (Jorge and Fonseca, 1999). Nevertheless, the problem remains open and inherently non-trivial due to the ambiguity and variability of hand-drawn inputs.

To manage the complex, interrelated issues involved, REFER structures its high-level vectorization process with the same core aim of balancing fidelity to the input with simplicity of the output. It is achieved by systematically addressing four critical aspects:

(1) Vectorization follows one of two distinct schedules based on timing: interactive (also known as "online") vectorization, and batch (or "offline") vectorization.
(2) Basic cues are extracted by analyzing various characteristics of individual strokes and their mutual relationships, helping to guide the vectorization process.
(3) REFER applies two distinct fitting strategies to recognize whether strokes correspond to straight or elliptical lines. Both strategies compute merits that quantify the quality of each fitting.
(4) Finally, the vectorization process is designed to interact with sketch refinement procedures. In the long term, REFER aims for fully concurrent vectorization and refinement. Current implementations approximate this through simpler, but effective, interactions between the two stages.

The user can configure the vectorization schedule, select the preferred fitting approach, and control the interaction with sketch refinements using the available options in the *2D Vectorization Configuration* dialog box, which can be accessed through the *2D Vectorization* toolbar.

3.5.5 *2D vectorization toolbar and configuration dialog*

The *2D vectorization* toolbar is located over the parent window and allows the configuration and management of the vectorization process (see Figure 3.8. Two commands control the vectorization schedule, and one command opens the dialog that configures the schedule, the fitting approach, and other sketch refinements.

The *2D vectorization configuration* command opens a dialog box with various settings to configure the vectorization process (see Figure 3.9).

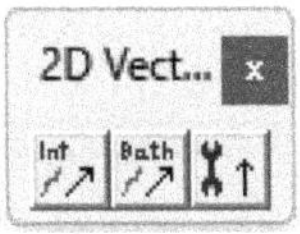

Figure 3.8. 2D Vectorization toolbar.

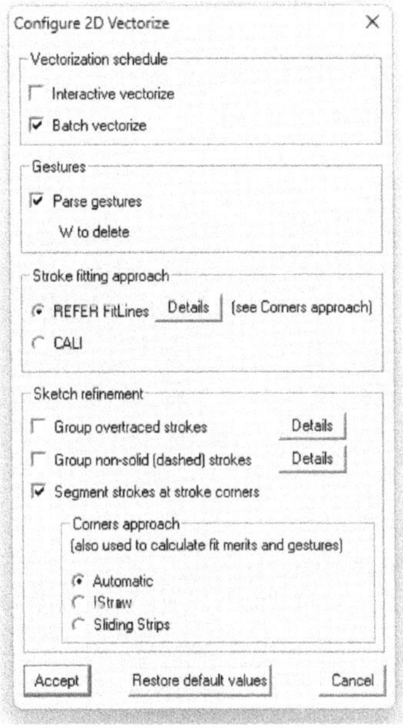

Figure 3.9. *2D vectorization configuration* dialog box.

The following aspects of the vectorization process can be configured:

- Vectorization schedules. Two options are available for processing sketches.
- Detection of gestures, which serve as an alternative interaction method to traditional command menus during sketching.
- Stroke fitting approach, which determines how strokes are interpreted as high-level primitives.
- Refinement operations applied to the sketch, focusing on grouping and segmentation of strokes.

Additional configuration options are available for specific processes through the corresponding "Details" buttons.

3.5.6 *Vectorization schedules*

REFER implements two different "schedules" to organize over time the sequence of actions required for sketch vectorization:

(1) Interactive (or online) vectorization.
(2) Batch (or offline) vectorization.

It is widely accepted that the online schedule is primarily intended for interactive reconstruction, where immediate feedback and continuous sketch interpretation are important. The offline schedule is more suitable for batch reconstruction workflows, where processing takes place once the sketching phase has been completed (Company *et al.*, 2009).

By default, REFER uses an offline vectorization schedule. This schedule includes both tasks that can be performed eagerly during sketching and tasks that are best deferred until the sketch is considered complete. However, consistent with REFER's philosophy as a testbed of ideas, users are welcome to experiment with either schedule.

REFER also implements the concept of concurrent vectorization, which is considered a promising direction for the future organization of the vectorization process. Concurrent vectorization addresses the intricate, interdependent relationships among vectorization tasks by allowing them to operate simultaneously rather than sequentially. Readers interested in deeper insights into concurrent approaches are encouraged to consult the work of Garcia Fernandez-Pacheco *et al.* (2012), who proposed a multi-agent system architecture to achieve concurrency in vectorization processes.

3.5.6.1 *Interactive vectorization*

In the interactive (or "online") vectorization schedule, REFER re-vectorizes the current sketch immediately after each new stroke is introduced. This mode is intended for exploring the feasibility of progressive online reconstruction, a technique designed to allow users to view a continuously evolving 3D model as they complete the sketch (Naya *et al.*, 2007). It is useful for this purpose provided that subsequent stages of the reconstruction process can also be triggered interactively and repeatedly.

This schedule, however, is somewhat controversial. Updating the line-drawing while the user is still sketching can disrupt the natural creative flow, diverting attention away from design thinking. Users may become overly focused on how the system interprets their input—worrying about minor or temporary misinterpretations by the vectorization engine—rather than concentrating on the intended form and function their sketch aims to convey.

An additional limitation of the interactive schedule is that vectorization and refinement processes dependent on broader context can hardly be performed reliably before the sketch is complete. For example, modifying lines to optimize junctions before all connected strokes have been drawn can introduce misalignments or inconsistencies in the final line-drawing. Premature decisions based on incomplete information often degrade the overall quality of reconstruction.

The result of the interactive vectorization process is displayed in the *2D Sketch* window as a line drawing rendered in green lines. This visualization can be toggled on or off using the *Display vectorized online Line Drawing* command available in the *Display 2D Sketch* toolbar.

When the *Interactive vectorization* command is deactivated, the application deletes the existing online vectorized line drawing. If reactivated, REFER recalculates and displays the current online vectorization based on the available strokes.

3.5.6.2 *Batch vectorization*

In the batch (or offline) vectorization schedule, REFER performs vectorization of the current sketch only during the designated stage of the reconstruction flow. This stage is initiated after the sketch is fully completed when the reconstruction command is executed.

The result of the batch vectorization process is displayed in the *2D Sketch* window as a line drawing rendered in black lines. This visualization can be toggled on or off using the *Display vectorized offline Line Drawing* command, available in the *Display 2D Sketch* toolbar.

The application deletes the existing offline vectorized line drawing when the batch vectorization command is deactivated. If reactivated, REFER recalculates and displays the updated offline vectorization based on the current sketch.

3.5.6.3 Concurrent vectorization

We advocate for a future vectorization approach that we call *Concurrent Vectorization* in which different vectorization sub-stages run in parallel while sharing common information. These sub-stages are not necessarily executed simultaneously. If true parallel computation is unavailable, they can proceed sequentially, where each sub-stage runs until it requires information from another, at which point it pauses until the necessary information becomes available.

The advantages of a concurrent vectorization approach can be illustrated through the segmentation problem. How should sequences of points be divided into strokes so that each stroke represents a single straight line or elliptical arc? It is not even clear when segmentation should occur. Fitting lines before segmentation can produce macro lines, where several distinct lines or curves are grouped together (as shown in Figure 3.10(b)). Conversely, segmenting strokes each time a candidate corner is detected can result in micro lines, as minor undulations or oscillations are misinterpreted as true corners (Figure 3.10(c)).

Humans tend to interpret scenes holistically, gaining general understanding before making detailed interpretations. For example, the strokes highlighted in red in Figure 3.11(a) are the exact same strokes shown in Figure 3.10. When analyzed without any context, the strokes are ambiguous when viewed independently. However, once the object is perceived as an extrusion (Figure 3.11(b)), it becomes clear that these strokes together represent an elliptical arc.

A similar example was proposed by Perkins (1968) to demonstrate that even "geometric" cues such as cubic corners are context dependent. The same corner, with identical angles, is clearly perceived as cubic in one context, but generally perceived as acute in another. The example is illustrated in Figure 3.12.

(a) (b) (c)

Figure 3.10. Sequence of points (a), macro lines (b), and micro lines (c).

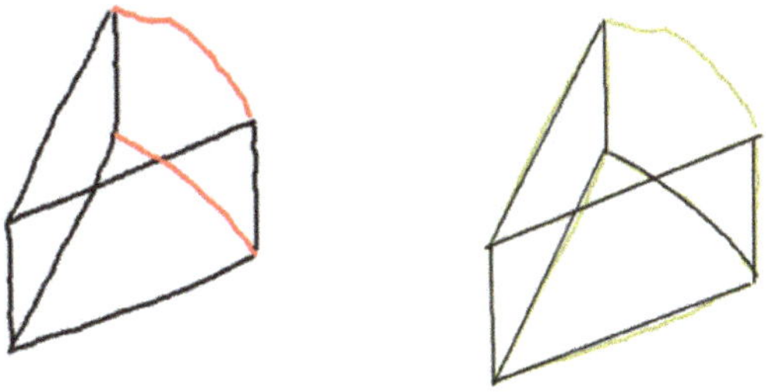

(a) (b)

Figure 3.11. The strokes highlighted in red (a) are ambiguous unless the object is perceived as a whole (b).

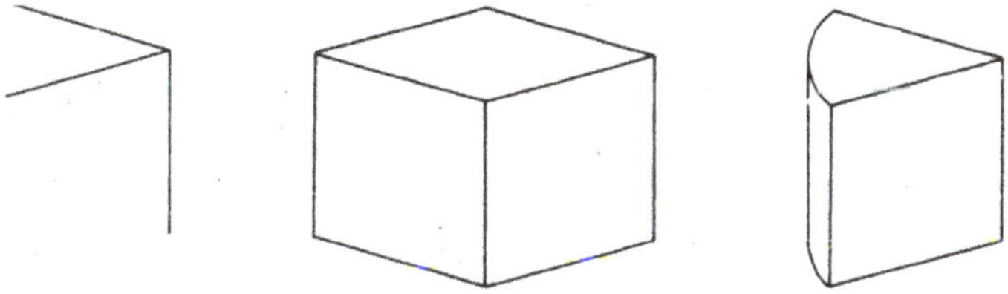

(a) (b) (c)

Figure 3.12. Perkins' example (1968). The same cue (a) can be interpreted as cubic (b) or acute (c) depending on the context.

3.5.7 *Vectorization cues*

3.5.7.1 *Introduction*

Effectively assigning meaning to a sketch is complex, as multiple interpretations are plausible. We can roughly emulate the human perceptual system by assuming that reasoning occurs based on rules. In particular, we accept that human beings assign meaning to the lines of a drawing depending on perceived regularities or characteristics. Consistent with this vision, REFER aims to artificially perceive cues that may help unambiguously interpret the lines implicit in the strokes of a sketch.

REFER currently detects one primary cue to aid in the process of fitting lines to strokes: points where strokes abruptly change direction, referred to as corners. By detecting these corners, a single stroke can be broken down into independent strokes that depict the consecutive lines of a polyline.

It is recommended to detect key cues and refine strokes prior to line fitting. While this approach deviates from the standard guideline–2D refinements are generally best applied after vectorization to preserve the original design intent–it can help improve accuracy by addressing noise and ambiguity early in the process. However, certain imperfections in strokes have no relation to design intent yet can significantly disrupt the vectorization process. As such, they are best addressed beforehand. In fact, operations like detecting stroke partitions and refining stroke tips fall under the category of sketch refinement, rather than drawing refinement.

3.5.7.2 *Stroke corners*

Attneave (1954) was the first to emphasize the significance of points where lines exhibit the most abrupt changes in direction. These dominant points, characterized by high curvature, were later termed "corners" (Rattarangsi and Chin, 1992). Corners are calculated and saved as cues to assist in the interpretation of each stroke.

Three different approaches are available in REFER for detecting corners, which can be selected in the *2D Vectorization Configuration* dialog box:

(1) Automatic, which uses IStraw (approach 2) when time sequence information is available, and switches to Sliding Strips (approach 3) when time data is unavailable.
(2) IStraw, proposed by Xiong and LaViola (2010), which relies on the time between consecutive points (but falls back to ShortStraw, also by the same authors, if time information is missing).
(3) Sliding Strips (Masood and Sarfraz, 2007; Sarfraz *et al.*, 2009; Swati *et al.*, 2009), which operates without requiring time sequence information.

The recommended approach is (1) automatic, as it dynamically selects the most suitable method based on the availability of time data.

In all cases, corner detection operates in a resampled space, since strokes are resampled to maintain a constant inter-point distance (Wolin *et al.*, 2008). After detection, corners are mapped back into the original stroke space. The output consists of a list of indices corresponding to the points in the stroke closest to each detected corner.

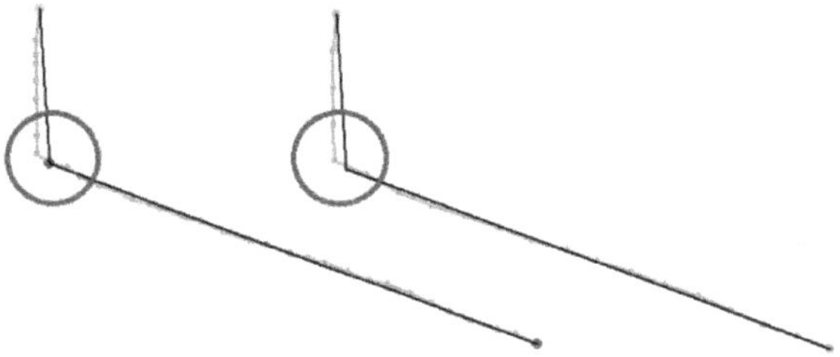

Figure 3.13. Unexpected result in the mapping process: the resampled node identified as a corner (right) lies between the original node perceived by humans as the corner and its immediate neighbor (left).

It is important to note that the mapping process can occasionally produce unexpected results, particularly when the resampled node identified as a corner (Figure 3.13-right) lies between the original node perceived by humans as the corner and its immediate neighbor (Figure 3.13-left).

3.5.8 *Vectorization fitting*

3.5.8.1 *Fitting approaches*

The task of extracting high-level semantic geometric primitives from sketches remains an open problem, even when the sketches are not densely packed (i.e., consisting of a discrete set of strokes) and are limited to basic line-drawings without additional annotations.

The practical process of finding these high-level semantic primitives involves fitting them to the sequences of points that define the strokes in the original sketch. These primitives may include straight lines or segments, circles, circular arcs, ellipses, and more (Wang *et al.*, 2013). Notably, a classical approach (Saga and Makino, 1993) uses fuzzy logic to address the inherent imprecisions in sketches while identifying seven types of geometric primitives: lines, circles, circular arcs, ellipses, elliptic arcs, closed free curves, and open free curves. However, this method treats positional ambiguity solely as a result of stroke speed, disregarding contextual factors that could indicate whether the user was drawing with care and precision or more freely.

REFER specifically requires fitting straight segments, as they represent the edges of polyhedral shapes, as well as elliptic arcs, which encompass circular arcs as a special case and are commonly used to represent contours of revolution shapes in pictorial views.

For readers interested in fitting splines or similar curves to sketched strokes, we recommend a recent approach to stroke classification (Gryaditskaya *et al.*, 2020). This method is compelling not only because it fits complex curves, but also because it facilitates the corresponding three-dimensional reconstructions. Notably, concurrent fitting and inflating these complex contours is enabled by the approach's prior detection and inflation of a polyhedral scaffold, which the authors currently restrict to the well-known problem of normalons.

In REFER, two distinct approaches are available for fitting primitives to sketches, which enable the detection of high-level semantic primitives within these representations:

(1) CALI Library (Jorge and Fonseca, 1999).
(2) *FitLines* Approach. This approach includes a procedure for perceptually based fitting of strokes into straight segments (Plumed *et al.*, 2015), as well as a procedure for fitting elliptic arcs (Company *et al.*, 2015b). Additionally, this approach calculates merits for each fitting.

Users can switch between these two approaches via the *2D Vectorization Configuration* dialog box. However, the implementation of CALI in REFER is limited to fitting straight segments, and it does not offer parameter tuning options, as it deterministically classifies whether a stroke represents a straight line. Conversely, the *FitLines* approach can detect both straight segments and elliptic arcs and provides configuration options for fine-tuning the fitting process via the fitting configuration dialog. *FitLines* introduces an innovative classification approach that categorizes well-formed strokes as either straight or elliptic, while labeling low-quality strokes as neither straight nor elliptic. For ambiguous cases, it assigns a merit score to quantify the level of uncertainty.

The argument supporting this approach is based on the fact that human sketches are inherently imprecise from a geometric perspective. However, human perception possesses the ability to filter out these imperfections and interpret the underlying geometry. Current recognition methods tend to focus on geometric or ultra-precise accuracy, often neglecting the perceptual aspects of sketching. We assert that algorithms should mimic human perception–accepting what humans accept, rejecting what humans reject, and remaining uncertain when humans doubt.

In our view, the fitting methods should prioritize speed, not only to provide real-time feedback, but also because higher-level interpretation processes may involve repeated use of the same methods, making excessive computation times disruptive to the user experience. While perceptual accuracy should take precedence over exact geometric precision, a minimum level of precision remains essential to produce reliable and meaningful outcomes. We also note that the approach by Tanaka *et al.* (2017), which has not yet been incorporated into REFER, is worth exploring, along with the recent contribution by Chatterjee and Bhowmick (2025).

3.5.8.2 *Fast vs. accurate fitting*

REFER's goal is to develop an approach where strokes are fitted to the most plausible types of lines, followed by a beautification process to arrange them in the most likely configuration. However, strokes representing polylines typically require segmentation, and lines fragmented into multiple strokes (such as dashed lines) need to be grouped together.

These constraints suggest that fitting is just one aspect of the broader and more complex problem of vectorization. In our view, fitting strokes separately from solving overtracing (Pusch *et al.*, 2007; Wang *et al.*, 2014) and grouping strokes (Xiong and LaViola, 2010) does not yield satisfactory results. Excessive grouping of overtraced strokes or fitting lines before segmentation leads to the creation of macrolines, where several actual lines or curves are incorrectly grouped together. Conversely, segmenting strokes each time a candidate corner is detected leads to microlines, as the segmenter mistakenly interprets oscillations as corners. To avoid these pitfalls, these processes should operate in parallel, passing information between each other until they converge on a solution. For this reason, we advocate for concurrent vectorization in the long term. In the current version of REFER, vectorization schedules introduce a form of primitive parallelization.

A crucial component of this approach is the need for fast fitting algorithms. Many current fitting techniques focus predominantly on geometric accuracy or even ultra-accuracy. However, recognition strategies in SBM must equally prioritize perceptual criteria. As Saund and Moran (1994) emphasized, "what you perceive is what you get" should be the goal, rather than "what you measure is what you get." This view is also central to the work of Fauvreau *et al.* (2016).

3.5.8.3 *Fitting details configuration*

REFER offers a range of tuning parameters via the *Details* button in the *2D Vectorization Configuration* dialog box, as shown in Figure 3.14. These parameters are tailored to the specific type of geometric primitive being fitted, allowing fine control over the vectorization process.

A detailed explanation of the principles underlying REFER's vectorization method—along with its tuning parameters—is provided in (Plumed *et al.*, 2015) and (Company *et al.*, 2015b). These sources emphasize the importance of understanding human perception thresholds, based on experimental data collected from representative user groups. Rather than aiming

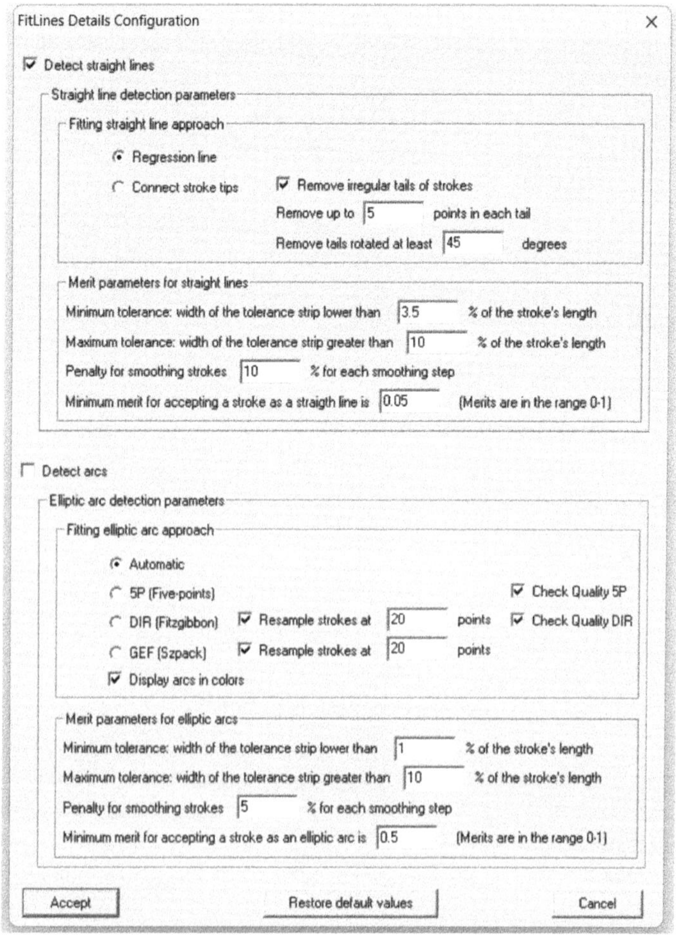

Figure 3.14. *FitLines Details Configuration* dialog box.

for ultra-precise geometric accuracy, REFER's algorithm is designed to align with human judgment– accepting what users typically accept, rejecting what they reject, and expressing uncertainty where users are unsure. The algorithm's default settings were calibrated to reflect this perceptually driven approach. For convenience, the main features of the straight line and elliptic arc fitting mechanisms used by the application are described next.

3.5.8.4 *Fitting straight line segments*

REFER implements two alternative approaches for fitting strokes into straight line segments. The first approach calculates the regression line determined by all the points in the stroke and then limits the line by the points closest to the minimum oriented bounding box of the stroke. This typical "line of best fit" algorithm considers the need to swap coordinates when the slope is greater than 1 to avoid the common issue of regression failure with vertical lines. An example is shown in Figure 3.15.

The approach provides the best local fitting but does not account for the global relationships between neighboring strokes, which are intended to depict shared vertices through junctions. Additionally, it tends to be more time-consuming than the second approach described below. Nevertheless, it is used in REFER because an accurate fit can sometimes make a significant difference, particularly in tasks such as grouping non-solid strokes.

The second approach for fitting straight segments in REFER is based on the perceptual assumption that straight lines typically depict edges that share common vertices with adjacent edges. Therefore, lines should run between these vertices, as shown in Figure 3.16.

The rationale is that the endpoints of strokes are the best estimates of the intended location of the vertices of the lines they represent, regardless of whether the stroke was accurately drawn. In other words, edges

Figure 3.15. Fitting straight line segments (Approach 1).

Figure 3.16. Fitting straight line segments (Approach 2).

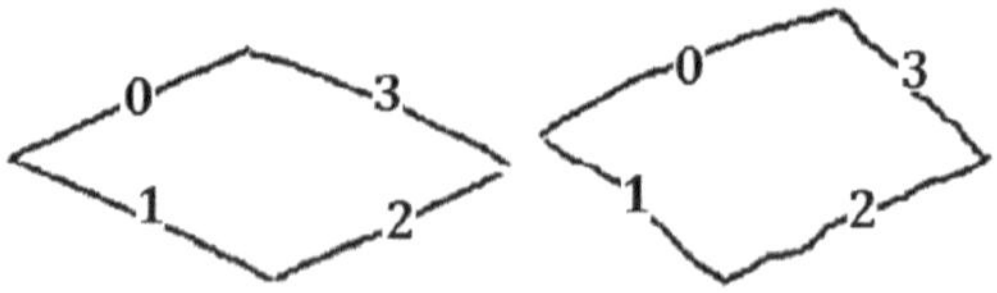

Figure 3.17. Perceptual relationships between neighboring strokes in a high quality (left) and low quality (right) sketch.

connect vertices, and the length and slope of these edges are secondary to this fundamental relationship. This concept aligns with the well-established observation that endpoint coincidence is the simplest and most common way for strokes to interact with other neighboring strokes (Igarashi *et al.*, 1997).

Other geometric constraints, such as horizontal or vertical alignment, parallelism, or perpendicularity, cannot practically be introduced reliably at this stage, as they depend on context that has yet to be defined. Even though isolated strokes are analyzed at this vectorization stage, these strokes are still part of a broader sketch. As a result, in the full recognition process, perceptual relationships with neighboring strokes will take precedence over pure geometrical straightness, position, or orientation. For example, in Figure 3.17, humans tend to perceive Stroke 0 as connected to Strokes 1 and 3, and parallel to Stroke 2, regardless of the position and orientation of any geometric fit, and even despite the quality of the sketch (high on the left, low on the right).

However, stroke terminations (referred to as tips) may be imprecise, as users may inadvertently draw unwanted "tails." By detecting and removing these tails, incorrect line fittings can be prevented based on inaccurate tip locations. The approach implemented in REFER for removing these tails is designed to be efficient in terms of time. However, it struggles with the complex and varied nature of undesired tails. As a result, this fitting approach is not recommended for use when dealing with such cases.

Once the fitting process is complete, the quality of the fit needs to be assessed. The main stages of the algorithm for calculating the fit's merit are described as follows:

(1) Define the segment by fitting to the regression line (although the straight line connecting both endpoints of the stroke is also available).
(2) Calculate the slope of the line and rotate the stroke to a horizontal position.

(3) Calculate the tolerance for the stroke.
(4) If the tolerance is below the minimum threshold, accept the stroke as a straight line with a merit figure of 1.0 (indicating certainty), and end the algorithm.
(5) Smooth out the stroke to prevent oscillations from affecting its evaluation as a straight line. Store the number of smoothing stages (NSS). For more details on the smoothing process, refer to the fitting straight segments merits section.
(6) Recalculate the tolerance for the smoothed stroke.
(7) Assign a merit figure of 1 for a tolerance below TolMin, 0 for a tolerance above TolMax, and a linearly decreasing value between 1 and 0 within this range to distinguish between good, average, and poor strokes.
(8) Reduce the merit figure for oscillating strokes based on a devaluation parameter (Penalty) and the number of smoothing stages (NSS): Merit = Merit – Penalty * NSS. Negative merits are not allowed.

The recommended values for the tuning parameters are provided in the *Fitting Details Configuration* dialog box.

3.5.8.5 *Fitting stroke tips*

In REFER, the endpoints of a stroke are referred to as tips. They can be easily determined as the first and last points of a stroke, since strokes are saved as ordered lists of consecutive points. Certain forms of stroke beautification have been shown to enhance the accuracy of tip detection, particularly when strokes are drawn imprecisely or with minimal care.

Unintentional hand movements during pen-down or pen-up actions can cause some users to produce unwanted "tails" at the end of strokes, as shown in Figure 3.18.

These tails can significantly affect the length, location, or orientation of lines fitted by certain fitting methods, particularly if the lines are derived by simply connecting the first and last points of the stroke, as illustrated in Figure 3.19.

To mitigate this issue, REFER incorporates a tip-repair mechanism that heuristically identifies and removes stroke-end segments exhibiting abrupt directional changes–features that typically signal the presence of unintended tails. Two strategies can be used to correct the tips:

Figure 3.18. Tails at the end of strokes.

Figure 3.19. Effect of tails in line fitting.

- If a corner is detected near the tip, the corner will replace the tip with the endpoint of the line.
- If no corner is found but neighboring points define segments that form steep angles with the line joining the vertices, then those neighboring points will replace the tip.

The procedure for correcting the tips is described as follows:

(1) The line segment connecting the first two points of the stroke is compared against the line connecting the second point and the last point of the stroke.
(2) If the angle between the two lines is smaller than a parametrically configured threshold (45 degrees, by default), the first point is accepted as the first endpoint of the vectorized line.
(3) If the angle is too large, the first point is replaced by the second point, and the procedure is repeated by comparing the line connecting points two and three with the line connecting points three and four.
(4) The procedure can be repeated up to a configurable number of points (five, by default).
(5) The same procedure is applied to the other endpoint of the stroke.

This approach is effective for removing tails that are reasonably straight and form a non-obtuse angle with the rest of the stroke, as shown in Figure 3.20.

The effectiveness of this approach, however, is highly dependent on its configuration parameters. In the example shown in Figure 3.20, the left-side tail was only partially removed due to a constraint limiting the

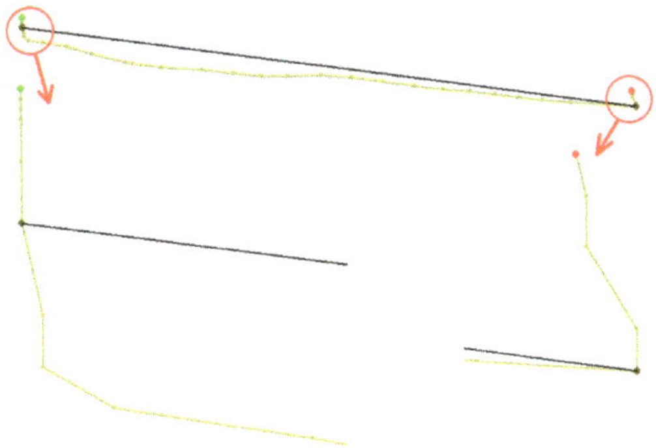

Figure 3.20. Example of removal of tails.

Figure 3.21. Example of tails that overlap the main stroke.

check to a maximum of five points. Moreover, the method faces challenges with irregular tails– particularly those that overlap and recede from the main stroke– making them more difficult to detect and eliminate, as shown in Figure 3.21 (the endpoint of the stroke, shown in red, does not define the end of the segment).

This tail refinement only affects the fitting approach that uses tips and does not modify the strokes themselves. Since these tails are generally regarded as unintended artifacts or failures, one could reasonably argue in favor of tidying up the strokes by detecting and removing them, thereby preventing their adverse impact on the vectorization process. However, beyond the general criticism that beautifying the strokes inevitably alters the original sketch—resulting in the irreversible loss of authentic user input –two further criticisms specifically apply to this type of sketch refinement. First, the heuristic procedures currently available in the literature for detecting and eliminating tails (including the method implemented here) operate without knowledge of the type of line being depicted. Second, these procedures lack awareness of the broader contextual relationships among strokes. As a result, tail detection is inherently limited by its reliance on purely local stroke information, leading to potential errors or inappropriate modifications depending on the true intent of the drawing.

Current procedures disregard the type of line being depicted, as this information is unavailable before the fitting process. However, the thresholds required for detecting tips on a stroke representing a straight segment may differ significantly from those needed for a stroke that represents an arc.

The approach implemented in REFER is specifically recommended for straight segment fittings, as tails are unlikely to affect the fitting of elliptic curves. In other words, the beautification process only occurs if a straight-line fitting is required, and if this particular approach (default is line regression) is selected.

Beautifying the tips of a stroke is, strictly speaking, a refinement of the sketch, not the line drawing. As this is a sketch refinement procedure, it should ideally be performed before line fitting to improve the results of the vectorization methods. Further exploration of concurrent vectorization might offer insights into how to incorporate such refinements in parallel with fitting processes. The key reason is that current methods rely on local information about the strokes, whereas handling all the variations of stroke tails would require broader context, which is not available before vectorization.

The option to activate or deactivate the tips beautification process is available in the *Fitting Details Configuration* dialog, along with a parameter that defines the extent of what is considered a "tail."

3.5.8.6 *Fitting straight line segments review*

Many approaches have been proposed for both stroke classification and the fitting of straight lines, ranging from the simplest—such as comparing chord lengths (Qin *et al.*, 2001)—to the most popular, such as the Hough Transform (Duda and Hart, 1972). Linear regression also plays a significant role, though it is known to produce identical results for visually distinct point distributions, as it considers only the overall cloud of points rather than the sequence in which they are arranged. This effect is famously illustrated by Anscombe's quartet (see Figure 3.22).

REFER includes a stroke fitting approach based on calculating the regression line of their points, generating a straight line that is then trimmed into a segment. The endpoints of the segment are determined as the intersections of the regression line with the stroke's minimum oriented bounding box. This method can be selected through the dialog box for configuring fitting details.

Beyond classical methods, other recognition strategies have been developed for or adapted to SBM contexts. Shpitalni and Lipson (1997),

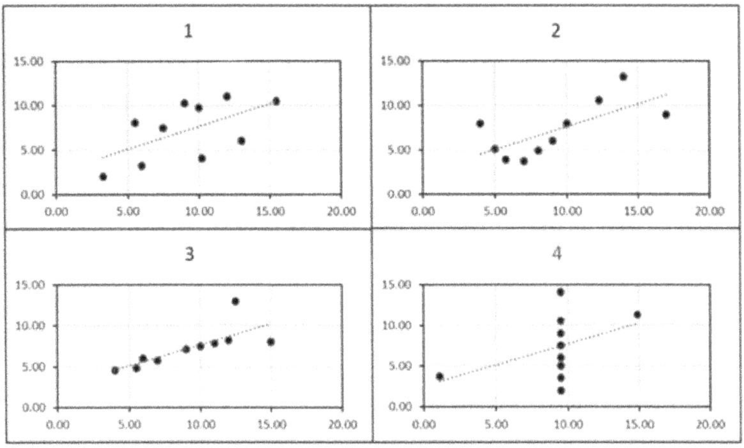

Figure 3.22. Anscombe's quartet.

for example, applied linear least squares fitting to a conic section equation. Qin (2005) proposed a classification method based on adaptive thresholds and fuzzy logic, considering the linearity and convexity of curves. Zhang *et al.* (2006) surveyed earlier techniques and introduced a seeded segment growing algorithm designed to extract graphical primitives from a stroke, refining their control parameters by incorporating relationships between primitives. Their method reportedly achieves high reliability in detecting straight segments. Additionally, Plumed *et al.* (2013b) introduced a variant of the Hough Transform tailored to SBM applications.

Two limitations emerge across these approaches. First, most assume from the outset that a stroke can and should be fit into a straight line. Consequently, the fitting process produces either a deterministic best-fit result or a rejection (failure) if a satisfactory line cannot be found. In contrast, REFER adopts a fundamentally different philosophy by seeking to assess the likelihood that a stroke represents a straight line, rather than forcing a binary (yes/no) classification. This probabilistic assessment supports more flexible, collaborative recognition flows, allowing subsequent stages of recognition to integrate degrees of certainty rather than absolute judgments.

Second, the filters used in SBM approaches are primarily designed to remove sketching noise—noise that is generally larger and more structurally complex than the optical or electronic noise found in camera-based image processing. Importantly, the threshold values applied by these algorithms are often set heuristically, based on the authors' intuition or isolated observations, without systematically correlating with human perception. A representative example is the work by Shpitalni and Lipson (1997), which adopted a minor-to-major axis ratio of 1:20 to distinguish between strokes depicting elliptic arcs and straight lines. Although effective in practice, this threshold was chosen subjectively, not validated against human studies. This case highlights a broader issue: many SBM algorithms are governed by thresholds that remain arbitrary rather than perceptually grounded. REFER's approach asserts that such thresholds must be revisited and systematically aligned with what humans intuitively recognize as lines or curves.

3.5.8.7 *Fitting straight line segments merits*

REFER determines the merits of strokes to guide vectorization. For users interested in controlling the merit calculation process, this section summarizes the procedure for evaluating merits when fitting strokes to straight segments (Plumed *et al.*, 2015).

REFER's approach builds on the concept of Tolerance—a well-established metric in Geometric Dimensioning and Tolerancing (ISO 1101-2017) for assessing the "straightness" of a line. It can be framed as a "minimum oriented bounding box" or "minimum-area enclosing rectangle" problem, for which the rotating caliper algorithm is the best-known solution. However, applying such exact calculations is unnecessarily time-consuming when evaluating strokes that are only approximately straight, and does not significantly outperform simpler methods when the goal is to reject strokes that are clearly poor candidates for straight-line fitting.

As a simpler and faster approach, REFER first computes the fitting line segment—either by linear regression or by connecting the tips of the stroke. This segment provides a reliable estimate of the overall slope for nearly straight strokes without extreme tails. The stroke is then rotated to align the fitting segment horizontally, establishing a local X-axis. A bounding box is computed relative to this orientation, defining an x-range and y-range, as illustrated in Figure 3.23.

The tolerance of straightness is then calculated as the absolute value of the y-range. The smaller this value, the straighter the stroke is considered. The study by Plumed *et al.* (2015) confirmed that this measurement aligns closely with human perception of straightness.

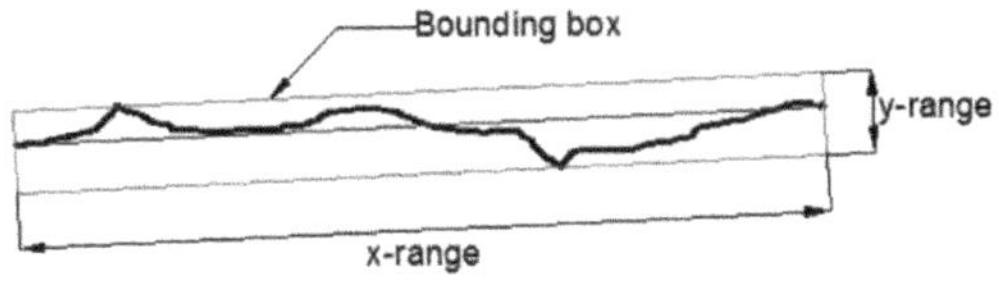

Figure 3.23. Bounding box as computed by REFER.

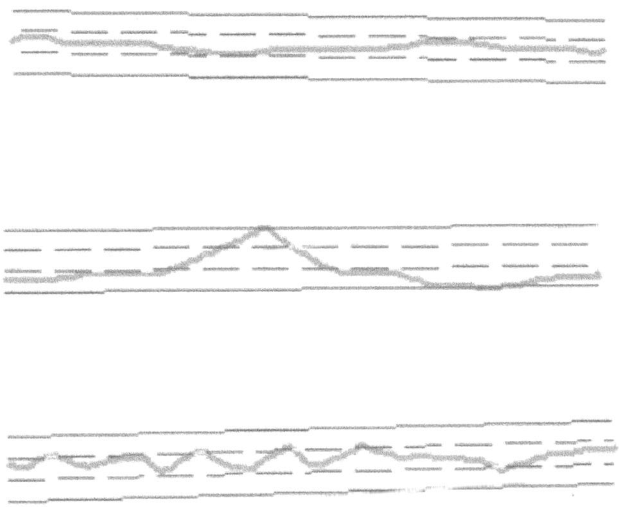

Figure 3.24. Good (top), average (middle), and poor (bottom) strokes.

Two tolerance bands are defined: a narrow (minimum) tolerance and a wide (maximum) tolerance. If the entire stroke remains within the narrow band, it receives maximum merit. If any point falls outside the wide band, the stroke receives no merit. Tolerances are defined proportionally to the length of the stroke. This system is effective in most cases, except for very short strokes (fewer than ten points) where even slight imperfections may prevent a stroke from being recognized as a straight segment. Within the intermediate range between the two bands, merits decrease linearly from 1 to 0, enabling finer distinctions among good, average, and poor strokes, as shown in Figure 3.24.

The command to display tolerance strips was designed to assist researchers in understanding how tolerances influence merit assignments. In addition to evaluating global straightness, two types of "irregularities" in strokes are further distinguished: long-wavelength shapes (called undulations) and short-wavelength features (called oscillations). These terms are drawn from signal processing, where undulations act as a carrier wave and oscillations represent higher-frequency modulations. Regardless of terminology, this distinction is important for replicating human perceptual behavior. Plumed *et al.* (2015) demonstrated that humans typically recognize oscillations as involuntary noise, whereas undulations are more likely interpreted as deliberate variations in intended shapes.

Various well-known techniques exist for removing high-frequency oscillations (Vaseghi, 2008), with moving averages (rolling averages) being particularly popular. However, correctly estimating parameters to distinguish noise from signal is critical (Durbin, 1959). Given REFER's focus on matching human perception, a simple, controllable smoothing strategy was adopted based on the two most perceptually salient parameters: corners and width.

In this approach, "micro-corners" are progressively removed by sequentially eliminating alternate points, provided that the smoothed stroke retains perceptible corners, and the width of the smoothed stroke remains similar to the original stroke width. Similarity is controlled via a threshold *TolSmooth*, which defines the maximum allowable variation in width. Points are only removed if their mutual distance is less than 2 × *TolSmooth*. The number of smoothing steps (NSS) applied to each stroke is recorded. The effect of smoothing is illustrated in Figure 3.25.

Corner detection uses one of three segmentation methods described in the vectorization segmenting section: *IStraw* (Xiong and LaViola, 2010)), *ShortStraw* (Wolin *et al.*, 2008), and *Sliding Strips* (Masood and Sarfraz, 2007)). The method is user-selectable, with IStraw set as the default option. Since IStraw requires timing data, if no timing information is available, the system automatically falls back to ShortStraw. Although

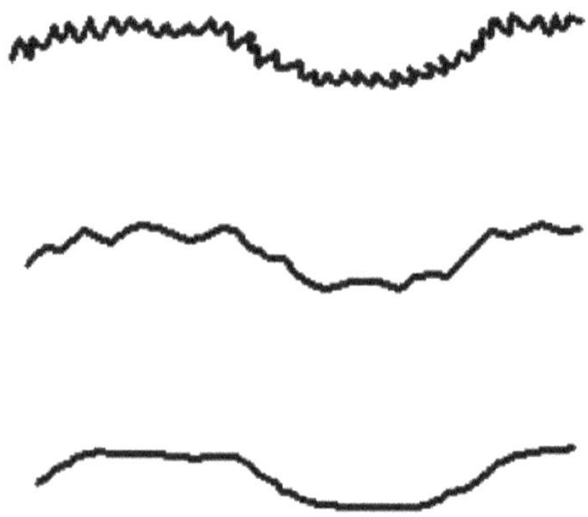

Figure 3.25. Effect of stroke smoothing.

corners are detected during this process, strokes are not immediately segmented. Instead, fitting is attempted first, and segmentation decisions are deferred until later stages.

Merits can be displayed as labels attached to the stroke numbers, activated via the *2D Modes* toolbar. However, these labels are only available after vectorization and provided that the display merits flag is enabled in the *Configure Display Options for Nodes and Corners of Strokes* menu in the same toolbar. Tolerance strips can also be displayed to better visualize how strokes are evaluated.

3.5.8.8 *Fitting elliptical arcs*

REFER employs an approach for fitting elliptical arcs that automatically combines existing algorithms to balance speed and precision. To measure precision, it introduces fast metrics grounded in perceptual criteria, ensuring tolerance toward sketching imperfections. Rather than producing a deterministic yes/no outcome, the algorithm returns a likelihood estimate based on these metrics, enabling its use within higher-level collaborative decision-making flows.

The input is a stroke—a time-ordered sequence of 2D points—captured by tracking the user's hand movements while sketching an ellipse. (In some systems, individual points may also be timestamped.) The output is an approximate, rather than exact, ellipse fit. To accommodate multiple interpretations, the result includes a merit figure that reflects the likelihood of the stroke being an ellipse. An example is illustrated in Figure 3.26.

The method is designed for computational efficiency, allowing for immediate feedback and enabling higher-level interpretation processes to invoke the fitting repeatedly as needed. An extensive explanation of the fundamentals of the fitting approach—along with its tuning parameters—can be found in (Company *et al.*, 2015b). Recommended default values for the tuning parameters are outlined in the *Fitting Details Configuration* dialog.

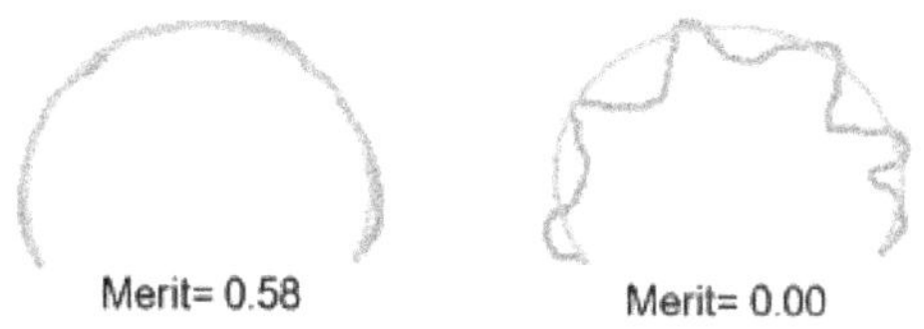

Figure 3.26. Example of merits for elliptical arcs.

3.5.8.9 *Fitting ellipses review*

This section summarizes the state of the art in fitting elliptic arcs to further clarify the method implemented in REFER. The most common approaches for fitting an ellipse with unknown parameters to a set of stroke points are based on minimizing the total fitting error, measured in various ways. A review of the field is provided by (Chernov *et al.*, 2014). A distinction is typically made between maximum likelihood, geometric, and algebraic fitting approaches (Yu *et al.*, 2013). Maximum likelihood methods are not considered viable here, since they tend to break down when the noise is large or when fitting highly eccentric ellipses.

Since an ellipse is uniquely determined by five points, the sequential ordering of points in a stroke can be leveraged by selecting regularly spaced five-tuples. Although Rosin (1993) introduced a method based on accumulating fits over many five-point combinations (5P method), each individual five-point fit is computationally inexpensive.

The Direct Ellipse Fit (DIR) algorithm of Fitzgibbon *et al.* (1999) is currently the most widely used algebraic method for fitting ellipses to scattered data. A more robust variant of DIR was later developed by Halir and Flusser (1998), with a C# implementation provided by Kotagiri (2010). DIR was first applied to sketch recognition by Shpitalni and Lipson (1997). While solving the eigenproblem for symmetric matrices is relatively simple, DIR involves non-symmetric matrices, making the computation more demanding.

As a representative geometric approach, the Guaranteed Ellipse Fitting (GEF) algorithm by Szpak *et al.* (2012) claims to outperform DIR when points are sampled from only a portion of an ellipse. GEF provides a fast and accurate approximation of the computationally intensive orthogonal-distance fitting method.

REFER includes Davis's (1996) implementation of the five-point (5P) algorithm, the Kotagiri C# implementation of DIR (Fitzgibbon *et al.*, 1999; Kotagiri, 2010), and a custom C++ implementation of GEF using Sampson distance correction (Szpak *et al.*, 2012). Company *et al.* (2015) identified typical failure cases for these methods, which may inform the configuration of the fitting process:

- 5P fails (1) for arcs close to 360 degrees, unless one endpoint is replaced by an additional intermediate midpoint; (2) for short (less than 180-degree angle) and flat arcs, or strokes with changes in curvature,

as these tend to produce elongated and inaccurate fits; and (3) for flat or undulating strokes, which typically do not yield valid ellipses.
- DIR fails for short and flat arcs due to its bias towards smaller ellipses when fitting only a portion of an ellipse. Numerical instability may also cause occasional failures.
- GEF does not fail, provided it receives a valid initial seed from either 5P or DIR. However, the execution times are too slow to be suitable for interactive applications.

The same authors also reported conclusions regarding the relative execution times:

- 5P is by far the fastest method (approximately 1,000 times faster than DIR and 100,000 times faster than GEF) with its speed advantage increasing as the number of points grows. If speed is a priority, 5P should always be tried first.
- DIR execution times may be acceptable for strokes with a few hundred points. When resampling to fewer than 100 points, DIR becomes a credible fast-fit alternative.
- Neither GEF/5P nor GEF/DIR combinations achieve acceptable performance for interactive applications, even when strokes are resampled to just 20 points. More specifically, GEF/5P is slower than GEF/DIR and thus not recommended.

3.5.8.10 *Fitting ellipses merits*

REFER evaluates stroke merits based on perceptual procedures which aim to replicate how humans perceive shapes embedded within sketched strokes. Design engineers are used to identifying ellipses in engineering sketches, but what criteria do they use when deciding whether and how to fit an ellipse to a curve? To understand the limits of human perception, it is necessary to ask humans directly. The only scientifically proven strategy is to conduct experiments with groups of participants and make their perceptions explicit (Company *et al.*, 2015c). To date, studies in the field of visual perception rarely provide sufficient detail to develop algorithmic approaches that replicate human perception.

Although some algorithms are described as perceptually oriented, we believe it is insufficient to merely cite human perception as an objective.

Instead, algorithms must be explicitly designed to emulate human perceptual processes and carefully tuned to align as closely as possible with human judgment.

Once human perception of strokes is better understood, the goal is to replicate this behavior by fine-tuning algorithms that assess the differences between the input stroke and the resulting arc. Since comparing differences based on perimeters would be too computationally intensive for interactive use, REFER employs fast, efficient metrics that provide a reasonable approximation of the perceptual discrepancy between the stroke and the fitted arc. For readers interested in modifying how merits are calculated, the procedure for evaluating the merits of fitted elliptical arcs is summarized below and fully described in (Company *et al.*, 2015).

The merit of a fitted ellipse is defined as 1 if the stroke corresponds to an ideal ellipse and 0 if it is highly non-elliptical. A stroke is accepted as representing an elliptical arc if it fits within a minimum tolerance band, or, alternatively, its smoothed version fits within a maximum (or "relaxed") tolerance band.

REFER uses merit values to control the fitting flow. First, it evaluates the merit of the stroke as a straight line. If the merit is less than 1, it then evaluates the stroke as an elliptical arc. The type with the higher merit is selected.

Merit values can be displayed as labels attached to stroke numbers. This feature can be activated through the *Display 2D Sketch* toolbar. Note that labels are only available after vectorizing the sketch and provided that the display merits flag is active in the *Configure display option for nodes and corners of strokes* settings in the *Display 2D Sketch* toolbar.

3.5.9 *Sketch refinement*

3.5.9.1 *Introduction*

Sketches managed in SBM applications must account for various 2D operating modes, which are typically treated as input options. However, some modes of operation are primarily linked to how a sketch is to be interpreted. In these cases, vectorization can contribute significantly by refining sketches and revealing the information embedded by users during different types of interactions with the sketch.

Although many types of refinement and interaction are conceivable, only a few have been systematically studied and implemented in the field of SBM. The following scenarios, which require sketch refinement, can be replicated in REFER by configuring the *2D Vectorization Configuration* dialog box (available in the *2D Vectorization* toolbar):

- Some users tend to redraw sketch lines by overlapping multiple strokes. This "line sweeping" action produces overtraced strokes, which require grouping these over traces to reveal the single stroke or line that was ultimately intended.
- Some users depict non-solid lines as sequences of consecutive strokes (e.g., drawing hidden edges with dashed lines). Grouping non-solid strokes simplifies vectorization and preserves higher-level information that may facilitate subsequent reconstruction.
- Some users do not lift the pen between strokes when a line ends, and another begins at the same point. Instead, they draw a single continuous stroke that represents multiple lines. This behavior produces poly-strokes, which must be segmented by an appropriate vectorization module to extract the individual geometric primitives contained within them.

REFER aims to run different vectorization tasks in parallel, allowing them to feed back into one another during progress. Extending this concurrency to both sketch refinement and line-drawing refinement, however, remains a challenge. In the current version of REFER, these tasks are executed sequentially.

Grouping tasks are performed immediately after fitting, since fitting information is necessary for effective grouping. Segmentation tasks are deferred until the end because segmenting overtraced or non-solid lines—often poor-quality strokes—tends to produce false segmentations. Furthermore, fitting generates merits that assist in deciding when to group or segment.

The current flow first attempts to fit strokes as lines. If the fitting merits are sufficiently high, the solution is accepted as final. If poorly fitted strokes remain, they are subjected to grouping or segmentation before fitting the resulting grouped or segmented strokes. The procedure currently ends after a single iteration, although the exploration of multiple iterations is considered a relevant area for future development.

In their recent book, Bonnici and Camilleri (2023) extensively reviewed current approaches to the simplification and vectorization of sketched drawings. Although primarily focused on sketches for artistic and graphic design purposes, the ideas detailed in the book are worth examining.

3.5.9.2 *Grouping overtraced strokes*

The mode for grouping overtraced lines is available via the *2D Vectorization Configuration* dialog box. Once activated, grouping is automatically performed during vectorization (grouping overtraced strokes is an automatic process but requires fine-tuning for optimal performance). The process is launched only during offline vectorization, as it makes no sense to group overtraced lines while their traces are still being drawn. Users can further configure the refinement of overtraced strokes by selecting the corresponding "Details" button and adjusting the parameters available in the *Overtraced Stroke Grouping Details Configuration* dialog.

The approach implemented in REFER for grouping overtraced strokes is inspired by the method proposed by Ku *et al.* (2006), although REFER cannot yet merge overtraced curves. The contributions by Bartolo *et al.* (2007) and Wang *et al.* (2012, 2014, 2015b) are also notable. An example of grouping of over-traced strokes is shown in Figure 3.27.

The procedure implemented in REFER is described as follows: candidate patterns are searched first. Only strokes with a positive merit as straight segments are considered. The largest valid and unvisited segment is selected as a candidate overtrace for a new overtraced line. The remaining segments are recursively parsed and added to the overtraced line if, and only if, they meet the following conditions:

- They have neither been previously visited nor already included in the pattern.
- They have a similar orientation to the current overtraced strokes.
- They overlap with the current chain of overtraced strokes.

The notion of "similar orientation" is more nuanced than it might first seem. Two segments, "a" and "b," are considered "nearly collinear" if they satisfy both angular and offset thresholds. Alignment is checked by ensuring that the angle between the two segments does not exceed a given threshold. Offset is measured between segments, i.e., the closer tip of one segment must not be significantly farther from the other segment than the allowed threshold. Distances between segment tips and their extensions are ignored. The concept is illustrated in Figure 3.28.

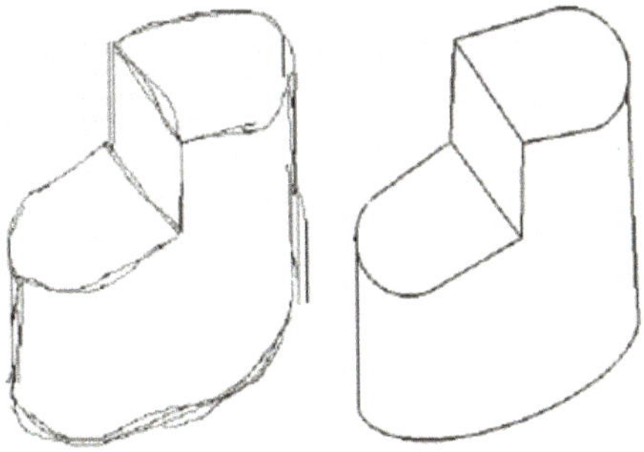

Figure 3.27. Sketch with overtraced strokes before (left) and after (right) grouping.

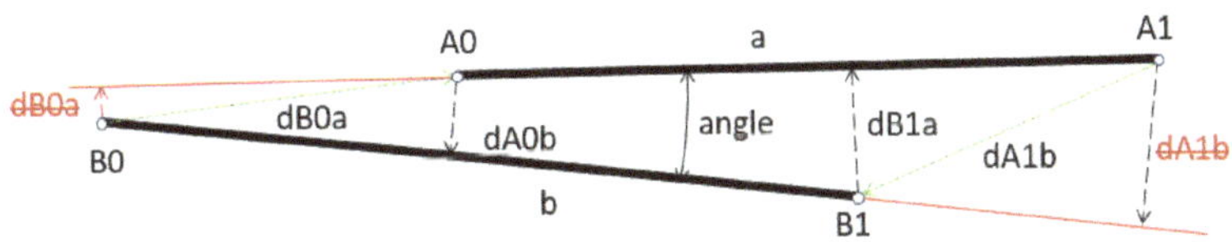

Figure 3.28. Similar orientation in overtraced strokes.

Once an overtraced line is detected, the sketch database is updated to group the overtraced strokes as follows:

- Cluster all overtraced strokes into a simplified new stroke.
- Replace the first stroke with the new clustered stroke.
- Combine all strokes that constitute the overtraced line into a single, unified stroke.
- Remove the original strokes containing the overtraces.

Clustering is performed in two sub-steps. First, all overtraced strokes are combined into a single, usually "zigzagging," stroke. In the example shown in Figure 3.29, the first tip (in green) of the rightmost stroke is selected as the starting point. Its endpoint (in red) is connected to the starting point (in green) of the second stroke, and the procedure is repeated for the remaining strokes until the last stroke is added (on the left).

If consecutive strokes are reversed, the connecting lines can produce a complex zigzagging shape. Note that if the first and/or last strokes are reversed, their start and endpoints may not be near the tips of the clustered stroke (see the green point on the left side of the example). To address this issue, the most extreme tips of the combined stroke are calculated and used to define the stroke. In the second sub-step, the clustered stroke is

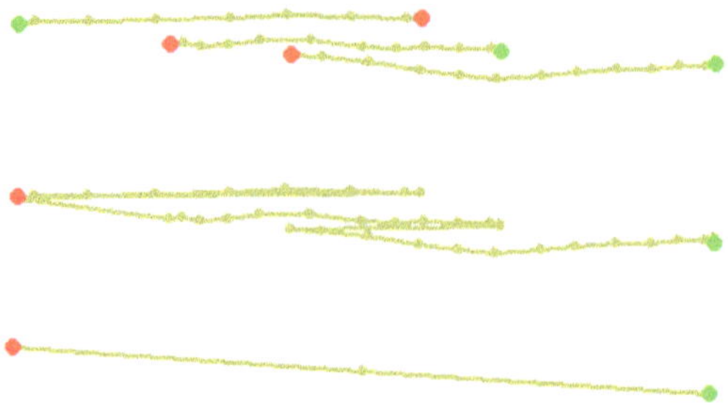

Figure 3.29. Stroke clustering process.

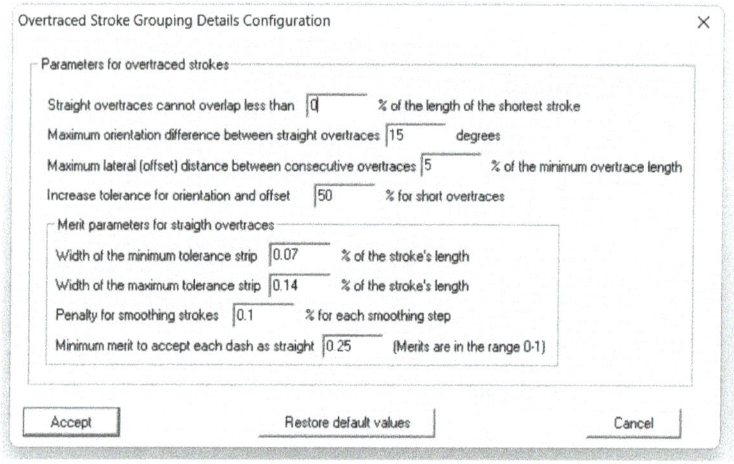

Figure 3.30. *Overtraced Stroke Grouping Details Configuration* dialog box.

fitted, and the tips and midpoint of the fit are used to define a simplified combined stroke.

3.5.9.3 *Overtraced stroke grouping details configuration*

The parameters that govern the behavior of the grouping of overtraced strokes can be edited in the *Stroke Grouping Details Configuration* dialog box. The recommended values are shown in Figure 3.30.

The minimum overlap required to consider two strokes as belonging to the same overtraced line is illustrated in Figure 3.31. Note that the default value is 0, as good sketchers tend to "elongate" strokes without overlapping them, as shown in Figure 3.32.

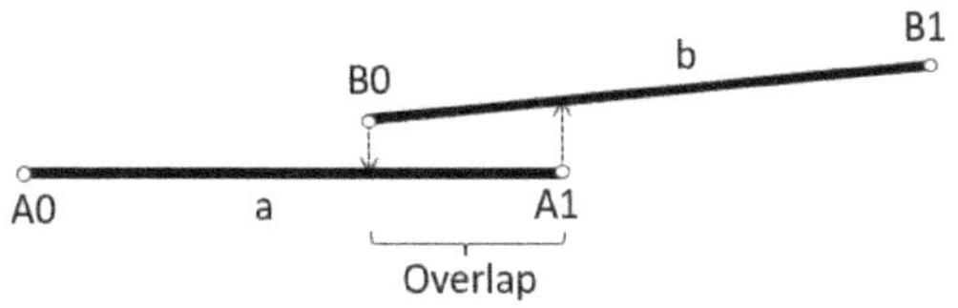

Figure 3.31. Minimum overlap required to be considered an overtraced line.

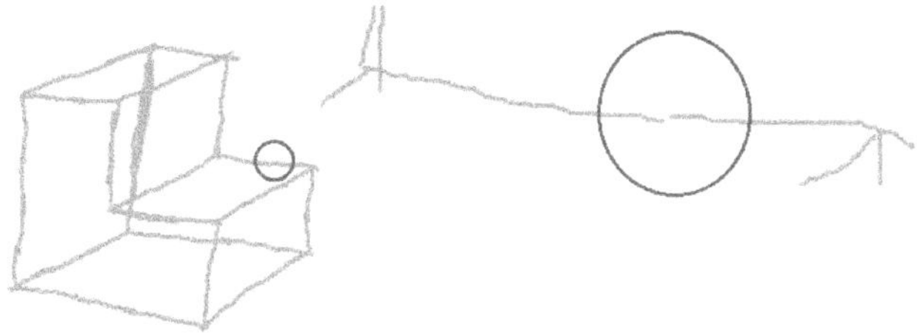

Figure 3.32. Elongated stroke without overlapping.

The approximate collinearity of segments is evaluated by checking two conditions:

- An angular threshold, with a default value of 15 degrees.
- An offset threshold equal to 5% of the length of the shortest stroke.

Both thresholds are increased for short strokes (up to 50% by default) based on the assumption that drawing errors are proportionally larger in shorter lines. Only straight strokes are considered candidates for merging into overtraced lines (see Figure 3.33). Straightness is evaluated using the merit parameter for straight lines, as defined in the dialog.

The justification for these specific parameters is that the required "straightness" of strokes in an overtraced line is typically lower than that of isolated strokes. It is assumed that during overtracing, users may draw with less precision. The default parameters perform well for simple overtraced drawings, such as the prismatic shape shown in Figure 3.34.

However, two common failure cases exist. The first occurs when lines converge at a small angle, like at the apex of the pyramid shown in Figure 3.35. Reducing the angular threshold (as shown in the pyramid on the right) improves the detection of lines converging at small angles. However, it also increases the risk of failing to group strokes on lines drawn with poorer quality (such as the right-side edge).

The second common failure occurs with nearly overlapping lines, each depicted by its own set of overtraced strokes, as illustrated by the bracket shown in Figure 3.36. Reducing the offset threshold can help to

Figure 3.33. Only straight strokes can be merged into overtraced lines.

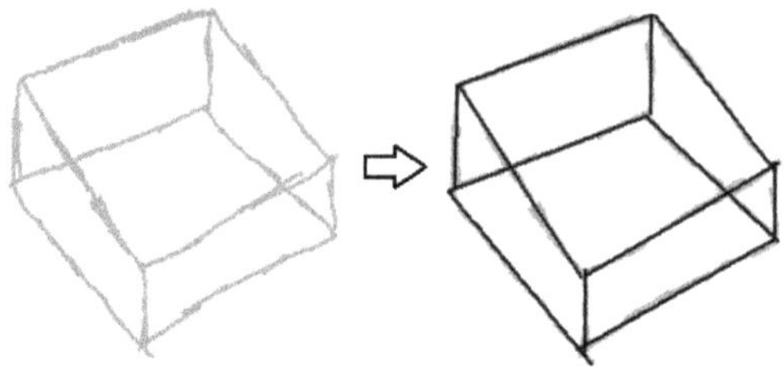

Figure 3.34. Successful grouping on strokes with default parameters.

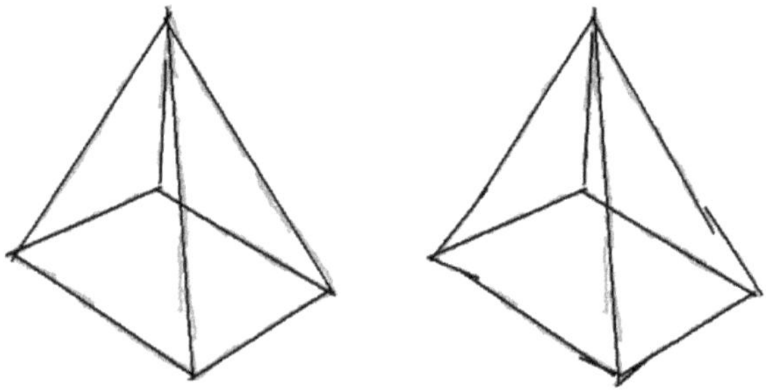

Figure 3.35. Unsuccessful grouping of strokes due to lines converging at a small angle.

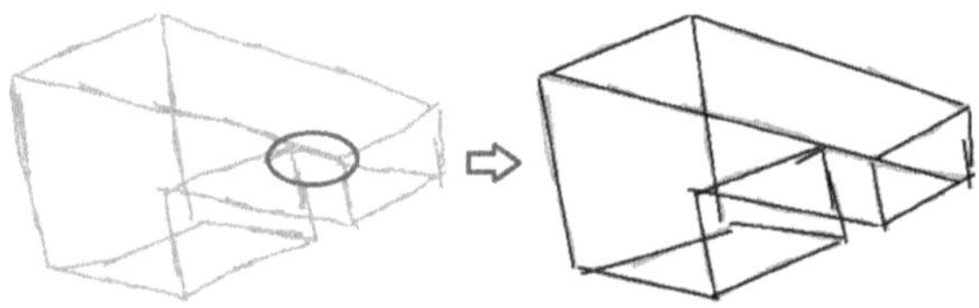

Figure 3.36. Unsuccessful grouping of strokes due to nearly overlapping lines.

separately identify the two lines, but at the cost of sometimes failing to group all the strokes correctly for each individual line.

In our view, properly detecting nearly overlapped lines will ultimately require a global approach. Strokes should be grouped not only based on proximity but also considering the broader structural meaning they contribute to the sketch.

3.5.9.4 *Grouping non-solid strokes*

The mode for grouping discontinuous lines is available in the *2D Vectorization Configuration* dialog box. Once activated, the grouping process is automatically launched during vectorization. Grouping is only triggered during offline vectorization, as it does not make sense to group lines while they are still being drawn.

To further configure the grouping refinement for non-solid strokes, users can click the corresponding "Details" button and adjust the parameters in the *Non-Solid Stroke Grouping Details Configuration* dialog box.

Various non-solid line styles should be supported, including dashed, dash-dot, and dotted lines. Each style serves a specific purpose in SBM, as such lines commonly represent meaningful elements in technical drawings (e.g., hidden edges, symmetry axes, construction lines, etc.). The current version of REFER supports only dashed lines.

The procedure detects isolated and consecutive straight strokes, which are presumed to be candidate dashed strokes. Once the grouping module is executed, the strokes defining dashed lines in the input sketch are grouped into a single stroke in the working sketch, which can then be displayed through display dashed lines. Additionally, these strokes are labeled as dashed in the database for easier subsequent analysis, such as finding hidden edges in the 2D line drawing. The process is illustrated in Figure 3.37.

Implausible strokes are marked as non-visitable before starting the recursive search:

- Very long strokes are discarded as non-candidate dashes (since dashed lines represent hidden edges, which are only obscured by faces delimited by edges larger than the dashes).
- Non-isolated strokes are also discarded (tips of candidate strokes must be distant from the tips of other strokes that are neighbors and non-collinear).

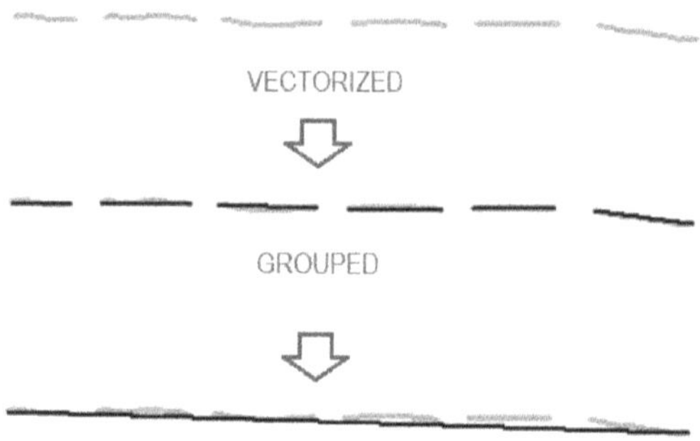

Figure 3.37. Parsing of dashed strokes in REFER.

Chains are calculated recursively by searching for consecutive dashes, which must meet the following criteria:

- The dashes should have consistent length (not too short or too long) to ensure uniformity and clarity.
- They either follow consecutively or precede the current dash in the sequence, and their spacing should be neither too close nor too far apart to maintain visual consistency.
- They are reasonably aligned with each other.

Once the dashed line is detected, the sketch database is updated by grouping the dashes as follows:

(1) Create a new stroke by connecting the tips and midpoints of all the individual dashes (reversing the direction of each stroke if necessary to ensure alignment with the overall direction of the dashed line).
(2) Group copies of all strokes that belong to the dashed line within the new stroke.
(3) Save the new stroke instead of the first dash.
(4) Remove the original strokes containing all dashes except the first one.

The grouped line is marked as "dashed' in the database, making it easier to detect hidden edges in later stages.

3.5.9.5 *Non-solid stroke grouping details configuration*

The parameters that govern the behavior of the grouping of non-solid strokes can be edited in the Stroke Grouping Details Configuration dialog box. The recommended values are shown in Figure 3.38.

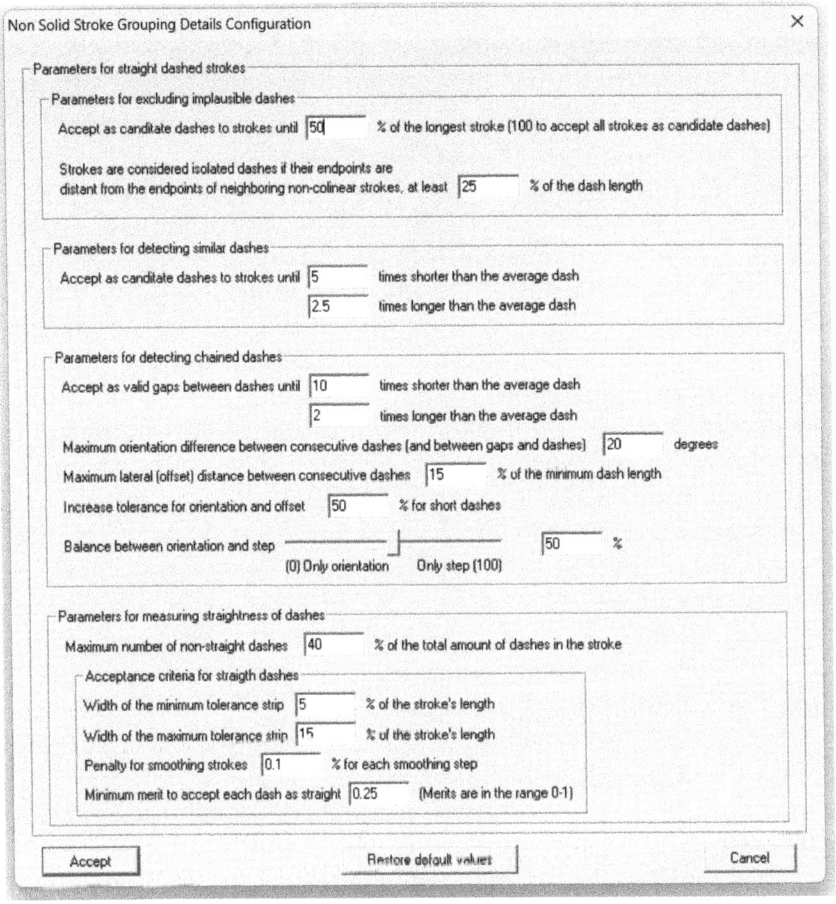

Figure 3.38. Non-Solid Stroke Grouping Details Configuration dialog box.

The first group of parameters controls the detection of implausible dashes. The first parameter determines how "isolated" the strokes must be to be considered candidate dashes. The second parameter controls the relative length between the longest strokes and those that can be considered candidate dashes. This setting is based on the assumption that longer strokes correspond to visible lines, which delimit faces that hide shorter strokes behind them.

The following four parameters determine the acceptable range of lengths for candidate dashes and their corresponding gaps. These parameters are used to quantify the assumption that dashes in a dashed line must be somewhat similar to each other.

The next three parameters fine-tune the acceptable differences in orientation and offset between dashes that may belong to the same dashed chain. Currently, these parameters are tuned to detect "nearly" straight chains of dashes. However, users are encouraged to experiment with relaxing these values to better detect curved dashed lines.

Similarly to the case of overtraced strokes, the notion of "similar orientation" in non-solid strokes must be defined. Two segments, "a" and "b," are checked to determine if they are "nearly" collinear with each other within a specified angular "threshold." In addition, this same threshold is used to check the angular alignment between each segment and the line defined by their midpoints. This approach assumes that poorly drawn consecutive dashes may not be perfectly collinear, but they should maintain a certain "global" collinearity, which can be locally inspected by ensuring that a global line passes close to their midpoints. The concept is illustrated in Figure 3.39.

Regarding the offset, REFER applies a strategy similar to the one described for overtraced lines. The main difference in this case lies in how the offset is measured. It is taken from the tips of each segment to the line containing the other segment, rather than between the segments themselves. This approach assumes that the segments do not overlap, rendering direct segment-to-segment distance irrelevant.

The final set of parameters in the dialog are used to evaluate the merit of each stroke as a candidate straight line. These are typically more relaxed than the general parameters used for evaluating solid straight lines, as short strokes tend to be less precise than longer strokes.

The example shown in Figure 3.40 illustrates how curved strokes are not detected as dashes (stroke 4) but are integrated into the dashed line. Strokes that are displaced laterally (stroke 5) are detected but not integrated. Finally, short strokes (stroke 7) are not detected as dashes, as explained in the section Fitting Straight Segments Merits. The figure demonstrates that relaxing the constraints enables the algorithm to identify a single solid line. However, this approach is only effective for

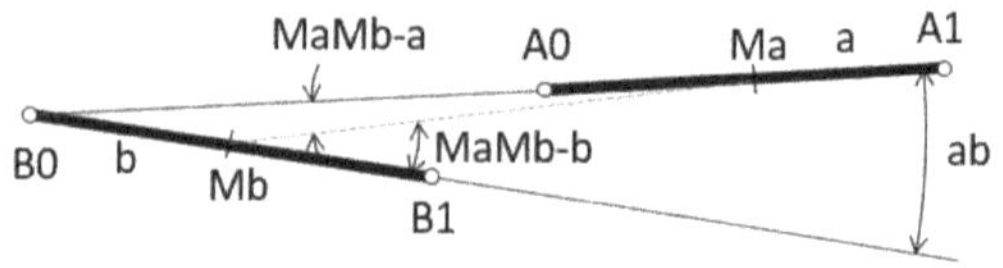

Figure 3.39. Similar orientation in non-solid strokes.

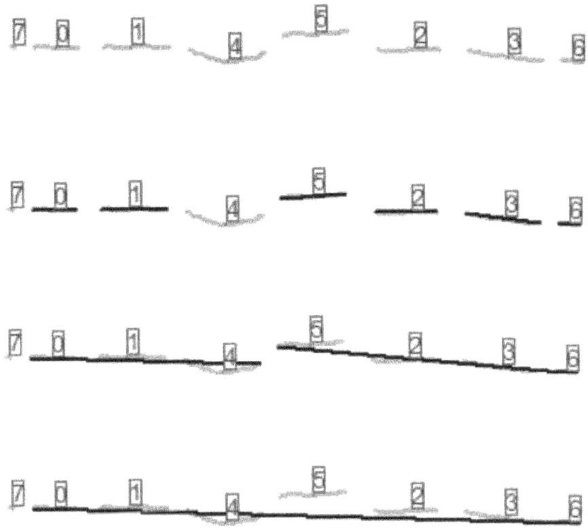

Figure 3.40. Example of the integration of curved strokes into a dashed line.

isolated dashed lines, where there is no risk of inadvertently combining dashes from neighboring lines.

It is important to note that users may produce widely varying styles of dashed lines. In most cases, the dashes are relatively uniform in length, with consistent spacing between them. However, inconsistently drawn dashed lines may include segments of significantly different lengths. Additionally, some users tend to create very "spaced out" or "dense" dashed lines, where the gaps are much larger or smaller than the dashes themselves, as illustrated in Figure 3.41.

REFER provides parameters that help adapt the detection approach to accommodate these variations, but certain dashed lines may still go undetected. For instance, in the sketch shown in Figure 3.42, the leftmost dash is positioned too close to the next one. Additionally, the long line present in the drawing ensures that the longer dashes are not misclassified as separate solid lines, in accordance with the filter controlled by the second parameter mentioned earlier.

3.5.9.6 *Segmenting PolyStrokes*

This type of refinement is aimed at splitting single strokes that represent multiple lines. Single strokes are identified in the sketch, and those that cannot fit straight segments or elliptical arcs are selected as candidates to

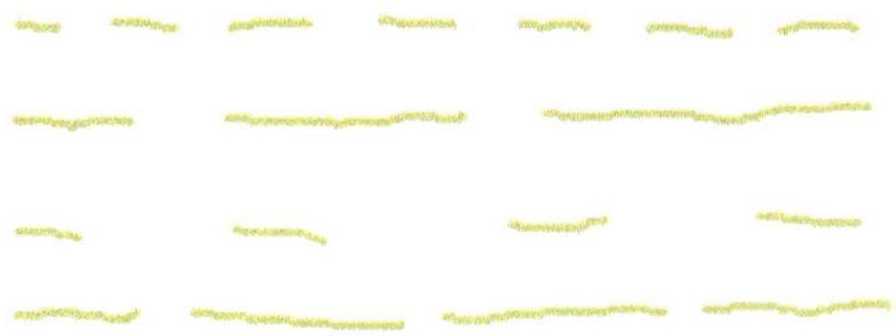

Figure 3.41. Different styles of dashed strokes.

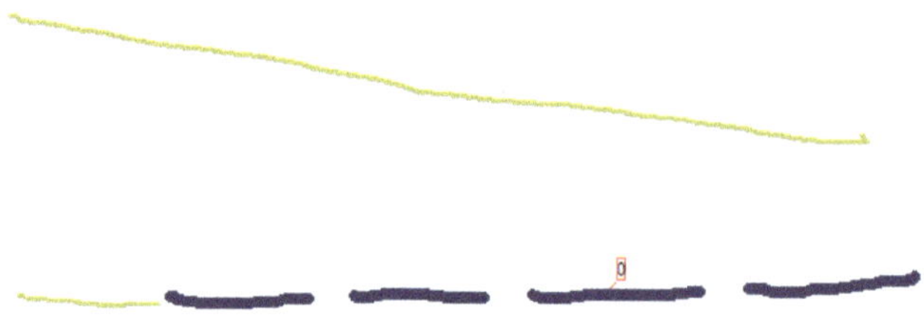

Figure 3.42. Undetected dash due to close proximity to the subsequent dash.

be trimmed into a set of simpler strokes, as illustrated in Figure 3.43. These simpler strokes are then fitted.

In the current implementation of REFER, segmentation is triggered during the fitting process. Fitting provides the segmenter with merits, which help determine which strokes are good candidates for segmentation, i.e., those that were poorly fitted before segmentation. In the segmentation module, every stroke is parsed to detect sub strokes. Detected sub strokes are segmented and stored separately to be subsequently fitted.

REFER's segmentation approach assumes that every corner is a candidate endpoint for two attached sub-strokes. Corners must be previously calculated and stored as cues for the stroke. They are saved as a list of positions of the points on the stroke that are closest to each corner. These positions are relative to the sequence of points on the stroke.

Most corner detection algorithms are commonly validated using imprecise or unreliable evaluation methods. Many developers incorrectly assume that the purpose of the algorithm is to detect the features they intended to depict, rather than those that users actually perceive. However, perception does not always align with intention. We advocate for a different approach that involves validating corner detection algorithms through human-centered experiments. By studying how people naturally interpret, and segment sketched drawings, it becomes possible to configure algorithms that more accurately replicate human perceptual processes.

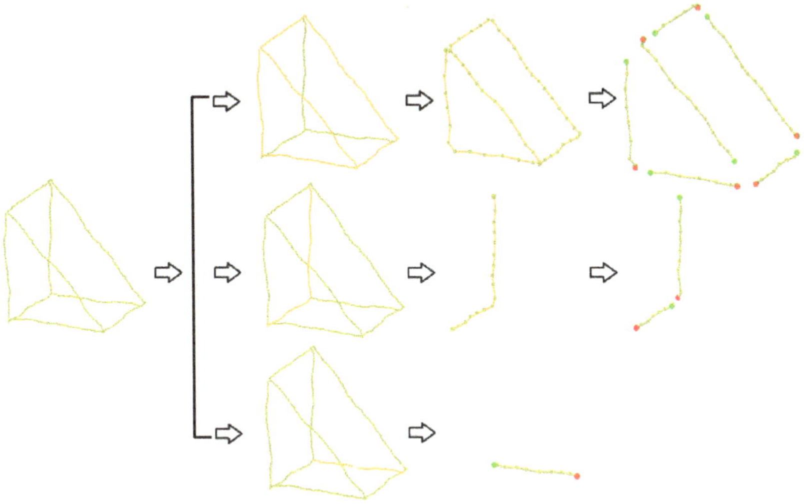

Figure 3.43. Segmentation of polystrokes.

A notable example of this methodology is the work conducted by Company *et al.* (2010) who carried out perceptual studies to better understand human segmentation behavior. In their study, the authors created a series of sketched drawings and asked participants to segment them according to their intuitive perceptions. This study allowed the researchers to gather empirical data on both the number and the locations of the segmentation points perceived by human observers, rather than relying solely on the designer's original intention. The findings from these experiments revealed an important insight: not all potential segmentation points are equally salient to human perception. Some corners or breakpoints are consistently identified by most participants, while others are vastly overlooked.

3.6 2D Refinement

3.6.1 *Introduction*

The vectorization process does not correct the inherent geometric imperfections typical of hand-drawn sketches. Therefore, several "post-vectorization" filters must be applied to further refine—or "tidy up"—the resulting vectorized line drawing.

This refinement stage is conceptually similar to what occurs in calligraphic interfaces, which dynamically process sketches in real-time—converting freehand input into straight segments, joining nearby segment endpoints together, enforcing geometric constraints such as parallelism, and more. However, when such refinements are applied "offline"—after the drawing is complete—they can be somewhat controversial. Some researchers argue that automatic refinements may unintentionally alter the original design intent embedded in the sketch.

To address this concern, REFER provides full control over the refinement process. Each filter can be independently activated or deactivated, allowing users to tailor the system's behavior to their specific needs or preferences.

The refinement filters implemented in REFER are based on techniques proposed in (Piquer, 2003), and (Company *et al.*, 2004, 2019). They operate during the "2D Refinement" stage, which is managed via the *2D Refinement* toolbar and configured through the *2D Refinement Configuration* dialog box.

REFER supports two operational modes for 2D refinement: interactive and batch. These modes can be selected via the 2D Refine Schedules, enabling users to apply filters either incrementally during the sketching process or collectively once the sketch is finished.

Currently, REFER includes six distinct filters for 2D refinement, each designed to address specific imperfections such as minor misalignments, unintended gaps, or irregular segment angles. These filters can be combined to achieve cleaner and more geometrically meaningful representations of the input sketch.

3.6.2 *2D refinement toolbar and configuration dialog*

The *2D Refinement* toolbar is located over the parent window. Two commands control the refinement schedule (interactive vs. batch), and one command opens the dialog that configures the filters, as shown in Figure 3.44.

The *2D refinement configuration* command opens a dialog box with various settings to configure filters in the 2D refinement process (see Figure 3.45.

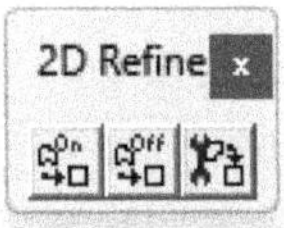

Figure 3.44. 2D Refinement toolbar.

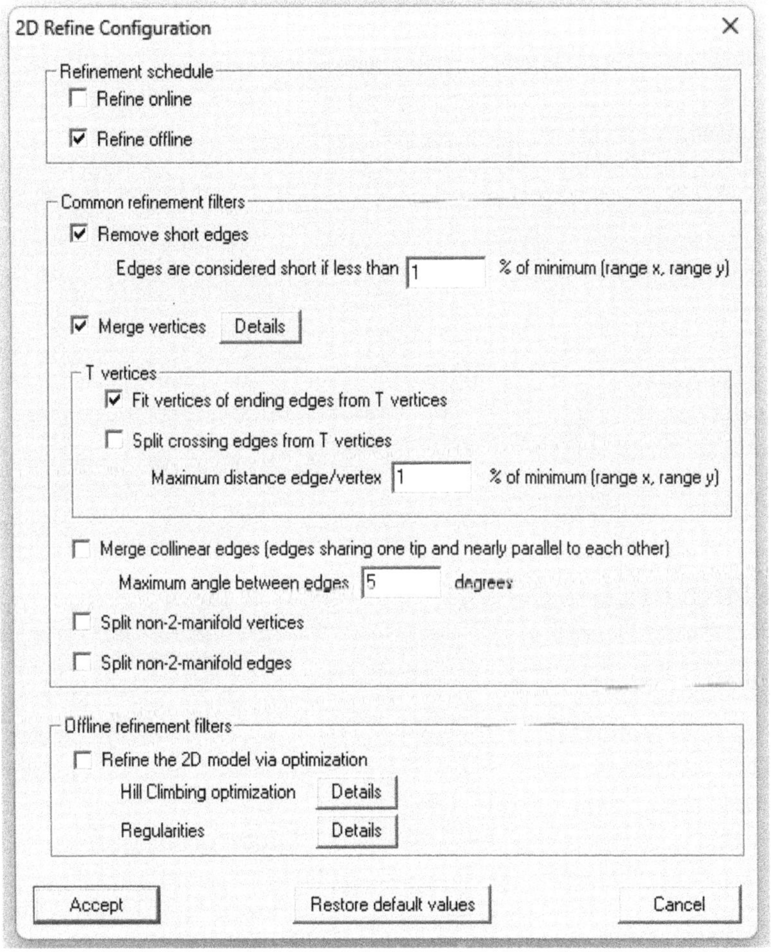

Figure 3.45. *2D Refinement Configuration* dialog box.

Users can configure both the activation of 2D refinement schedules, and the specific filters applied during the refinement of the 2D line drawing. For each group of refinement options, a "Details" button is provided, which opens the corresponding dialog with advanced configuration parameters:

- *Merge Vertices*: This dialog allows users to adjust the behavior of the vertex merging filter, which merges nearby endpoints in the sketch to eliminate unintended gaps or overlaps.
- *Hill Climbing Optimization*: This dialog provides access to the parameters used by the coordinate optimization filter, which applies hill climbing to refine vertex positions and improve the geometric coherence of the sketch.
- *Regularities*: This dialog allows users to configure the regularities (e.g., parallelism, orthogonality, etc.) used by the same coordinate optimization filter to further enhance the structural consistency of the line drawing.

These advanced settings provide users with fine-grained control over the behavior of each filter, making it possible to customize the refinement process to match specific design intentions.

3.6.3 *Merge vertices details dialog*

The *Merge Vertices Details* dialog box—shown in Figure 3.46 and available in the *2D Refinement Configuration* dialog—enables the configuration of the specific parameters associated with the merge close vertices filter.

REFER implements a simple yet effective algorithm for detecting trihedral corners—which is not recommended for vertices of higher valence. The method identifies edge tips that are not yet connected to any other edges. These tips must satisfy two conditions: (1) they are spatially very close to one another, and (2) exactly three such tips are found. The system considers three tips to be "close" if circles centered on them intersect each other pairwise. A circle intersection algorithm is used for this purpose.

The radius of each circle is measured in screen pixels (typically 6), which ensures that only groups of tips perceived to be very close are

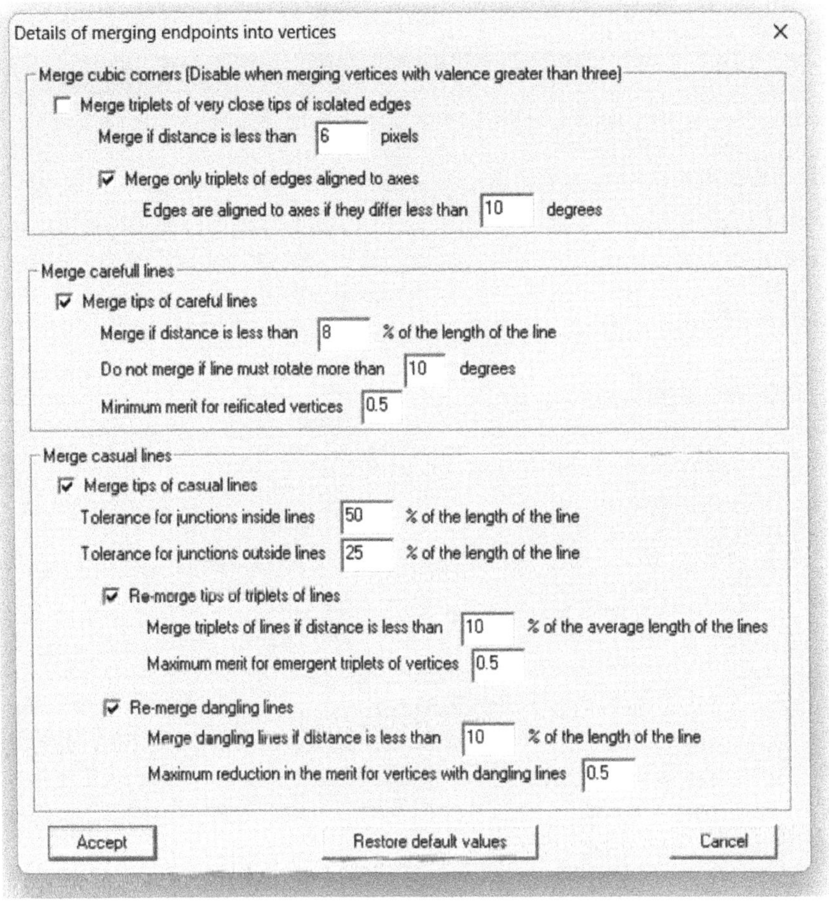

Figure 3.46. *Merge Vertices Details* dialog box.

considered candidates for junctions. Since the detection threshold is defined in pixels, the actual distance in world coordinates changes with zoom level. When zoomed in, tips appear farther apart in pixel space, making junction detection stricter. This mechanism gives users indirect control over geometric tolerance, as described in the section on using zoom as a gesture.

Users can choose whether to detect only trihedral corners aligned with the primary axes or to allow arbitrary orientations. By default, the filter enforces axis alignment, discarding triplets not parallel to any main axis. This behavior can be reconfigured through the dialog interface. The application estimates main axes beforehand to facilitate this filtering, although this estimation is tentative, given that the final geometry of the model is not yet defined. A default angular divergence threshold of 5 degrees is used to assess whether edges are aligned with these main axes.

Disabling the requirement for orthogonality allows the detection of trihedral junctions regardless of their orientation. Low-quality junctions may or may not be merged depending on the threshold settings. This trihedral detection mechanism is particularly effective as a pre-filter to identify obvious junctions before applying more advanced merging algorithms.

The main merging algorithm implemented in REFER is based on the method proposed in (Company *et al.*, 2019). To understand the algorithm's behavior and the design of its configuration interface, two key concepts must be considered:

(1) *Sketch quality continuum*: Sketches can be broadly categorized along a continuum of quality, ranging from very precise to highly imprecise. Engineers and designers—typically skilled sketching experts—often produce two distinct types of sketches: careful sketches, intended to convey detailed or finished ideas; and casual sketches, used to capture quick or rough ideas, often incomplete. To accommodate both sketch types, the dialog presents two blocks of parameters that provide separate configuration for "careful" and "casual" sketches. These sketch types are illustrated in Figure 3.47.

(2) *Mixed-mode sketching and perceptual scoring*: Although it is possible to detect whether a user is sketching carefully or casually, most sketches contain a combination of both. REFER uses a two-stage

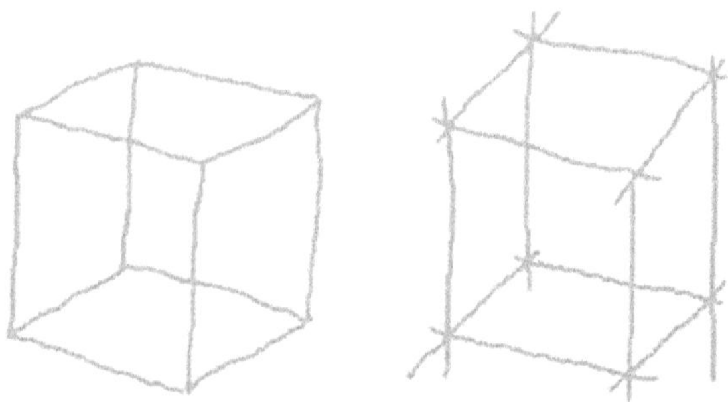

Figure 3.47. Example of careful (left) and casual (right) sketches.

approach to handle this mixture. Rather than binary decisions ("this is a vertex" vs. "this is not a vertex"), the algorithm assigns merits—numerical scores that represent how likely a junction is to be valid. This strategy aligns with the philosophy of vectorization merits and reflects a perceptual rather than strictly geometric interpretation. The idea is not to assert definitively but to suggest "this junction is more or less likely to be valid." The approach supports future extensions of REFER aimed at integrating global perceptual reasoning. For instance, a local set of strokes that barely suggests a junction might be more confidently interpreted as a vertex when considered in the broader context of the full scene. This situation is based on the hypothesis that humans naturally interpret shapes in context, which influences their perception of structure. This perceptual scoring system adds a layer of semantic flexibility, enabling REFER to better mimic the way human designers interpret line drawings.

3.6.4 *2D refine regularities details dialog*

The *2D Refine Regularities Details* dialog box—shown in Figure 3.48 and accessible from the *2D Refinement Configuration* dialog—enables the configuration of the regularities used to optimize the coordinates of vertices in the 2D line-drawing refinement process.

Users can enable or disable any of the three regularity-based cues—perceived structural features of the sketch presumed to reflect intentional geometric regularities. These cues influence how the vertex coordinates

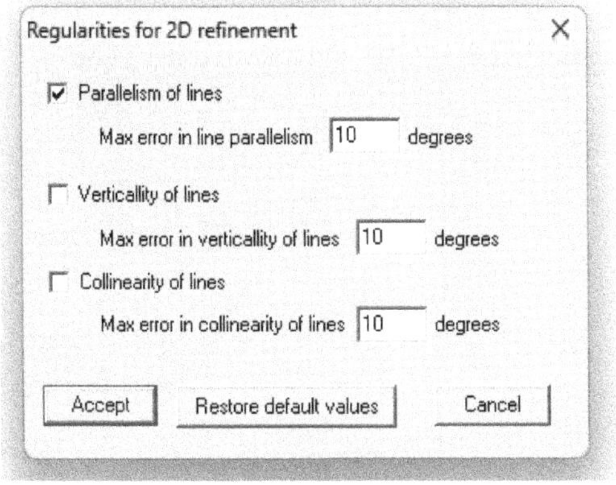

Figure 3.48. *2D Refine Regularities Details* dialog box.

are optimized. Each cue is governed by user-defined thresholds or snapping tolerances that determine whether a candidate geometric feature (such as two nearly parallel lines) should be adjusted to enhance regularity.

The parallelism regularity assumes that lines which are nearly parallel should be adjusted to become as parallel as possible. During optimization, the algorithm adjusts the x and y coordinates of the involved vertices to reduce angular deviation between lines, giving preference to line pairs that are initially more closely aligned. In the example shown in Figure 3.49, a cube drawn with slight perspective distortion is transformed into an axonometric form by enforcing parallelism (e.g., between edges 0 and 11, or 2 and 4).

The collinearity regularity assumes that lines which are nearly collinear should be adjusted to become as collinear as possible. In the broad sense, collinearity overlaps with parallelism. However, strict collinearity applies only to consecutive edges—lines that share a vertex. While the option to refine collinearity remains available for legacy reasons, in practice it is often redundant, as the merging of collinear edges typically provides a more robust solution. Nevertheless, as shown in Figure 3.50, applying a large tolerance (e.g., 35 degrees) for "nearly collinear" lines can simplify irregular shapes, such as converting an irregular octagon into a nearly rectangular form.

The verticality regularity assumes that lines which are approximately vertical should be adjusted to become strictly vertical. However, since REFER can already detect the three main axes, and subsequently refine edges that align with any of these directions, enforcing verticality as a separate process has limited practical value. It is generally more effective

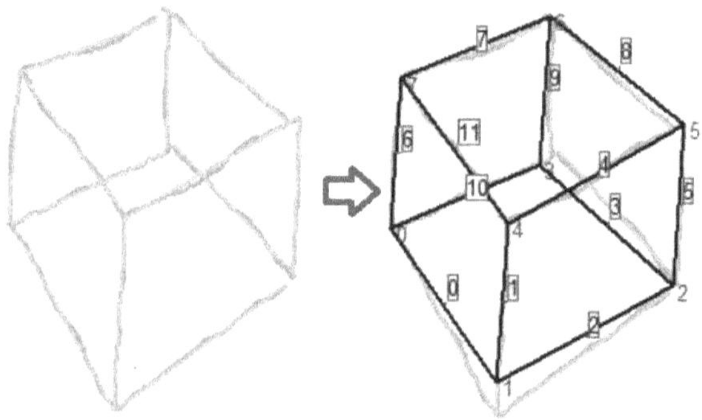

Figure 3.49. Cube with slight perspective distortion transformed into an axonometric form by enforcing parallelism.

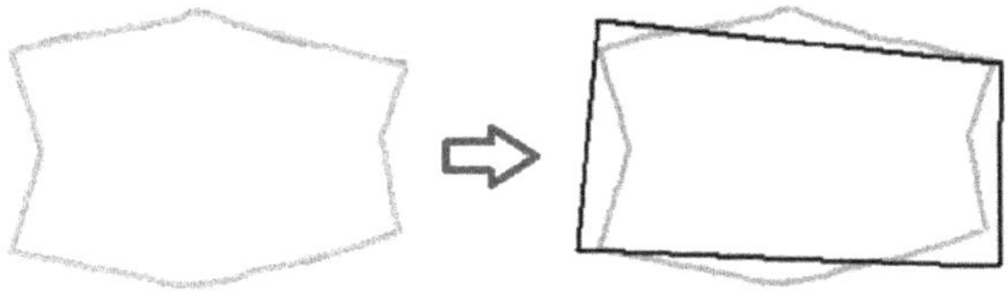

Figure 3.50. Simplification of an irregular shape by applying a large tolerance for "nearly collinear."

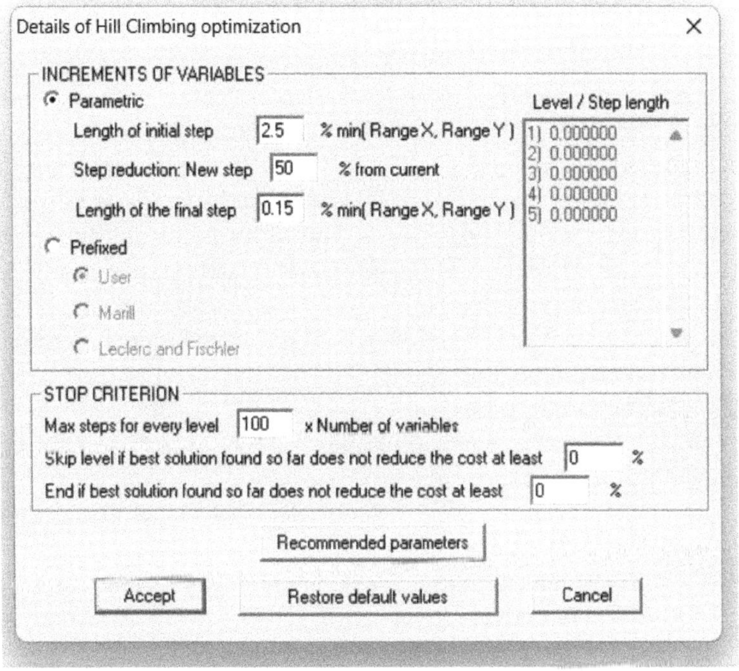

Figure 3.51. *2D Hill Climbing Details* dialog box.

to refine lines based on their alignment with all main directions rather than privileging verticality alone.

3.6.5 *2D hill climbing details dialog*

The *2D Hill Climbing Details* dialog box—shown in Figure 3.51 and accessible from the *2D Refinement Configuration* dialog—enables the configuration of the parameters that govern the Hill Climbing optimization algorithm. Users can adjust both the step increments for variable updates and the stopping criteria for the algorithm. These controls allow fine-tuning of the optimization process.

Hill Climbing is an iterative descent algorithm that explores the search space exhaustively by evaluating all possible directions at each step and selecting the one that yields the steepest descent in the objective function. Its key advantages are its straightforward implementation, robustness, and repeatability—since the same search path is followed consistently for a given input. Given an objective function

$$F = F(x, y) \tag{1.1}$$

where x and y are two vectors of "n" variables (in our case, the coordinates of the line-drawing vertices), the algorithm begins from an initial value of the vectors. It then evaluates F at 4n neighboring positions obtained by applying positive and negative offsets of a given step length S to each of the variables in x and y. The offset producing the greatest decrease in F (i.e., the best fit to the desired regularities) becomes the new current solution.

The process is repeated iteratively until one of the following stopping conditions is met: (1) the decrease in the objective function falls below a predefined minimum threshold, or (2) the number of iterations exceeds a maximum allowed value. When a stopping criterion is triggered, the cycle restarts with a reduced step size "S" to enable finer adjustments. The implementation allows for both parametric and fixed strategies to reduce the step size across iterations.

By default, the algorithm uses the parametric length adjustment strategy proposed by Conesa (2001), which balances performance and precision. Alternatively, users can define custom fixed step lengths or select from historically recommended values, such as: Marill's sequence: 1.0, 0.5, 0.1 (Marill, 1991) or Leclerc and Fischler's sequence: 0.125, 0.0625, 0.03125, 0.015, 0.007 (Leclerc and Fischler, 1992). Recommended settings for improving the speed, precision, and robustness of the optimization process are summarized in the dialog accessible via the *Recommended Parameters* button, as shown in Figure 3.52.

3.6.6 *2D refinement schedules*

REFER supports two distinct refinement schedules for processing vectorized line-drawings: interactive (online) refinement and batch (offline) refinement. The online schedule is typically associated with interactive

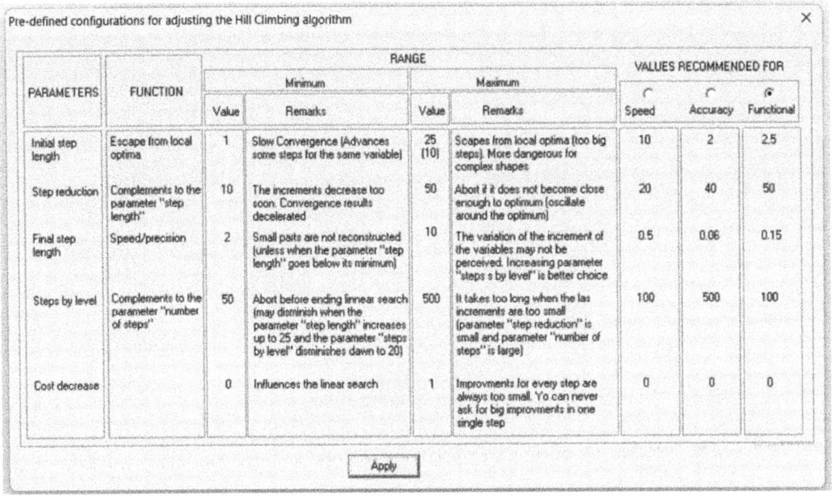

Pre-defined configurations for adjusting the Hill Climbing algorithm

PARAMETERS	FUNCTION	RANGE				VALUES RECOMMENDED FOR		
		Minimum		Maximum				
		Value	Remarks	Value	Remarks	Speed	Accuracy	Functional
Initial step length	Escape from local optima	1	Slow Convergence (Advances some steps for the same variable)	25 (10)	Scapes from local optima (too big steps). More dangerous for complex shapes	10	2	2.5
Step reduction	Complements to the parameter "step length"	10	The increments decrease too soon. Convergence results decelerated	50	Abort if it does not become close enough to optimum (oscillate around the optimum)	20	40	50
Final step length	Speed/precision	2	Small parts are not reconstructed (unless when the parameter "step length" goes below its minimum)	10	The variation of the increment of the variables may not be perceived. Increasing parameter "steps s by level" is better choice	0.5	0.06	0.15
Steps by level	Complements to the parameter "number of steps"	50	Abort before ending linnear search (may diminish when the parameter "step length" increases up to 25 and the parameter "steps by level" diminishes dawn to 20)	500	It takes too long when the las increments are too small (parameter "step reduction" is small and parameter "number of steps" is large)	100	500	100
Cost decrease		0	Influences the linear search	1	Improvments for every step are always too small. Yo can never ask for big improvments in one single step	0	0	0

Apply

Figure 3.52. Recommended parameters for Hill Climbing algorithm.

reconstruction workflows, whereas the offline schedule is used for batch-oriented reconstruction tasks.

By default, REFER employs the offline refinement schedule, which combines both immediate refinement tasks and those more appropriately applied after the sketch has been completed. Users are free to enable or disable either schedule via the *2D Refinement* Toolbar to explore alternative refinement flows.

3.6.7 *2D refinement filters*

3.6.7.1 *Short edges*

During the sketching process, users may inadvertently create very short strokes while moving the device that controls the cursor from one location to another—similar to how non-expert designers might leave unintended pencil marks when handling an actual pencil. In fact, it is uncommon to draw these types of small details intentionally during freehand sketching.

When this filter is active, edges that are shorter than a specified tolerance are detected and removed, as they are considered unintentional or extraneous lines. By default, the tolerance is set to 1% of the size of the 2D line-drawing. The value can be adjusted interactively through the *2D Refinement Configuration* dialog.

3.6.7.2 *Merge vertices*

A key stage in the refinement of 2D drawings is the detection of junctions. Vertices of the depicted 3D model—points where edges meet—are represented in the drawing by junctions, which are the locations where the tips of lines converge. Junctions are defined as X, Y coordinate pairs corresponding to shared endpoints of lines in the line-drawing that likely represent vertices of the 3D object. Many computer-based reconstruction approaches assume that junctions are carefully drawn in sketches. However, junctions are not always explicitly present in the original sketch, nor are they automatically generated during vectorization.

Detecting vertices requires merging the dangling tips of vectorized line segments in order to identify valid junctions. Shpitalni and Lipson proposed an endpoint clustering scheme that uses adaptive tolerances across different parts of a sketch (1997). Bonnici and Camilleri developed a vectorization algorithm based on multiple concentric families of circles placed over a dense grid in the image space (2015). Their method demonstrated the effectiveness of concentric circle samplers in detecting off-centered junctions within the sampled region. However, their experiments were limited to natural drawings.

Company *et al.* (2019) proposed an algorithm for identifying vertices in sketched drawings of polyhedral shapes, grounded in two central assumptions: (a) detecting junctions in vectorized sketches involves both geometric and perceptual considerations, and (b) experienced users may draw in either a precise or a casual manner, or both.

REFER also features a straightforward yet effective algorithm for detecting trihedral corners formed by exactly three edges whose tips converge within a small area. Based on the common occurrence of well-drawn trihedral (and often cubic) corners, this method is recommended as a pre-filter to identify the most evident vertices. It performs reliably in "box-like" prismatic structures, where thin walls produce closely spaced vertices that should not be incorrectly merged, as illustrated in Figure 3.53.

The approach by Company *et al.* (2019) is implemented in REFER as a 2D refinement filter. The objective is to first merge junctions that are clearly perceptible, while providing a second opportunity for less accurately drawn junctions to be identified in subsequent steps. This local perception approach is rooted in two complementary principles. Following the Gestalt law of proximity, carefully drawn tips that lie close to one

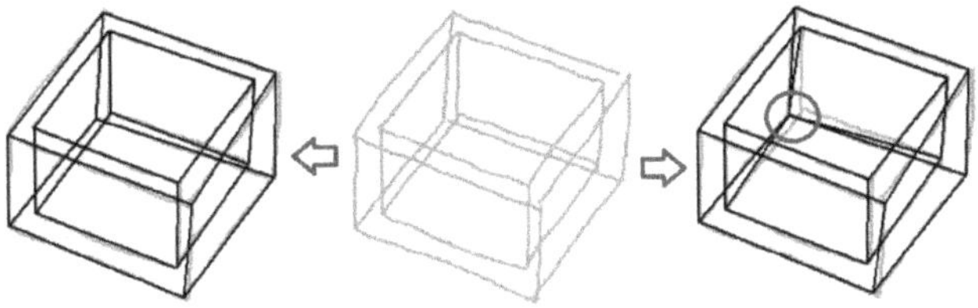

Figure 3.53. Correct (left) vs. incorrect (right) merging of vertices.

another are perceived as a single junction. This idea aligns with Hoffman's (2000) rules, particularly rule #2: "If the tips of two lines coincide in an image, then always interpret them as coinciding in 3D," and rule #4: "Interpret elements nearby in an image as nearby in 3D." Conversely, if the sketch is presumed to be casual, tips are considered imprecise, and junctions should instead emerge at the intersections of the lines or their extensions. In this case, the law of closure takes precedence, in line with Hoffman's rule #13: "If three or more curves intersect at a common point in an image, interpret them as intersecting at a common point in space."

Based on these principles, the approach proceeds in three practical stages:

(1) *Stage 1*: Merge neighboring tips, if they fall within a distance and orientation threshold.
(2) *Stage 2*: Merge triplets of dangling lines, if their intersections are near each other.
(3) *Stage 3*: Merge isolated dangling lines that pass near an existing junction.

The first stage employs a nearest-neighbor approach, based on the assumption that tips of carefully drawn lines corresponding to the same vertex are likely to be adjacent in a Delaunay triangulation of all tip points. This triangulation method is adapted from (Bourke, 1989). The algorithm sequentially merges the best-fitting pairs of tips according to the following conditions:

- The tips must be neighbors (no closer tip exists) and must belong to different lines. Neighbors are identified via the Delaunay triangulation, and after merging, the neighbor list is updated to include connections from previously merged tips.

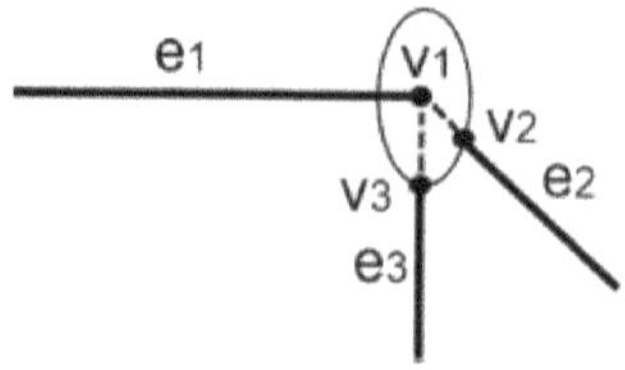

Figure 3.54. Distance threshold (*maxDistance*) between line tips.

- The tips must be closer than a distance threshold *maxDistance*, which is adjusted based on the angle between the lines. A trigonometric weighting scheme prioritizes perpendicular lines (200% allowance) over collinear ones (100%). The default *maxDistance* is 8% of the line's length. This threshold is illustrated in Figure 3.54.
- The angular difference between the lines must be less than *maxRotation*, typically set at 10 degrees.

Lines are processed in order from longest to shortest, under the assumption that shorter lines are more likely to be auxiliary. For nearly collinear lines (i.e., when their angle is below the *maxRotation* threshold), the common junction is not placed at the intersection point—which may lie far beyond the tips due to sketching imperfections—but at the midpoint between the two tips.

Initially, all vectorized line tips are disconnected, meaning each tip has a valence (i.e., number of incident lines) of 1. However, as merging proceeds, tips may become part of existing clusters. In such cases, the algorithm prioritizes junctions with higher valence. The rules are:

(1) If the valences differ, the tip with higher valence remains stationary, and the other tip is moved to the cluster.
(2) If the valences are equal (and both greater than 1), both tips are merged into the midpoint between them.

To ensure structural consistency, a merge is only accepted if the redirection of all lines connected to the moving tip requires less than *maxRotation* (10 degrees) of angular adjustment.

In the second stage, the algorithm identifies junctions by detecting clusters of three dangling lines that intersect near their tips. These intersections are valid only if the pairwise distances between them are less than 10% of the longest line in the triplet. Additionally:

- If the intersection lies inside a segment, its distance from the tip must be less than 50% (*inTol*) of the average length of the three lines.
- If it lies outside (i.e., on an extension), the maximum allowed distance is reduced to 25% (*outTol*) of the average length.

This asymmetry (*inTol* > *outTol*) favors intersections within the segments, which were found to more accurately match perceptual behavior. The resulting junction is placed at the centroid of the intersection triangle formed by the three lines. To avoid incorrect mergers, such intersections must be closest to the line tips.

Finally, in the third stage, remaining dangling lines may be merged to nearby existing junctions. A dangling line is clustered to a junction (of valence greater than or equal to 2) if the perpendicular distance from the line to the junction (e.g., distance *d3* in Figure 3.55) is less than *maxDistDangling*, which is set at 10% of the length of the dangling line. This strategy allows the inclusion of lines that would otherwise require excessive rotation to meet their original neighbor tips, thus relaxing the geometric constraints.

To quantify the likelihood that a junction is perceptually valid, a figure of merit is assigned to each one. This merit ranges from 0 to 1:

- 1 indicates that the junction is almost certainly perceived by human observers.
- 0 suggests that the junction is very unlikely to be perceived.

For carefully drawn junctions, the merit is 1 when the line tips fit perfectly. It decreases to a minimum value (default: 0.5) as the tips reach the maximum allowed separation. For casually drawn junctions, the maximum merit is capped at 0.5, decreasing to 0 when tips reach the distance

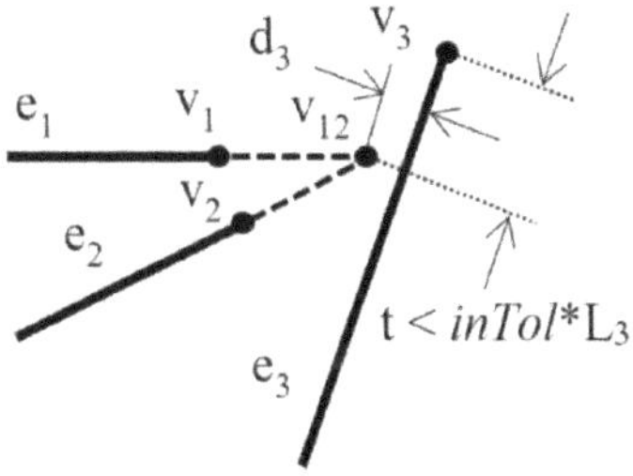

Figure 3.55. Merging dangling lines to existing junctions.

threshold. This scheme ensures that such "emerged" junctions have only average to low perceptual significance. When merging a dangling line to an existing junction, the merit is the minimum of two components:

- It is 1 when the line passes directly over the junction and decreases to 0 as the distance approaches *maxDistDangling*.
- It is 1 when the tip is the closest point on the line to the junction and decreases linearly if another point along the line is closer, up to the limit of the valid range.

Dangling lines with negative merit are not merged. The final merit of a junction that incorporates a dangling line is the lesser of the junction's current merit and the line's merging merit.

The default parameters were tuned to match the perceptual behavior of a test group composed mostly of individuals with technical backgrounds. This novel approach involved analyzing how participants perceived junctions in both carefully and casually drawn technical sketches (Company *et al.*, 2019). The aim is to match human perception (not necessarily the original intention of the sketch author). For this reason, developers should avoid validating perceptual algorithms solely against their own examples. Figure 3.56 shows how junctions were perceived in casual sketches. The data was collected from an experiment in which participants (mostly engineers) marked the tips of lines they perceived as forming junctions.

This process mirrors the behavior that the REFER algorithm emulates through its merit assignment system, capturing the essence of what is meant by a "perceptually driven" approach. In general, the algorithm is effective in replicating human perception. However, some casual intersections that are perceptually obvious to humans are not detected when using the default parameters, which are conservatively set to avoid false positives. Increasing parameter values can recover these missed junctions but also increases the risk of false mergers.

It is important to emphasize the local nature of the algorithmic approach, which focuses solely on the neighborhood of each junction, intentionally avoiding influence from global shape perception. This focus allows the formulation of reliable, isolated rules before exploring their integration into larger systems. Although machine learning (ML) approaches are likely to offer improved results, they inherently depend on global perception. Training an ML model using abstract sketches lacking

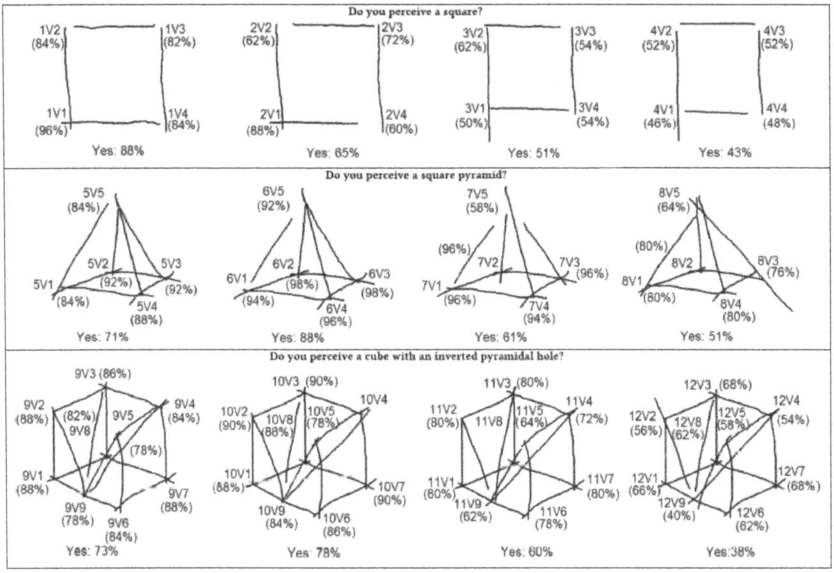

Figure 3.56. Sample data that shows how junctions are perceived in casual sketches.

any recognizable form would offer minimal insight. Conversely, training on meaningful sketches would unavoidably introduce human bias based on global shape understanding. These challenges undermine the goal of extracting the local perceptual rules that humans unconsciously apply when detecting vertices. Therefore, a strong foundation of local heuristics is essential before meaningful integration with ML-based methods can occur. Once established, synergies between well-tuned local strategies and modern AI techniques are worth pursuing.

A notable contribution in this direction is the work of Yin *et al.* (2022), who focused on repairing unintended gaps between strokes in artistic drawings. Their model uses local context cues to train classifiers that assess whether stroke pairs form end-to-end or T-junctions. The authors reported success in detecting intentional end-to-end junctions in wireframe polyhedral drawings. Their method seems to be effective for careful sketches—where strokes may fall just short of connecting—but struggles with casual sketches that contain longer-than-intended strokes. Such cases may result in false mergers. Furthermore, their model may not handle closely spaced junctions that correspond to distinct vertices at different depths, especially when they nearly occlude each other in the 2D projection. Despite these limitations, the use of machine learning in their approach is compelling and merits further exploration.

Regardless of whether traditional or AI-driven methods are employed, the most difficult and persistent challenge occurs when "local cues" are obscured or disrupted by "visual noise." This issue arises when neighboring vertices are separated at distances comparable to typical sketching inaccuracies, as shown in Figure 3.57. A path toward replicating the full spectrum of human perceptual abilities lies in effectively combining current "local strategies" with appropriate "global strategies" such as identifying vertices through face detection and vice versa.

3.6.7.3 *T-vertices*

Line labeling is an approach used to identify each line in a drawing as convex, concave, or occluding. The task requires a preliminary step that involves identifying the types of junctions in the drawing. One of the three basic types of three-line junctions is the *T-junction* (the other two are the *Y-junction* and the *arrow-junction*). A T-junction occurs when two of the three lines are collinear. Line labeling and junction classification are discussed in detail in the section on Cue Labels. While refining 2D line-drawings, special processing of T-junctions is often necessary. All T-junctions represent occlusion in trihedral objects. As such, they do not appear in wireframe drawings, but only in natural, perceptually accurate sketches, as shown in Figure 3.58.

In theory, T-junctions would be processed like any other junction. However, a unique issue arises in practice: people often draw collinear segments as a single continuous stroke, which hinders proper detection of the junction. In such cases, only one tip appears to intersect between the two tips of the other line, effectively obscuring the presence of a vertex.

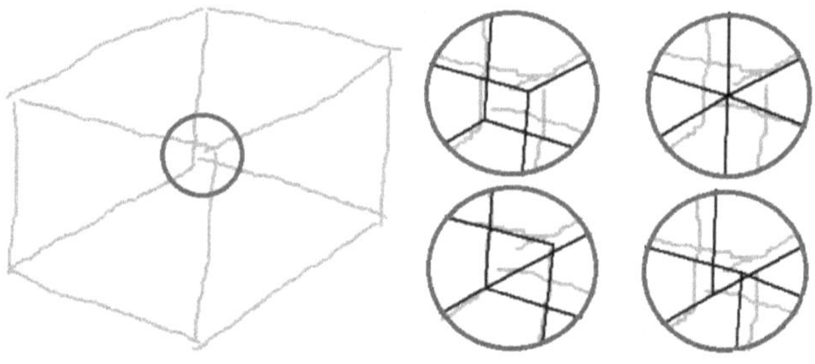

Figure 3.57. Example of local cues disrupted by visual noise.

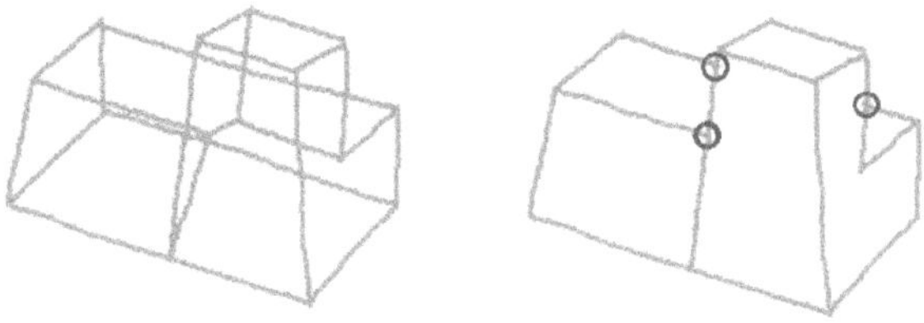

Figure 3.58. T-Junctions in a natural sketch.

This function addresses such issues by correcting T-junctions– adjusting their geometry and, when necessary, splitting the intersecting edge. The algorithm identifies a T-junction when one edge has a tip located within a specified tolerance of another edge, as long as neither of the two edges is already connected to additional segments. The algorithm iterates over all edges in the line drawing, searching for dangling edge tips that are close to another edge. When a candidate T-junction is found, the following steps are applied to both components of the T:

(1) The tip of the terminating edge (the lower part of the *T*) is moved to lie exactly at the intersection point of both lines.
(2) The crossing edge (the top bar of the T) is split into two new edges that share the newly adjusted vertex. This operation is only performed if the *Split* flag is active in the *2D Refinement Configuration* dialog.

3.6.7.4 *Collinear edges*

Overtracing (or redrawing) is a natural process by which designers iteratively translate their mental designs into refined sketches (Company and Varley, 2009). Overtracing can be categorized into different types: decoration (adding shadows, textures, etc.), thinking over the line (continuing to draw while mentally evaluating the design), and auto-correction (real-time correction of perceived drawing errors). Overtracing for decoration conveys meaningful visual information (e.g., curvature), while overtracing due to thinking or auto-correction should typically be interpreted as a simple line. Detecting overtraced lines is a complex task. The process involves at least three aspects:

(1) *Context*: the human ability to interpret 2D drawings as representing 3D shapes.
(2) *Formal code*: official drawing standards (e.g., ISO, ASME, EN).
(3) *Informal convention*: drawing practices passed down through education and training.

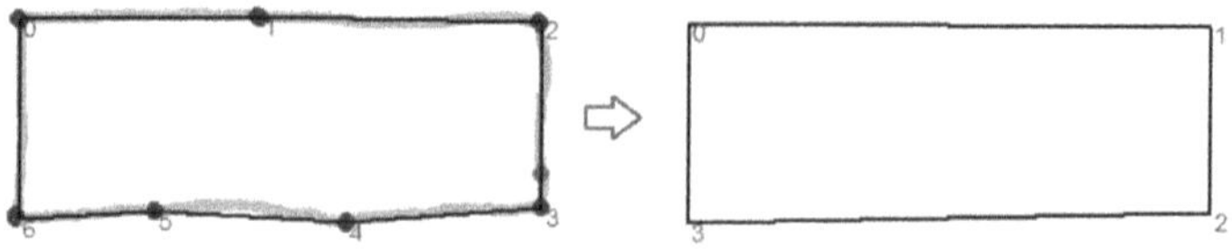

Figure 3.59. Refinement of collinear edges.

Reliable automated detection remains an open problem. This refinement filter addresses a basic but relatively frequent case. In brief, two edges that share one tip and are nearly parallel (within 5 degrees by default) are merged into a single edge, and the shared vertex is removed. The result is shown in Figure 3.59.

To avoid unintended remerging of edges involved in T-junctions (possibly created by the T-Vertices filter), collinearization occurs only when the two candidate edges are the only ones connected to the shared vertex. Note that this filter only affects edges that are already connected through a shared vertex. As a result, the vertex-merging filter must first identify and merge nearly parallel edges with closely situated tips before this filter can be applied. Once this step has been completed, the filter can effectively finalize the merging of collinear edges—even in casual sketches, as illustrated in Figure 3.60.

In future developments, this method could be combined with algorithms for merging dashed lines (perhaps by distinguishing between short and long strokes, and short and long gaps) to address one of the remaining challenges in vectorization grouping. Collinearity between edges that should remain separate can be managed using the *2D Refinement Optimization Coordinates* filter, which treats collinearity as one of several geometric regularities.

3.6.7.5 *Non-two-manifold vertices*

A two-manifold (or “2-manifold”) is defined as a body in which, at any point on its surface, a disk of arbitrarily small size can be constructed that is topologically equivalent to a disk on a plane (Hoffmann, 1989). A 2-manifold polyhedron is shown in Figure 3.61(a). A polytope is displayed in Figure 3.61(b), which is not a solid body, as indicated by the presence of a single hidden line. This structure is manifold-less since no disk equivalent to a planar disk can be constructed at the marked point. Finally, a non-2-manifold polyhedron is shown in Figure 3.61(c), where

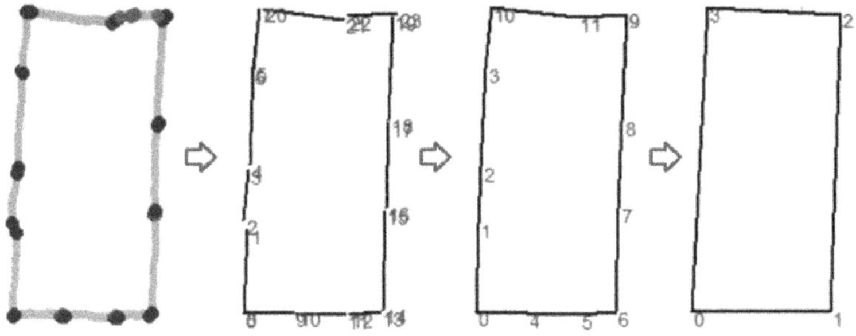

Figure 3.60. Merging of collinear edges.

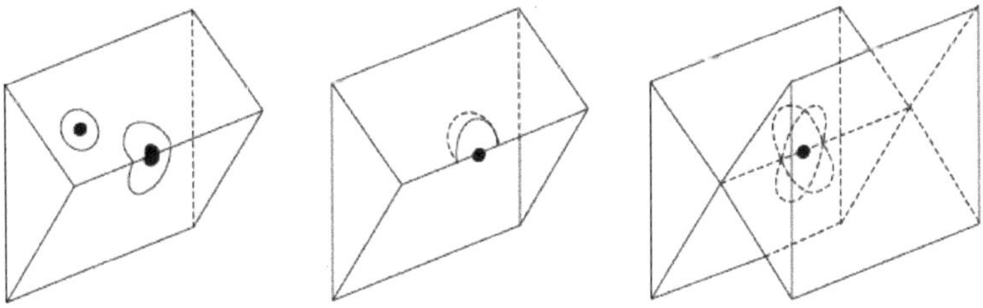

(a) (b) (c)

Figure 3.61. A 2-manifold polyhedron (a), a polytope (b), and a non-2-manifold polyhedron (c).

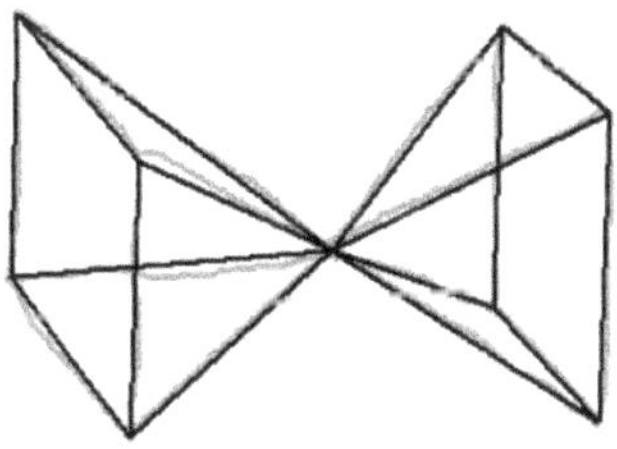

Figure 3.62. A non-2-manifold in REFER.

more than one such disk can be formed on different intersecting planes at the same point. Non-2-manifold shapes, like the one shown in Figure 3.62, are incompatible with most of the algorithms implemented in REFER.

To address this issue, REFER includes a filter for detecting and splitting non-2-manifold vertices to generate a set of manifold-compliant shapes. This filter only applies when the *Split non two-manifold vertices* option is enabled in the *2D Refinement Configuration* dialog. A vertex is considered non-2-manifold if:

(1) Its valence is 6 or greater (i.e., it is shared by six or more edges).
(2) Splitting it into separate (yet overlapping) vertices leads to a multi-graph drawing.

The procedure for handling such vertices is as follows:

(1) Each edge tip converging at the non-2-manifold vertex is reassigned to a new, individual vertex located at the same position.
(2) Next, the non-2-manifold vertex is reconstructed by merging only those edge tips that belong to the same subgraph (as determined after temporarily removing the original vertex).

This filter splits endpoints of edges that may have been previously merged by the *2D Refinement Merge Vertices* filter. While this might appear redundant, the distinction lies in the following criteria:

- The *Merge* operation uses a local criterion, typically based on proximity.
- The *Split* operation uses a global criterion, based on subgraph connectivity.

Both steps are essential. The merge filter would underperform if global constraints were introduced prematurely, while the split filter cannot function properly unless the endpoints have first been maximally merged. Together, they enable a more robust handling of complex sketch topologies.

3.6.7.6 *Non-two-manifold edges*

A non-two-manifold edge prevents most face-detection algorithms from working correctly and may even lead to catastrophic errors. To mitigate this issue, REFER includes a filter that identifies and splits non-two-manifold edges, producing a set of manifold-compliant shapes. The filter works by detecting such edges and splitting them into a pair of overlapping edges, each associated with one of the two distinct parts of the polyhedral structure. This filter only applies when the *Split non-two-manifold edges* option is enabled in the *2D Refinement Configuration* Dialog.

An edge is considered non-two-manifold if:

(1) Its valence is 5 or greater (i.e., it is connected to five or more faces).
(2) Splitting it into two (distinct but overlapping) edges results in a multigraph drawing.

The steps of the algorithm are described below:

(1) Both endpoints of the candidate non-two-manifold edge are split, which means that every edge tip converging at one of the endpoints is reassigned to a new, individual vertex at the same location—until the endpoints are isolated from all other edges.
(2) The subgraphs of the resulting graph are computed. Then, the subgraph defined by the non-two-manifold edge is temporarily removed.
(3) If and only if the remaining graph splits into exactly two disconnected subgraphs (thus confirming that the edge is non-two-manifold), the previously split vertices are selectively remerged. Specifically, all edge tips initially connected to each endpoint of the original edge and now belonging to the same subgraph are grouped into a single vertex.
(4) A new copy of the non-two-manifold edge is created for each subgraph, connecting the corresponding remerged vertices. Finally, the original (now decoupled) edge is removed.

3.6.7.7 *Optimize coordinates*

When the starting drawing is a rough sketch with imprecise (x, y) coordinates, the resulting 3D inflation often leads to distorted or even unbuildable shapes. To address this issue, one effective strategy is a process known as 2D reconstruction by optimization, which aims to output a geometrically consistent line drawing by refining vertex positions. This optimization process is implemented as a filter in REFER and is only meaningful when applied as part of offline refinement in a predefined refinement schedule.

Accordingly, REFER incorporates a dedicated reconstruction stage that performs an optimization-based offline refinement of the 2D drawing.

This approach adjusts the X and Y coordinates of the vertices to better align with up to three key regularities commonly found in line drawings (Piquer, 2003; Company *et al.*, 2004):

- Parallelism. This regularity assumes that many lines in a drawing are meant to be parallel. The algorithm identifies pairs of nearly parallel lines and penalizes their angular deviation in the objective function. The penalty is inversely proportional to the angle of convergence between each pair, encouraging those lines to become more parallel.
- Verticality. Based on the observation that designers tend to choose viewpoints in which the vertical dimension of objects aligns with the vertical axis of the sketch, this regularity assumes that almost-vertical lines are intended to be perfectly vertical.
- Collinearity. Derived from Hoffman's Rule 3 (2000), which states that lines appearing collinear in a drawing are typically interpreted as collinear in the 3D object, this regularity encourages such visual alignments to be preserved during optimization.

The activation, deactivation, and threshold configuration for each of these regularities are controlled via the *2D Refinement Regularities Details* dialog.

The optimization itself is carried out using a Hill-Climbing algorithm, an iterative method that begins with an initial configuration and applies small, incremental changes to search for an improved solution. After each iteration, the process continues from the improved configuration until no further enhancements can be made. While hill-climbing performs well for correcting small deviations in vertex coordinates, it is sensitive to poor starting conditions and can become trapped in local minima. The algorithm's behavior is configurable via the *2D Hill Climbing Details* dialog.

The 2D optimization filter typically yields high-quality results, greatly enhancing the accuracy and realism of the final 3D model. However, it comes with a computational cost. While simple sketches can be processed in seconds, more complex drawings may take several minutes, which limits the practicality of this method for interactive applications unless further performance improvements are made.

There are also theoretical limitations to this approach. Parallelism and collinearity are broad assumptions that describe common, but not universal, design patterns. In cases involving unusual geometries or accidental viewpoints, these regularities may introduce false positives. For example,

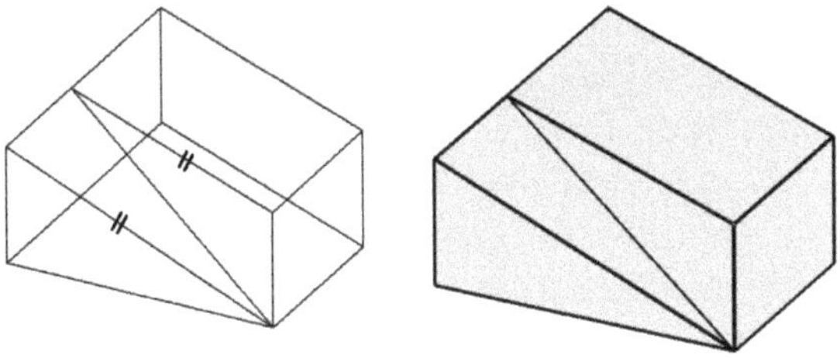

Figure 3.63. The two edges marked with the symbol "||" are parallel in the sketch (left) but not parallel in 3D.

two edges that are not parallel in 3D may appear parallel in 2D due to projection effects (see Figure 3.63), leading to incorrect refinement.

To visualize the impact of this refinement process, the *Display Optimization Process* view can be used, which shows the modifications made to the 2D drawing as a result of coordinate optimization. Note that the data is replaced once the 3D model has been inflated. Therefore, the visualization is only available when the drawing has been refined but the 3D inflation has not yet occurred.

3.7 Detection

3.7.1 *Introduction*

Detection is the process of identifying the key information necessary to reconstruct the 3D shape that is implicitly described within a 2D line drawing. This process follows an inductive approach, which progresses from specific visual observations to more abstract generalizations. Both geometric analysis and perceptual inference are employed to extract the embedded semantics that are present in the drawing.

Semantics are defined as the relationship between 2D signifiers, such as points, edges, and line arrangements, and what they represent–the intended 3D functionality or structural purpose of a designed object. Detection thus involves uncovering the technical and design-oriented meanings behind these visual cues, interpreting the drawing not just as geometry, but as an expression of a conceptual design solution.

Semantic information can be broadly classified into two categories: low-level semantics, simple geometrical regularities or arrangements that suggest fundamental structural patterns; and high-level semantics, more complex features or combinations of elements that reveal the overall intent or functionality of a design component. These levels are typically interrelated, forming a loosely hierarchical structure, where low-level semantic elements often serve as foundational clues for deducing more abstract high-level interpretations.

The general workflow for configuring and executing the detection process is divided into two primary detection modes, both accessible from the *Detection* toolbar:

(1) *Detection of cues*: Cues are localized geometric patterns or regularities that convey low-level semantic meaning. These include configurations like parallelism, symmetry, or vertex valence. Their detection is managed via the *Detect Cues Configuration* dialog box.
(2) *Detection of features*: Features are larger, more structured geometric elements that relate to high-level semantics, such as handles, holes, protrusions, or specific functional shapes. These are configured through the *Detect Features Configuration* dialog box.

REFER implements a *cascade* detection strategy. It first identifies simple cues such as vertex valences, which then serve as the basis for discovering more complex patterns and features. Furthermore, REFER does not aim to extract all possible information from the drawing. Instead, it focuses on identifying only the data necessary for reconstruction and interpretation, in contrast to traditional labeling approaches that attempt to fully annotate every element. Nevertheless, labeling capabilities are included in REFER, acknowledging its historical significance in the field and its potential value for future research. As REFER serves as an experimental platform for exploring diverse modeling strategies, such legacy techniques remain accessible and may inspire or complement modern detection pipelines.

3.7.2 *Detection toolbar*

The *Detection* toolbar is located over the parent window (see Figure 3.64).

The following commands are provided:

- *Toggle detection*: A switch button that enables or disables the automatic detection of cues and features. When enabled, detection will automatically be triggered during the reconstruction process, as initiated through the *Reconstruction* Toolbar.

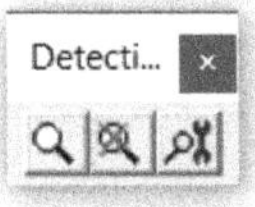

Figure 3.64. Detection toolbar.

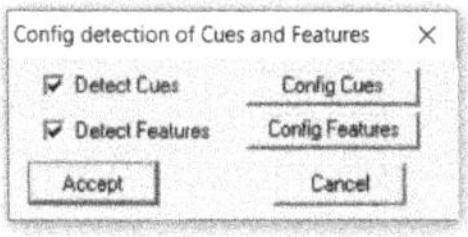

Figure 3.65. Configure Detection of Cues and Features dialog box.

- *Clear detected data*: Deletes all previously detected cues and features from the internal database. This operation is automatically performed at the beginning of each full reconstruction. It is particularly useful in manual workflows, where partial rebuilds are being performed and previous results need to be explicitly cleared.
- *Configure detection settings*: Opens configuration dialogs to adjust the detection parameters for cues and features. The *Configure Detection of Cues and Features* dialog box is used to enable or disable the detection process and, more importantly, to access two specialized configuration panels. These panels allow the user to specify which cues and features should be detected, as well as to define the procedures used for their identification. These panels can be opened by clicking the corresponding buttons: *Configure Cues,* to adjust settings related to cue detection, and *Configure Features,* to adjust settings related to feature detection, as shown in Figure 3.65.

3.7.3 *Detect cues configuration dialog*

The *Detect Cues Configuration* dialog box can be accessed from the *Detection* Toolbar, which manages both cue and feature detection, or from the *Cues Configuration* toolbar. The *Detect Cues Configuration* dialog allows users to enable or disable detection for each specific cue and configure the respective detection parameters. The cues are organized hierarchically in the dialog to show their parent/child relationships. The *Details* buttons in the dialog box open additional dialog boxes to control the specific parameters of different cues. The *Detect Cues Configuration* dialog is shown in Figure 3.66.

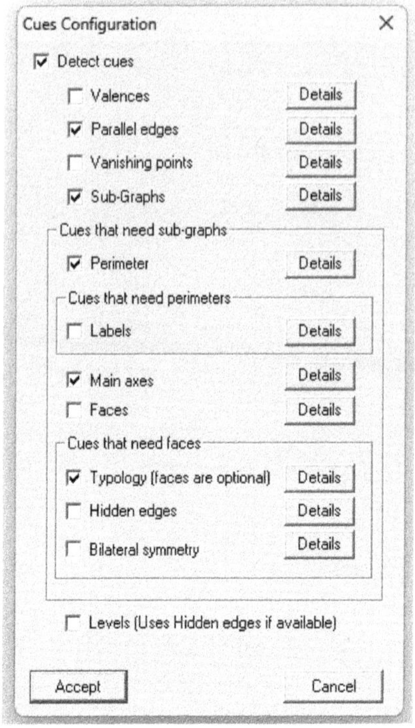

Figure 3.66. *Detect Cues Configuration* dialog.

3.7.4 *Detect features configuration dialog*

The *Detect Features Configuration* can be accessed from the *Detection Toolbar or the Feature Configuration* toolbar. The dialog allows users to individually enable or disable the detection of specific features. More importantly, it provides access to detailed configuration options for each feature through dedicated *Details* buttons. The Feature Configuration dialog box is shown in Figure 3.67.

Because the order in which features are detected can significantly affect both accuracy and efficiency, the system includes a default detection sequence that has been optimized through extensive testing. Users can modify this sequence to experiment with potentially more effective alternatives. It is important to note that some features may depend on others, so the order of detection can be critical to successful identification. Additionally, detecting certain features together can be advantageous. This strategy is especially effective when features share similar topological structures or are revealed by similar cues in the 2D sketch.

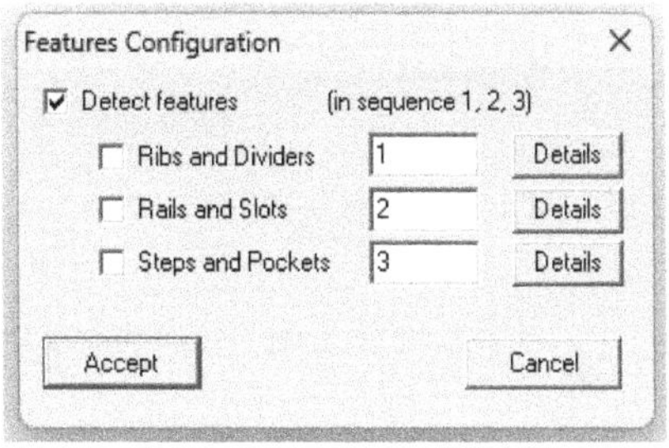

Figure 3.67. *Feature Configuration* dialog.

The *Details* buttons in the main dialog open individual configuration dialogs, where users can fine-tune the parameters specific to each type of feature. For example, the algorithm for detecting ribs and dividers has been calibrated based on the following thresholds:

- (T1) Lines are considered short if their length is less than 20% of the minimum side of the sketch's bounding box.
- (T2) A line is considered long within a quadrilateral if its length exceeds twice the longer of the two short sides of the quadrilateral.
- (T3) Lines are considered collinear if their angular deviation is less than 9 degrees.

Thresholds for detecting similar faces are:

- (T4) Vertex pairs are allowed if their distances differ by less than 20% (relative to the median).
- (T5) The orientation of segments connected to paired vertices must differ by less than 15 degrees.

The minimum edge-length threshold for feature detection is defined relative to the shortest side of the sketch's overall bounding box. This global reference is used in place of evaluating the lengths of individual lines.

While this may seem restrictive—particularly for identifying thin, isolated walls—it aligns with the goal of detecting thin features that are functionally embedded within larger shapes. The thinness is assessed in proportion to the overall object, not as an absolute dimension. Similarly, maximum thresholds are defined relative to the minimum size, not to the full object size. The dedicated dialog for configuring ribs and dividers is shown in Figure 3.68.

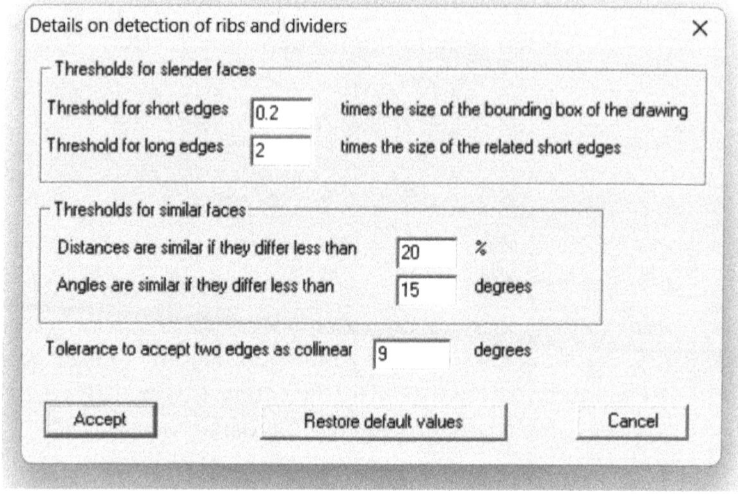

Figure 3.68. *Details on detection of ribs and dividers* dialog.

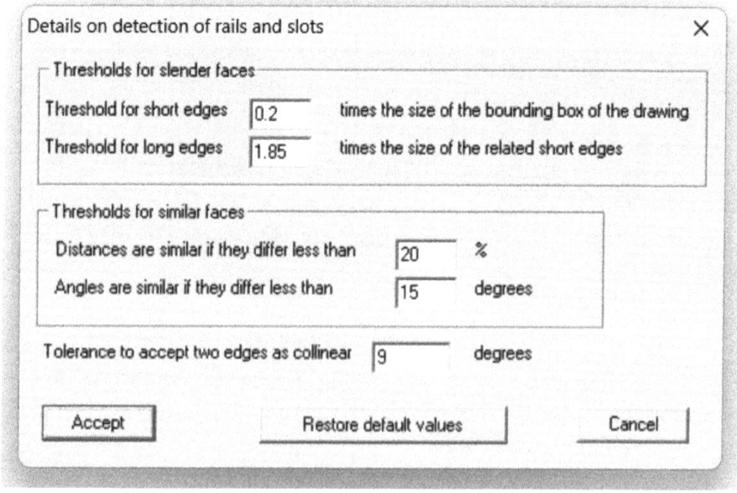

Figure 3.69. *Details on detection of rails and slots* dialog box.

The configuration for detecting rails and slots is similar, though with some variations to account for their specific characteristics, as shown in Figure 3.69. The recommended settings for detecting steps and pockets are illustrated in Figure 3.70.

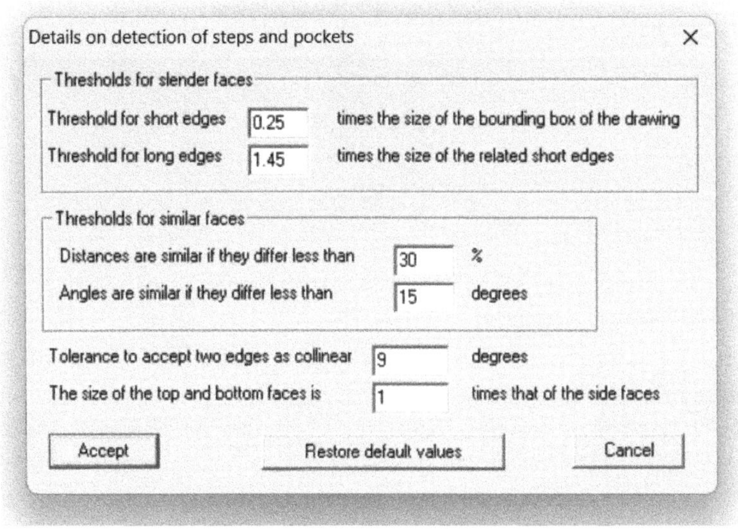

Figure 3.70. *Details on detection of steps and pockets* dialog box.

Feature detection is highly sensitive to the values of the configuration parameters, and small changes can significantly alter the detection results. For example, in Figure 3.71, the algorithm fails to recognize a slot due to projection effects: the top and bottom faces, which are large in the actual 3D model, appear very small due to the oblique viewpoint. While human perception compensates for this effect, geometric measurements do not. To resolve it, the final parameter in the slot detection configuration must be adjusted—specifically, the base-to-side surface area ratio must be reduced to 0.7 to allow correct detection.

3.7.5 *Cues*

3.7.5.1 *Introduction*

Cues (also known historically as clues or regularities) are defined as properties or patterns in a 2D sketch or drawing that indirectly reveal structural or functional characteristics of the 3D object being depicted. These visual hints help bridge the gap between the two-dimensional input and the three-dimensional shape it implies. The concept of cue is explored in more depth in the following section.

The idea of cue was first introduced by Marill (1991), who proposed that certain patterns in line drawings could suggest three-dimensional

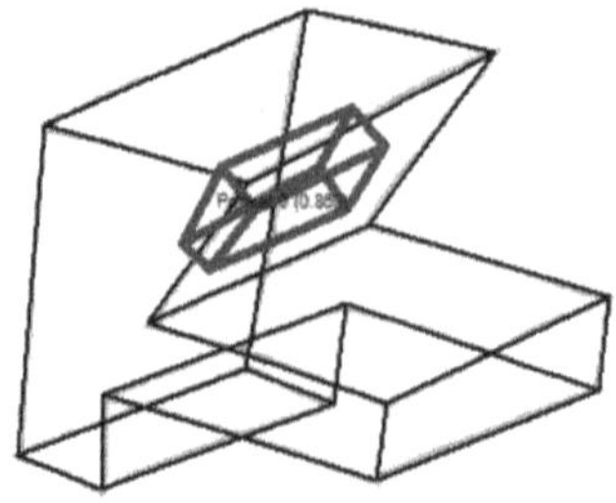

Figure 3.71. Slot detection failure due to projection effects.

structure. Leclerc and Fischler (1992) later formalized some of these visual regularities through analytical formulations, providing a theoretical basis for cue-based detection. Lipson and Shpitalni (1996) expanded on this approach, delivering consistent and practical formulations for a broader set of regularities. Yuan *et al.* (2008) took a step further by attempting to automatically select the most relevant regularities for a given sketch through a process they termed *Automatic Relevance Detection.*

While the automated relevance detection method proposed by Yuan *et al.* (2008) is technically interesting, its underlying premise is problematic. Regularity selection, in our view, should not be based purely on statistical or algebraic optimization. Rather, it should be grounded in perceptual and cognitive criteria—guided by how humans naturally interpret sketches. Nonetheless, we recognize the value of ongoing research in this area and welcome continued contributions that refine or challenge current methodologies.

REFER implements a perceptually motivated approach to cue detection. A regularity is assumed to be present if it is one the user is likely to perceive. This approach reflects the cognitive paradigm that underpins REFER's development: if the user sees it, then it should be considered to exist.

Perceptual cues enable the interpretation of information that is not explicitly encoded in the drawing. The pioneering work of Suwa and Tversky (1997) helped clarify how perception operates in the context of sketches, debunking several common misconceptions. Additional contributions include Palmer's comprehensive study of visual perception (Palmer, 1999) and Hoffman's algorithmic exploration of perceptual organization (Hoffman, 2000), which links perception to computational processes.

Cues are also used beyond traditional engineering drawing interpretation. For instance, *Miao et al.* (2015) leveraged symmetry cues to guide the inflation of free-form 3D shapes from sketches, illustrating the broad applicability of cue-based reasoning in computer graphics and shape processing.

To work effectively with cues in REFER, users should consider the following:

(1) The calculation of cues is managed through the *Cues Configuration* toolbar.
(2) Detected cues are displayed in the *2D Sketch Window*, which visualizes the drawing and its interpreted structures.
(3) Tools for controlling the visualization of cues can be found in the *Display Cues* toolbar.

3.7.5.2 *Cue definition*

Assigning meaning to what the eyes see is not a deterministic or immutable process. Visual perception is inherently ambiguous—each scene or sketch admits countless interpretations. To resolve this ambiguity, the perceptual system relies on principles—fundamental propositions that serve as the foundation for reasoning and interpretation.

Some researchers refer to these perceptual propositions as "principles" precisely because they are not as strong or empirically rigid as scientific laws (Palmer, 1999). From these principles, more concrete rules—defined as prescribed guides for action—can be derived to support algorithmic interpretation of visual input. These rules often resemble pseudocode in structure, acting as procedural steps that govern how the visual system (or a visual algorithm) infers structure from appearance.

The term *cue* is used to denote a visual stimulus that, when perceived, provides information about a characteristic or feature of the depicted object. Historically, these cues were called regularities, reflecting the observation that early cues corresponded closely to properties that define geometric regularities. In formal geometry, regular shapes are those that are edge-transitive, vertex-transitive, and face-transitive—meaning there exists a symmetry that maps any edge, vertex, or face onto any other of its kind. This concept of geometric uniformity inspired early computational approaches to cue detection.

In a broader cognitive and psychological context, cues may be understood as natural or unintended signs. The American Psychological Association defines a *perceptual cue* as "a feature of a stimulus that is perceived and used by an organism in a particular situation or setting to identify and make judgments about that stimulus and its properties." (American Psychological Association, n.d.).

Clearly, the number of possible visual cues is infinite. Nevertheless, psychologists have made notable progress in identifying and formalizing a core set of reliable principles. Of particular relevance is the algorithmically-oriented framework proposed by Hoffman (2000). Meanwhile, artists, architects, and geometers developed principles of graphical projection long before computational modeling existed. These include projection invariants, which are properties of 3D shapes that must be preserved in any geometrically valid 2D representation. These invariants provide essential cues for model reconstruction and REFER is designed to algorithmically identify as many of them as possible.

Once a sketch has been vectorized and refined, REFER enters the cue detection stage. During this phase, the system attempts to detect as many significant cues (as configured by the user) as possible before proceeding with 3D reconstruction.

Cues are organized in a fixed priority hierarchy. Some cues depend on others and must be computed at the right time. For example, detection of perimeters relies on prior detection of subgraphs. The *Cues Configuration* toolbar allows users to manage cue detection, including enabling or disabling specific cues and understanding their dependencies. The cues currently supported and detected by REFER include:

- Hidden edges (identified as dashed strokes during the non-solid stroke grouping phase).
- Valences of vertices.
- Parallel edges.
- Vanishing points.
- Subgraphs.
- Perimeters.
- Labels for junctions and lines based on their typology.
- Main axes.
- Faces.
- Typology.
- Hidden edges (further refined post-vectorization).

- Bilateral symmetry.
- Levels.

Historical development also plays a role in cue arrangement. For example, the algorithm to detect candidate lines for parallel edges was developed before the vanishing point detection algorithm. The latter not only identifies vanishing points but also clusters lines that converge toward them, effectively superseding the earlier method in some cases. However, both algorithms are maintained for two key reasons: (1) each method has strengths and weaknesses and may fail under different conditions, and (2) REFER is designed as a research and experimentation platform, giving users flexibility to explore and compare alternative interpretations.

The principles and rules used to interpret sketches are not "hard" constraints. Rather, they function as soft or probabilistic guides. They are often referred to as ceteris paribus (all else equal) rules, which means they hold true in general but may conflict under certain conditions. In such cases, the perceptual system resolves the ambiguity by assigning weights or figures of merit to each cue, assessing the likelihood that each one accurately reflects a real feature.

The concept of figures of merit was already used in the earlier vectorization stage. In cue detection, plausible cues are automatically assigned high merit, while ambiguous ones receive a merit score reflecting their credibility. An important early contribution to managing sketch imprecision through merit-based reasoning is the ADG framework (Lipson and Shpitalni, 1996), which inspired our current approach to computing main axes. Varley (2002) also provided valuable insights into assigning merit scores, although the computational approaches remain heuristic. Continued research is needed to formalize and refine these strategies.

3.7.5.3 *Cue configuration toolbar*

The *Cue Configuration toolbar* (shown in Figure 3.72 is located over the parent window.

The toolbar contains the following commands:

- Detect valences
- Detect parallel edges
- Detect vanishing points

- Detect subgraphs
- Detect perimeters
- Detect labels
- Detect main axes
- Detect faces
- Detect typology
- Detect hidden edges
- Detect bilateral symmetry
- Detect levels
- Detect Cues Configuration dialog box

The commands to initiate the detection of cues are organized in a parent-child sequence. Child cues can only be accessed after the corresponding parent cues have been activated. For example, the *face detection* command activates hidden edges and bilateral symmetry.

3.7.5.4 *Cue valences*

The valence of a vertex is defined as the number of edges connected to it. In the context of 2D sketches, valence involves counting the number of line tips that are so spatially close to one another that they are assumed to form a single junction, implying a shared vertex in the corresponding 3D model. The concept is illustrated in Figure 3.73.

Valence information is useful in various processes, including the detection of main axes, determining model typology, and executing normalon inflation.

Figure 3.72. *Cue Configuration* toolbar.

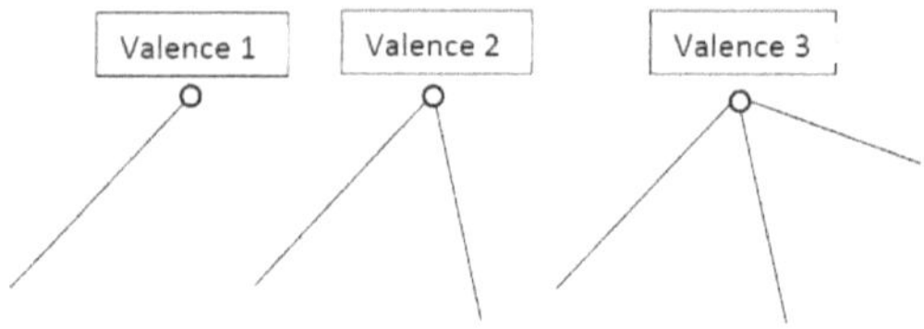

Figure 3.73. Cue valences.

This cue is not activated by default. It is computed and saved only upon user request, and solely for visualization purposes (see *Display Valences of Vertices*). Examining valence values is a common strategy to detect edge tips that have been incorrectly grouped, which can lead to the incorrect definition of vertex structures in the sketch, as shown in Figure 3.74.

Throughout the application, valence values are recomputed on-the-fly as needed by calling the *Get Size Edges In Vertex* operation from the *Data Base* module. Since the database changes frequently, recalculating the values ensures accuracy and consistency, and is preferred over persistent storage. When requested by the user, the system can treat collinear edges as a single edge when calculating valences, as shown in Figure 3.75. This setting is available in the *Valence Settings* dialog.

The configuration dialog for cue valences can be accessed by clicking the corresponding *Details* button in the *Detect Cues Configuration* dialog box. This dialog allows the user to choose how valences are calculated—either by treating collinear edges as a single edge or by counting each edge individually. This setting affects how the application interprets junction complexity.

When collinear edges are treated as a single edge, vertex valences tend to be lower, as illustrated in Figure 3.76. This approach can be useful for reducing noise in highly regular or linear structures. However, collinearity is subject to a detection threshold, i.e., edges must fall within a defined angular margin to be considered collinear. This scenario is evident in vertex V4 in the example shown in Figure 3.76.

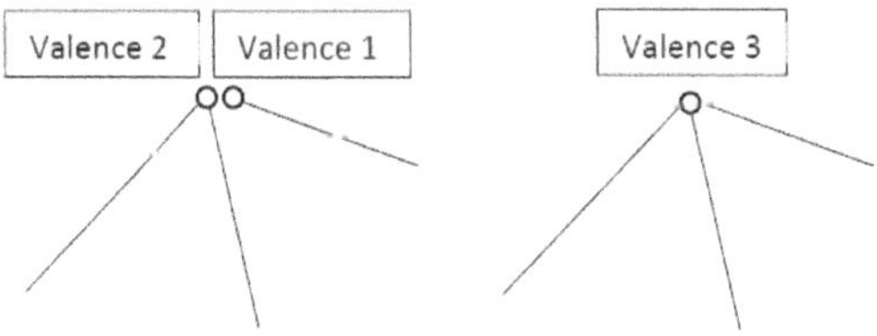

Figure 3.74. Incorrect (left) and correct (right) definition of vertex structures.

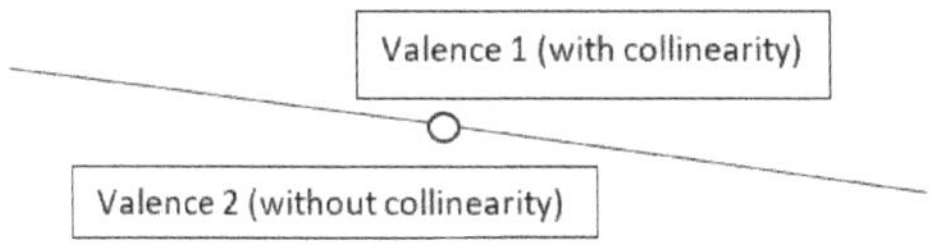

Figure 3.75. Valences in the treatment of collinear edge.

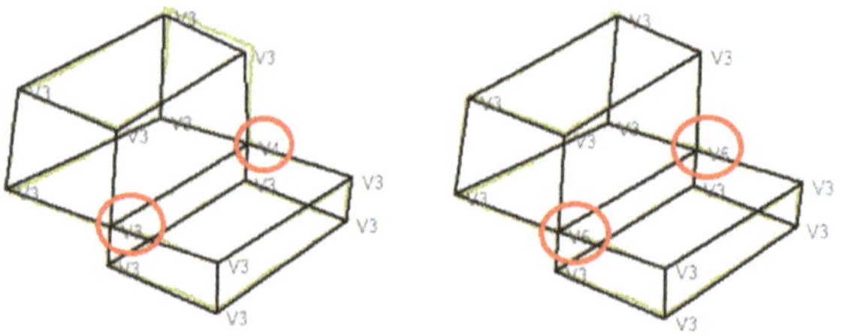

Figure 3.76. Lower valences in collinear edges when treated as a single edge.

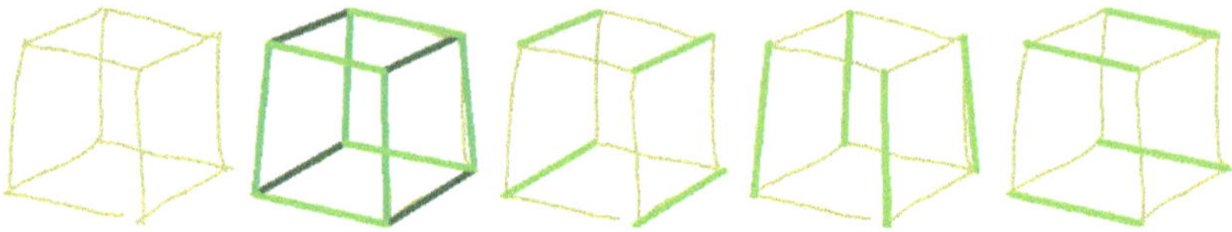

Figure 3.77. Cue Parallel edges.

Choosing whether to group collinear edges depends on the context of the sketch and the desired granularity of vertex connectivity. In cases where precise control over geometry is needed, such as when distinguishing between slight angular deviations, disabling collinearity merging can provide more accurate results.

3.7.5.5 *Cue parallel edges*

This feature returns the lines of the drawing grouped into bundles, each representing sets of edges that are intended to be parallel in the 3D model. These grouped lines provide insight into the spatial structure of the sketch and can be visualized through the *Display Parallel Edges* option. An example is show in Figure 3.77.

REFER implements an automatic detection algorithm to identify parallel edges which adapts its method depending on the characteristics of the sketch. For shapes drawn in parallel projection and exhibiting regular angular structure–referred to as *normalon* shapes–the algorithm leverages angular distribution analysis. In other cases, particularly those drawn with perspective distortion, it switches to the vanishing points method.

A shape is classified as *normalon* if it satisfies two criteria: (1) there are exactly three dominant angles present in the sketch, and (2) all edges align with one of the three principal directions. The angular distribution method for detecting parallelism in such shapes is based on the technique proposed by Lipson and Shpitalni (1996). The method involves constructing an angular histogram by computing the orientation of each line and recording the frequency of each angle. The histogram is refined using a smoothing technique:

Each angle contributes to the histogram through a Gaussian distribution centered on that angle, with a default standard deviation of 7 degrees ($\sigma = 7$), to accommodate sketching inaccuracies. These Gaussian curves are superimposed and weighed appropriately. Once all angles have been processed, the combined graph is normalized so that its peak reaches a value of 1.0.

The resulting Angular Distribution Graph (ADG) displays dominant angle groupings as noticeable humps or peaks. A peak is considered valid if its crest exceeds a predefined frequency threshold, known as *MinCrest*. These peaks correspond to the prevailing directions of edges in the sketch and form the basis for grouping lines into parallel bundles.

The same ADG-based method is also applied in REFER to compute the Main Axes of the sketch. Readers interested in a comprehensive explanation of the technique can refer to this section.

Various tolerance parameters can be configured to account for imperfections and variations in hand-drawn sketches. These settings can be adjusted in the *Parallel Edges* dialog, providing flexibility in edge detection sensitivity.

The dialog box for configuring parameters related to the *Cue Parallel Edges* feature can be accessed by clicking the corresponding *Details* button in the *Detect Cues Configuration* dialog box. The dialog is shown in Figure 3.78.

The set of tolerance parameters available for adjustment varies depending on the type of shape being analyzed:

- For normalon shapes, only a general threshold is required to group lines that are "nearly" parallel. This threshold helps accommodate the inherent inaccuracies of freehand sketches.
- In the specific case of cabinet views, the Axonometric Inflation process may encounter failures due to incompatible angle relationships. To resolve this issue, two parameters are provided that allow users to manually substitute conflicting angles with custom-defined values.
- The detection of pyramids assumes that a majority of edges converge at a single vertex. The algorithm uses this assumption to properly identify bundles of parallel edges.

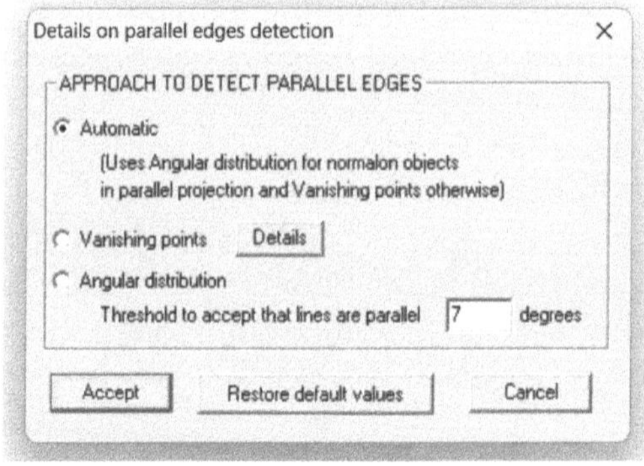

Figure 3.78. *Details on parallel edges detection* dialog.

- For prisms, detection is based on the assumption that a significant number of edges run parallel to one another. The system uses this pattern as the basis for identifying groups.
- Undefined shapes–those that do not clearly fall into the categories above—are labeled as type *D* (Dominant) when a substantial number of their edges are parallel to the main axes.

3.7.5.6 *Cue vanishing points*

For SBM systems to emerge as a viable alternative to both traditional WIMP-based CAD tools and hand-drawn conceptual sketching, they must support the entire range of conceptual design sketches. Although most engineering sketches leverage orthographic projection (Plumed *et al.*, 2010), it is also essential to support perspective projection—particularly when designers sketch informally. Perspective projection, in turn, requires the accurate detection of vanishing points (VPs).

Several vanishing point detection algorithms have been developed that align with human visual interpretation and can be fine-tuned to approximate human perception (McLean and Kotturi, 1995; Rother, 2002). However, none of these methods perform reliably in the presence of the imperfections commonly found in freehand sketches.

Most existing VP detection methods were originally developed for processing 2D camera images. The types of errors these algorithms address—such as lens distortion or noise from line segment extraction—are typically much smaller in magnitude than those found in manually drawn sketches. Among these approaches, the clustering strategy introduced by McLean and Kotturi (1995) is notable for its relative robustness to noisy input data.

REFER focuses exclusively on methods that take vectorized lines as input. Bitmap-based techniques, such as those by Barnard (1983) and Magee and Aggarwal (1984), are not considered. Other methods like Varley's (2003) which rely on or infer higher-level semantic information from the input are also excluded. However, the method proposed by Tardif (2009) is particularly relevant. It addresses the challenge of identifying vanishing points from sparse edge data by returning both a set of candidate VPs and a classification for each input edge—either associated with a VP or labeled as an outlier. Tardif's use of J-Linkage clustering improves on earlier approaches but introduces two limitations: (a) it is computationally expensive, and (b) it is non-deterministic, i.e., results vary with each execution and reflect likelihood rather than certainty.

REFER's objective is to detect the vanishing points that are implicit in conceptual engineering sketches, particularly for polyhedral objects. Adapting image-based computer vision techniques proves inadequate for this purpose, as they cannot handle the sketch inaccuracies that are commonly encountered. For example, a rectangular grid—often used in perspective analysis—fails when sketch lines deviate slightly from expected angles, as illustrated in Figure 3.79.

While these types of deviations may confuse algorithms, they do not typically affect human interpretation. In this regard, REFER implements a vanishing point detection algorithm that emulates human perceptual

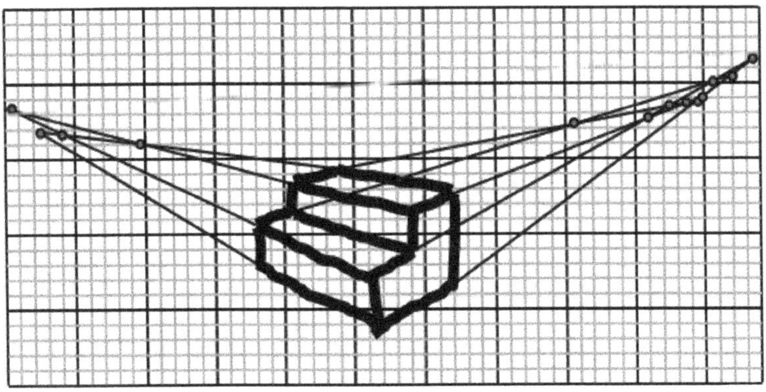

Figure 3.79. Rectangular grids fail when sketch lines deviate slightly from expected angles.

mechanisms. The algorithm was validated using perspective sketches derived from engineering drawings of polyhedral objects.

The algorithm takes as input a set of vertices and edges and returns groups of lines—bundles that are assumed to represent sets of edges that are parallel in the corresponding 3D model. REFER's approach is influenced by the exhaustive candidate exploration proposed by Rother (2002), as well as the clustering techniques of McLean and Kotturi (1995) and Tardif (2009). Two core ideas distinguish REFER's method:

(1) Direct clustering of candidate vanishing points. Unlike prior work that clusters lines, REFER forms clusters from the candidate VPs themselves. This strategy reduces computational complexity by eliminating the need to repeatedly compute intricate distances (e.g., voting distances (Rother, 2002) or probabilistic estimators (Tardif, 2009)). It also increases resilience to noise in the input data.
(2) Using polar coordinates centered at the sketch centroid: In the method implemented by REFER, similarity between vanishing point candidates is evaluated based on angular difference alone. Since sketching inaccuracies typically cause more distortion in position than in orientation (Plumed *et al.*, 2013), this polar representation is more tolerant to uncertainty in VP locations while preserving directional accuracy.

Despite these improvements, imperfect clustering may still occur. To address this issue, REFER applies several filtering and correction stages. The vanishing point detection algorithm implemented in REFER consists of the following eight stages (Company *et al.*, 2014):

(1) Candidate VP enumeration. All plausible vanishing points are computed. To mirror human perception, a valid region is defined based on the sketch size. Candidates must lie between an inner radius (*Ri*) of 1.5 times the sketch's size and an outer radius (*Ro*) of 5 times its size.
(2) Clustering VPs. Candidate points are grouped based on angular proximity in polar coordinates.
(3) Filtering intersecting edges. Lines that intersect at inappropriate angles or positions are removed.
(4) Filtering repeated edges. Redundant or overlapping lines that may skew clustering are eliminated.
(5) Filtering parallel edge groups. If the centroid of a VP cluster lies beyond the outer radius (*Ro*), the associated lines are likely to be parallel rather than convergent.

(6) Filtering divergent lines. Identifies and removes lines that diverge rather than converge toward any VP.
(7) Filtering opposite convergence. Removes groups that converge in opposing directions. This stage also includes:
(8) Resampling isolated parallel edges (edges not yet classified but potentially belonging to valid VP groups).

The results of this process—including detected vanishing points and grouped edges—can be visualized through the *Display Vanishing Points* feature.

To manage sketch variability, the algorithm allows several tolerance parameters to be configured via the *Vanishing Points* dialog, which is available via the corresponding *Details* button in the *Detect Cues Configuration* dialog box. The dialog is shown in Figure 3.80.

The vanishing point detection algorithm is highly sensitive to a small set of tolerance parameters, which can be adjusted in this dialog to fine-tune performance and handle sketch imperfections. It is recommended to review the detailed analysis provided by Company *et al.* (2014) before modifying these parameters, as inappropriate adjustments may negatively affect detection accuracy.

3.7.5.7 *Cue subgraphs*

Line drawings are commonly represented using graph structures in both discrete mathematics and computer graphics. This cue identifies the subgraphs present within the graph that corresponds to the current drawing. In the context of sketches depicting the edges and vertices of polyhedral shapes, a drawing is formally modeled as a graph $(G = (V, E))$, where the

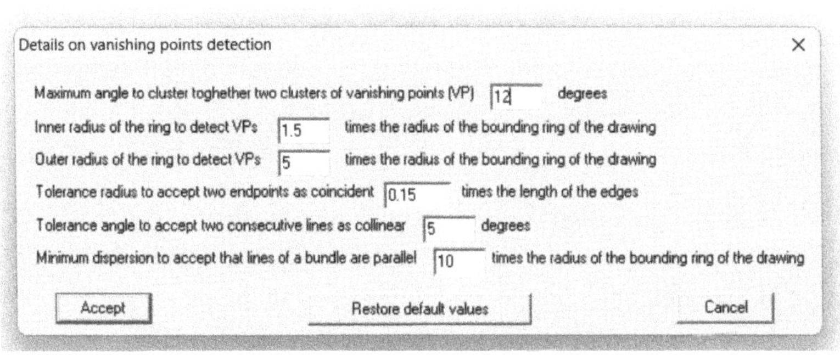

Figure 3.80. *Details on vanishing points detection* dialog.

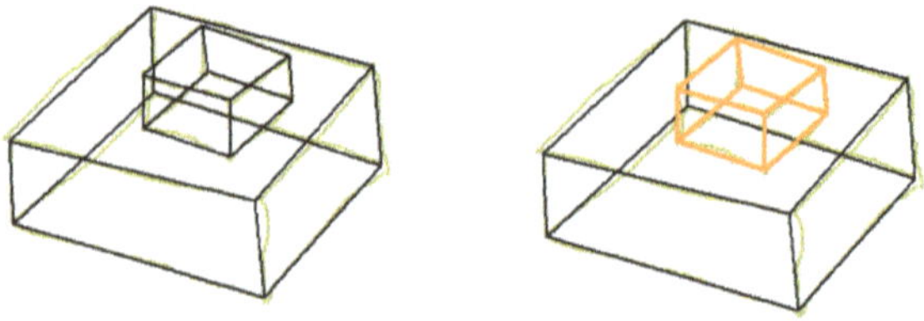

Figure 3.81. Original sketch (left) and subgraph highlighted in orange (right).

vertices (V) represent the endpoints of lines in the drawing, and the edges (E) represent the lines themselves. A graph is *connected* if a path of edges exists between any pair of vertices Vi and Vj. Otherwise, it is disconnected. A subset of the graph that is internally connected but has no connection to the rest of the graph is referred to as a disconnected subgraph or simply a subgraph. An example is shown in Figure 3.81.

The identification of disconnected subgraphs provides valuable insight when analyzing sketches or line drawings of polyhedral objects. This information is particularly important in scenarios where 3D reconstruction is performed using propagation-based methods, such as those employed in Axonometric Inflation.

Two alternative, but functionally similar, algorithms to identify subgraphs in REFER can be selected and configured in the Subgraphs Dialog:

- Varley's approach (2003): This implementation is optimized for compactness and uses an iterative search strategy to identify subgraphs:
  - Begin by selecting a vertex that has not yet been visited and assign it to a new subgraph.
  - Iteratively assign to this subgraph any edge (and its adjacent vertices) that connects to any already-assigned vertex.
  - Once no further connections can be made, begin a new subgraph and repeat the process.
- Company's Approach: This method prioritizes clarity and maintainability and is based on a breadth-first search. It operates as follows:
  - Select an unvisited vertex and perform a breadth-first traversal to visit all reachable vertices, marking them as part of the current subgraph.
  - Repeat the traversal for any remaining unvisited vertex, creating a new subgraph with each pass.
  - Continue until all vertices have been assigned to a subgraph.

Both methods output the total number of identified subgraphs and assign labels to each vertex and edge, specifying the corresponding subgraph to which they belong. The identified subgraphs can be visualized interactively using the *Display Subgraphs in Sequence* tool.

The dialog box for configuring parameters related to the *Cue Subgraphs* feature can be accessed via the corresponding *Details* button in the *Detect Cues Configuration* dialog box. At the user's request, subgraphs can be computed using one of two available approaches, as described in the *Subgraphs* section.

3.7.5.8 *Cue perimeters*

It has been stated that the number of contour edges in polyhedral projections is small, and the number of their intersections is even more limited (Kettner and Welzl, 1997). Therefore, detecting the perimeter in sketched drawings of polyhedral shapes is a valuable goal—particularly when this information can be extracted prior to searching for higher-level semantic cues. In contrast to other well-known approaches (Pop *et al.*, 2001), REFER's method does not assume prior knowledge about faces or their orientations during perimeter detection. Instead, the goal is to obtain the perimeter first, so it can be used to guide subsequent detection of visible and occluded faces.

It is important to distinguish between the terms perimeter and silhouette. The latter is typically defined as a set of successive points—or sometimes an external polyline—that outlines a region of interest and serves as a cue for figure-ground distinction (Isenberg *et al.*, 2003). Thus, the silhouette output from many existing methods applied to sketches of polyhedral shapes is not a subset of the original lines and junctions but rather an overlaid graphical entity (Qin *et al.*, 2007).

The earliest work directly addressing perimeter detection in polyhedral drawings is Roberts's study on the perception of 3D objects from line drawings, particularly the section on polygon recognition (Roberts, 1963, 1965). Roberts's approach is simple and efficient: at each junction, the connected lines are sorted based on their orientation. Polygon tracing then proceeds by selecting an initial line at random and, at every subsequent junction, following the next line in the sorted sequence. This process continues until the starting point is reached and a closed cycle is formed. This method can identify regions in natural drawings (e.g., faces or part-faces of the object) and detect the exterior boundary. However, it does not work well with wireframe drawings.

Labeling methods such as those proposed by Huffman (1971), Clowes (1971), and Waltz (1972), and extended by Varley and Martin (2000a, 2000b) can also infer perimeters. Nevertheless, perimeter detection is not the main objective of these methods, which rely on junction label catalogs. To date, only trihedral and tetrahedral junctions have been comprehensively cataloged; extending this classification to 5-hedral or higher-order junctions is considered impractical.

REFER includes an envelope method to detect object perimeters (Company *et al.*, 2017). Unlike digital image contour extraction methods, which locate the contour without identifying which elements of the drawing it consists of, REFER's approach identifies the actual lines and junctions (intersections and T-junctions) that define the perimeter in the original drawing. The algorithm takes as input a set of vertices and edges, forming one or more subgraphs (previously calculated). It then outputs a list of vectors representing the sequences of vertices and edges that form the perimeters of each subgraph and of the overall line drawing.

The method identifies closed circuits (perimeters) by analyzing the 2D connectivity of the drawing and is compatible with both wireframe and natural line drawings, as illustrated in Figure 3.82.

Some lines may only partially belong to the perimeter. For example, in wireframes, visible parts of the occluded lines may terminate at intersections, while in natural drawings they may terminate at T-junctions. REFE"s method detects and includes such junctions in the perimeter sequence, as shown in Figure 3.83.

Because the approach handles intersections, it can also compute the perimeter of multigraph line drawings (Figure 3.84), where no single path of edges visits all vertices. It determines the perimeter of each subgraph, in addition to the global perimeter of the full line drawing. This cue requires subgraphs to be calculated beforehand.

The algorithm's tuning parameters can be configured in the *Perimeters* Dialog, which is available via the corresponding *Details* button in the

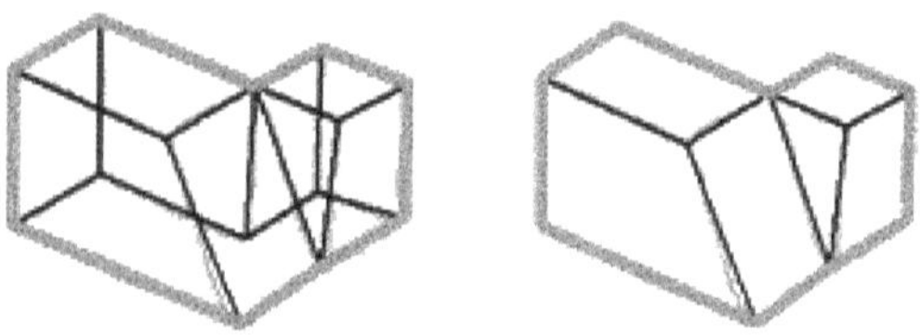

Figure 3.82. Identification of perimeters in wireframe (left) and natural (right) drawings.

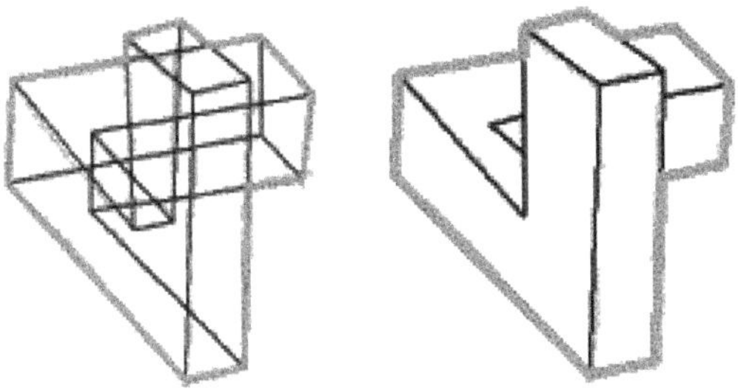

Figure 3.83. Perimeter detection in wireframe (left) and natural (right) sketches with occluded lines.

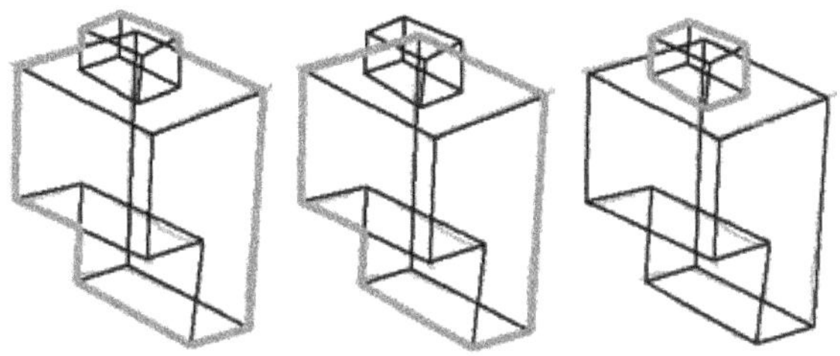

Figure 3.84. Perimeter detection in multigraph line drawings.

Detect Cues Configuration dialog box. Detected perimeters can be interactively visualized via the *Display Perimeters in Sequence* tool. When requested by the user, perimeter detection can be configured to:

(1) Include or exclude chains of dangling edges, which may appear even in very simple sketches.
(2) Apply a threshold to accept imperfect intersections during the crossing test. This threshold is critical for identifying intersections and T-junctions. Each time a new line is added to the perimeter, the algorithm tests whether any other lines cross the current perimeter line, or terminate at it (i.e., form a T-junction). If either condition is met, the algorithm inspects the intersection points. The closest intersection to the previous junction is added to the perimeter as if it were a corner. Then, a left turn along the intersecting line is added as the outgoing line of the perimeter. This threshold may typically be set to 1% of the total drawing size if the line-drawing has been refined. However, it should be increased if the option *Merge edge and vertex in T-vertex* is not enabled in the *2D Refinement* stage.

3.7.5.9 *Cue labels*

The line-labeling technique was first introduced in the field of Machine Vision to help narrow down potential 3D interpretations of line drawings by classifying each line as convex (+), concave (–), or occluding (→). Line labeling is preceded by the task of junction type identification. The first attempt at junction classification is attributed to Guzman (1968), but Clowes (1971) and Huffman (1971) independently provided the first systematic catalog for polyhedra with trihedral vertices, as shown in Figure 3.85. Varley (2003) extended the catalog to include tetrahedral vertices, emphasizing the exponential growth in complexity as vertex valence increases.

Using line labels to reconstruct models from sketches was soon discarded as a general method. Its use declined because it is limited to natural line-drawings (with hidden lines removed) and extending it to complex polyhedral shapes proved difficult. The method was also criticized for providing limited value beyond identifying which T-junctions are occluding or not: "the only essential output from line labeling for frontal geometry is the determination of which T-junctions in a drawing are occluding and which are non-occluding" (Varley *et al.*, 2005).

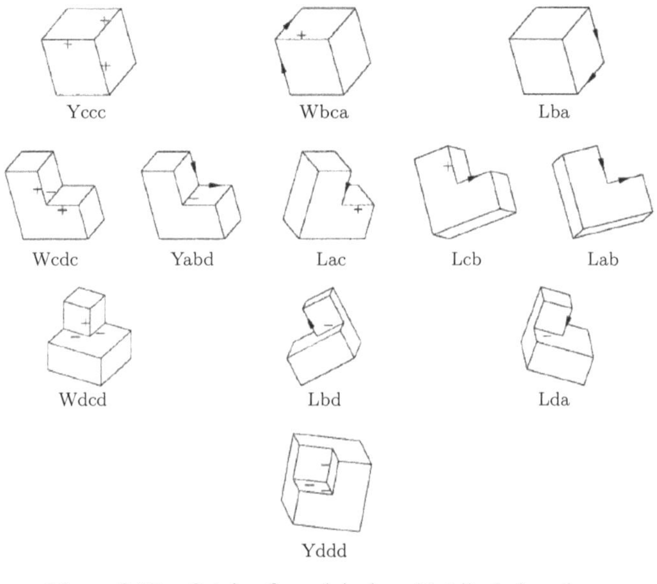

Figure 3.85. Catalog for polyhedra with trihedral vertices.

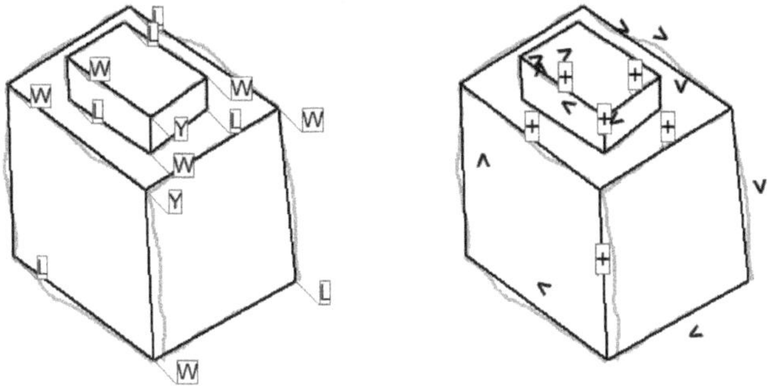

(a) (b)

Figure 3.86. Visualization of vertex (a) and edge (b) labels.

Labeling is available in REFER via an implementation based on RIBALD (Varley, 2003). Users can activate both the detection and visualization of vertex labels and edge labels, as shown in Figure 3.86.

The labeling algorithm in REFER operates independently of perimeter detection. Because REFER employs its own method for perimeter detection, the perimeter must be computed beforehand. The labeling process assumes the sketch represents a single, complete object, with all outer boundary lines treated as occluding. This assumption streamlines the computation and improves efficiency.

The labeling algorithm supports subgraphs, but prior subgraph detection is required. However, it does not handle implications for subgraphs representing related shapes (e.g., a step on top of a cube).

According to Varley (2003), "combinations of labelled lines meeting at a junction which do not produce a junction label listed in the catalogue can be rejected. The task is translated into a discrete constraint satisfaction problem, with the constraints that each line must have the same label throughout its length, and each junction must be allocated a labeling listed in the catalogue. Several effective algorithms for this discrete constraint satisfaction problem have been proposed."

Three line-labeling algorithms are implemented in REFER. The behavior of these algorithms can be configured via the *Labels* dialog.

(1) Kanatani's combinatorial labeling
(2) Relaxation labeling
(3) Lamb-Bandopadhyay geometric labeling

According to Varley (2003): "The best-established method of labelling line drawings is by means of a catalogue of permissible junction labels—any combination of labelled lines meeting at a junction which does not produce a junction label listed in the catalogue can be rejected. The Clowes-Huffman catalogue for line drawings of trihedral polyhedra is well-established, although the limitation to trihedral vertices is somewhat restrictive."

The Kanatani method (1990) is a deterministic algorithm built upon the Clowes-Huffman framework. It modifies the labeling process by decomposing non-trihedral vertices into trihedral configurations, introducing a zero-length line when necessary to achieve this structure. Bonnici and Camilleri (2023) have recently revisited the labeling approach.

In set-intersection labeling, only discrete information is propagated between junctions and neighboring edges. In contrast, relaxation labeling propagates probabilistic information. Relaxation is faster but potentially less reliable.

The labeling module in REFER is primarily intended for demonstration, as labeling is not utilized in the reconstruction process. A variety of sample sketches can be found in the Ribald Files input folder. The primary reason for excluding labeling from the core process is the frequent absence of a unique solution.

Additionally, the need for ever-larger junction catalogs makes the method non-scalable. As Varley (2003) puts it, "the labeling problem remains non-trivial, especially when non-trihedral vertices are allowed, and no perfect solution has been found."

Although REFER does not currently use labels for reconstruction (focusing instead on wireframe representations), it does implement a specific form of inflation by levels which draws inspiration from junction cataloging, assigning relative depths based on junction types.

The parameters to configure the cue labels can be accessed through the corresponding *Details* button in the *Detect Cues Configuration* dialog box. The labels can be calculated using the three available approaches, selectable by the user.

3.7.5.10 *Cue main axes*

In sketch-based modeling from single-view drawings, aligning the sketch with the projection of a 3D orthogonal reference system prior to reconstruction is often beneficial. While some reconstruction methods require users to manually specify the projected reference axes, several automated techniques have also been proposed in the literature.

A simple strategy for determining the main axes is to assume that the first three edges drawn are parallel to the three principal axes. However, experiments evaluating the validity of this hypothesis concluded that it lacks perceptual basis—most people do not draw sketches this way. Naturally, if all edges are aligned in just three directions (i.e., the object is a normalon, a polyhedron where all adjacent edges and faces meet at right angles), these directions can be considered the main axes. However, the inaccuracy inherent to sketches must be considered.

Lipson and Shpitalni (1996) proposed a heuristic strategy to identify prevailing angles and main directions despite imperfections. The Angular Distribution Graph (ADG) enables automatic detection of the main axes. In their words, "it is possible to identify the main axis directions of the intended object from this graph."

The ADG is built by sampling the angle of every line segment in the drawing and plotting the results in an angular distribution histogram. The graph is smoothed out using a Gaussian distribution ($\sigma = 7$ degrees), then normalized so that its maximum is 1.0. The resulting graph highlights the dominant angles in the drawing. In orthogonal objects (normalons), three dominant angles typically emerge, as shown in Figure 3.87.

REFER's implementation of the ADG algorithm does not apply a fixed 7-degree threshold. Instead, it uses it as a minimum value and the threshold is increased up to the minimum angle between edges connected to the same vertex. These settings ensure that edges sharing a vertex (even the most acute) are not considered parallel. Below this threshold, lines can be safely treated as parallel, reducing the number of candidate directions. The three most frequent directions are then selected as the main directions, though this assumption is not always valid. The example in

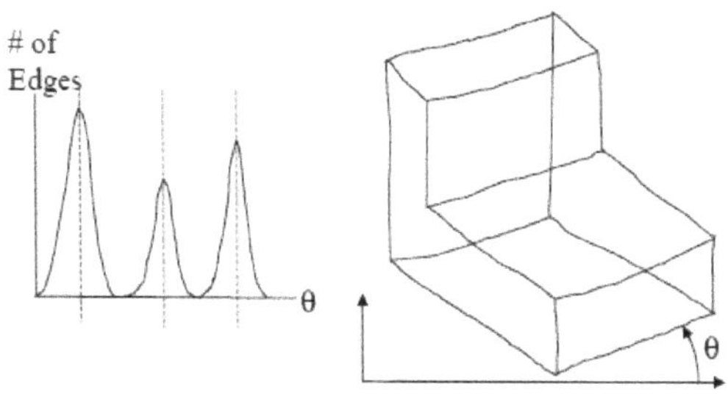

Figure 3.87. Histogram of dominant angles (left) and corresponding normalon (right).

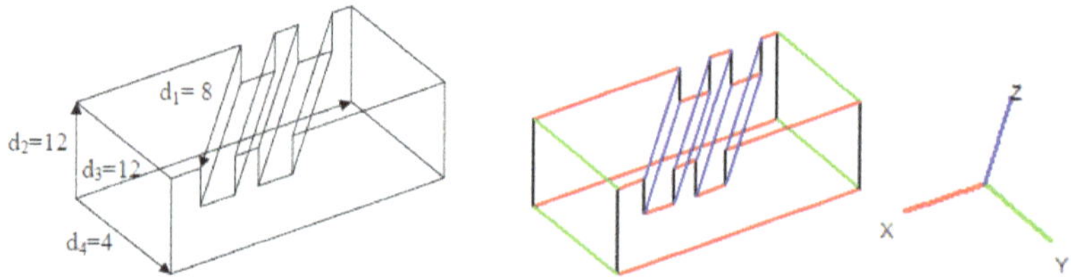

Figure 3.88. Directions and frequencies for a sketch.

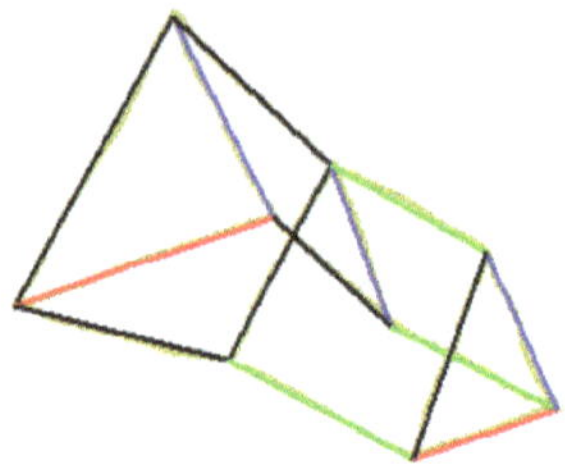

Figure 3.89. Failure to detect the main vertical direction.

Figure 3.88 shows a model from Lipson and Shpitalni (1996) where four directions and their frequencies are evident.

An additional check can verify whether these three directions coexist in at least one vertex. If not, their likelihood of representing the main axes is reduced. Note that this criterion assumes vectorization refinement has been performed—specifically, that vertices have been merged.

In addition, the resulting candidate axes should be validated using the Cubic's corner criterion by Perkins (1968) to ensure they depict actual normalons, not false configurations. For example, Figure 3.89 shows a failure to detect the main vertical direction, which is visually obvious, but undetectable if only edges are considered.

For normalons in perspective-like projections, as well as for other object types, it is advisable to compute vanishing points and identify the three most frequently populated ones as the principal axes. As a further validation step, Perkins' cubic corner criterion (1968) should also be applied.

REFER implements a *projective* approach (Conesa, 2001), which proceeds as follows: parallel edges are grouped using the ADG; next, all triplets formed by combinations of these parallel groups are evaluated using a set of criteria. Triplets are filtered using the following sequence of criteria:

(1) Projective consistency of angles.
(2) Matching corners (triplets must occur at least at one vertex).
(3) Preference for isometric axonometry.
(4) Frequency of directions in the sketch.
(5) Priority given to the first drawn edges.

The first criterion is binary: triplets failing it are discarded. The remaining criteria increase the "merit" of surviving triplets. The valid triplet with the highest merit is ultimately selected.

The first two criteria are based on the principle that human perception rejects geometries that violate projection rules. The first criterion reflects the theory that perceptual rules are constrained by projective geometry (Perkins, 1968; Attneave and Frost, 1969). Specifically, to represent a solid rectangular corner, either all three angles around it must be greater than 90 degrees, or two angles—each less than 90 degrees—must sum to more than 90 degrees. This criterion applies to both axonometric and perspective projections (as long as the view direction is approximately perpendicular to the drawing plane).

Figure 3.90 illustrates the grouping process for a line drawing. The triplets $d_1d_2d_3$, $d_1d_2d_4$, $d_1d_3d_4$, and $d_2d_3d_4$ should be checked, and only $d_1d_3d_4$ and $d_2d_3d_4$ should pass this first criterion. After the second criterion, triplet $d_1d_3d_4$ would be rejected. Since only the triplet $d_2d_3d_4$ would remain, the process would finish.

The isometric preference criterion favors configurations closer to isometric projection. It helps disambiguate certain shapes but can mislead in cases like the one shown in Figure 3.91, where vertex 2 gives the correct interpretation, whereas vertex 1 produces a perceptually implausible result.

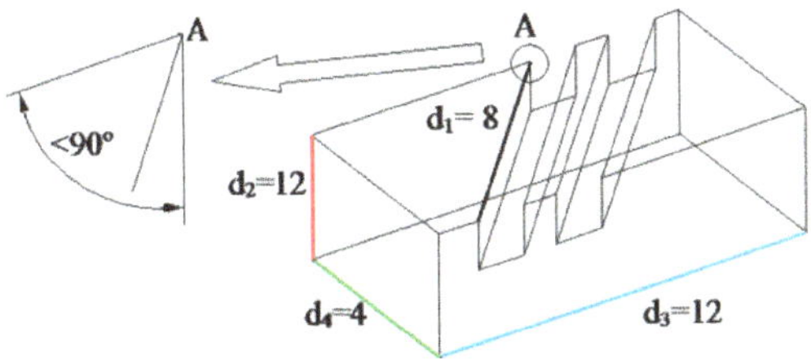

Figure 3.90. Grouping process for a line drawing.

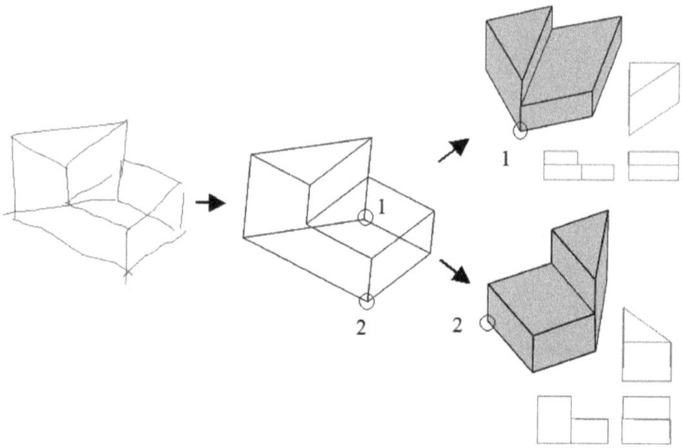

Figure 3.91. Perceptually implausible result in the application of the isometric preference criterion.

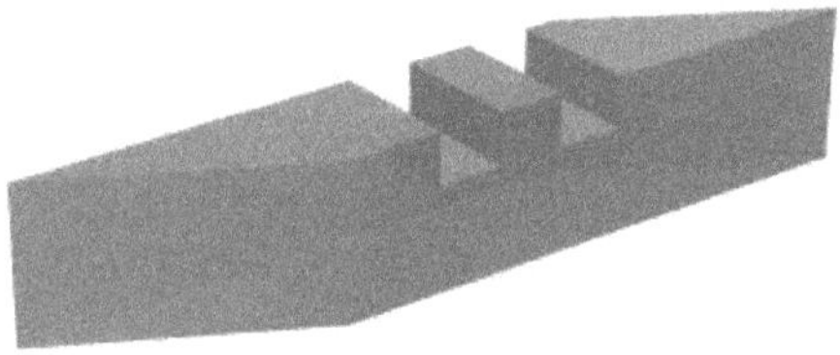

Figure 3.92. Reconstructed model after the application of criterion 4.

The fourth criterion considers triplet frequency: the most common is preferred. Crucially, this step occurs after applying the projective rules. Otherwise, the triplet $d_1d_2d_3$ would be selected in the example shown in Figure 3.91, since their frequencies are greater than the ones given by $d_1d_3d_4$, and both fit the second criterion. Consequently, the reconstructed model would be the one shown in Figure 3.92.

As noted, a general optimal solution is not yet available, and sub-optimal solutions require case-specific tuning. All parameters for this approach can be configured via the main axes dialog. After the first four criteria, ambiguity may still exist between valid triplets. In these cases, the triplet containing the earliest drawn edges is selected.

Although alternative methods have been proposed (Kang *et al.*, 2004), the problem remains unresolved. Current approaches are unable to consistently group lines that correspond precisely to 3D-parallel edges. A key challenge lies in the absence of an optimal angular "bucket" size–what works well for one sketch may be too narrow or too broad for another.

3.7.5.11 *Cue main axes - Varley*

Varley *et al.* (2004b) proposed an algorithm for identifying the main directions in single-view line drawings. The goal of this algorithm is to "inflate" the sketch by assigning provisional depth coordinates, which can then support subsequent inferences such as edge labeling and 3D structure recovery. The underlying assumption is that the object depicted has three dominant axes (denoted by i, j, and k), each corresponding to a visually identifiable cluster of approximately parallel lines in the 2D drawing.

To determine these axes, the authors evaluated seven different selection strategies or variants, each using a different heuristic combination of geometric or statistical criteria. These strategies are summarized as follows:

(1) Standard isometric projection. Assumes a fixed isometric frame where the $+k$ axis points vertically downward (aligned with the 2D y-axis), and the $+i$ and $+j$ axes are oriented at ± 120 degrees from it.
(2) Most populous groups. Selects the three most frequent line orientations (by angular histogram) to define i, j, and k.
(3) Hybrid of 1 and 2. Uses Method 1 to fix the k axis vertically and selects i and j from the three most populous groups, excluding the one closest to vertical.
(4) Geometrically longest groups. Chooses the three orientation groups associated with the longest total edge lengths.
(5) Hybrid of 1 and 4. Uses Method 1 for k and selects i and j from the three longest groups, again excluding the group nearest vertical.
(6) Bisector-based from population. Chooses i and j from the two most populous non-vertical groups, and sets k to be the group closest to their angular bisector.
(7) Bisector-based from length. Same as Variant 6 but uses line length instead of population for selecting i and j.

Among the tested variants, Variant 6 consistently performed best across the evaluated datasets. Variant 3 also demonstrated notable effectiveness and had some theoretical appeal due to its alignment with isometric projection assumptions. In light of this ambiguity, the authors introduced a tuning parameter that allows interpolation between the outputs of Variants 3 and 6, providing flexibility for adapting to different input conditions.

Several of these variants, especially those relying on frequency or length metrics, depend on the ability to group lines that likely represent parallel edges in the underlying 3D structure. To address this issue, Varley (2003) introduced a line bundling algorithm that partitions drawing edges into sets—termed bundles—where each bundle ideally contains lines that correspond to parallel 3D edges. The bundling algorithm operates iteratively as follows:

- Repeat until all edges are assigned to bundles.
- If any unbundled edge is necessarily parallel to a bundled edge, add it to the same bundle.
- If any unbundled edge is nearly parallel to a bundled edge, add it to the bundle if possible.
- For each face containing unbundled edges:
    - If an unbundled edge is close to a bundled direction and topologically expected to be parallel, include it if possible.
- If no new assignments were made during the iteration and unbundled edges remain, assign any remaining edge to a new bundle.

The process continues until every edge is placed into a bundle. The bundling strategy combines angular similarity with topological context, leveraging adjacency and face structure to make more informed grouping decisions. This hybrid of geometric and structural cues increases robustness across varied drawing styles and levels of noise.

Although the bundling approach and direction-selection variants perform well in many cases, some limitations persist. Even the basic isometric assumption used in Variant 1 can yield axes that do not align with human intuition, particularly for visually simple or symmetric objects. One such case is illustrated in the sketch of the sheet metal part shown Figure 3.93. The example underscores the ongoing challenge in ensuring perceptual alignment between algorithmically determined axes and human expectations in 3D shape interpretation.

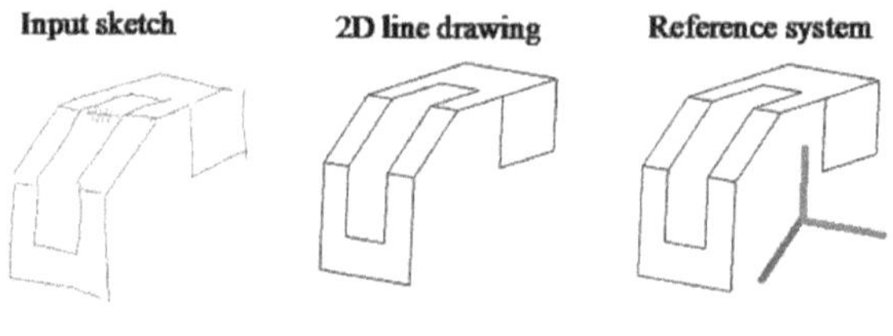

Figure 3.93. Sketch of a sheet metal part.

3.7.5.12 *Cue main axes dialog*

The parameters for configuring the *cue main axes* can be accessed via the corresponding *Details* button in the *Cue Detection Configuration* dialog box. The main axes can be computed using one of three available methods:

- Rule based. The rules that govern axis detection can be customized through the *Main Axes Rules Dialog*. The rules are classified into two types: rules used to filter (i.e., retain or discard) candidate main axes, and rules used to assign relative merits among valid candidates.

  Each rule can be independently enabled or disabled by the user, providing flexible control over the axis selection process.

  The *projective rule* acts as a strict geometric criterion. It discards any candidate triplet of directions that could not plausibly result from an orthographic projection of a 3D object. This rule functions as a binary pass/fail filter and is applied prior to evaluating perceptual merit. The remaining rules contribute to scoring candidate axis triplets based on how well they conform to perceptual or heuristic criteria. Each rule contributes a tunable "merit" value to a global score, enabling the system to prioritize those triplets that are deemed perceptually stronger or more likely to represent the true main axes. It is important to note that the matching corner rule depends on the completion of vertex merging. As a result, this rule cannot be applied if the sketch has not undergone 2D refinement.

  The dialog also enables the configuration of threshold angles used in two key areas: (1) to eliminate redundant candidate triplets whose directions are too similar (i.e., nearly overlapping), and (2) to determine how closely a candidate direction must align with specific reference directions—such as the edges of a corner, standard isometric directions, or the vertical/horizontal axes—to be considered a valid match. These thresholds provide fine-grained control over the sensitivity of the axis detection algorithm. The *Main Axes Rules Dialog* is shown in Figure 3.94.

- First three edges. The system will automatically assign the first three strokes drawn in the sketch as corresponding to edges parallel to the three main axes (i, j, k). These strokes are interpreted in the order they were made during the sketching process.
- User-selected edges. Users can manually select three strokes of the input sketch via the selection mode. These strokes will then be used to infer the main axes once cue detection is initiated. Another option,

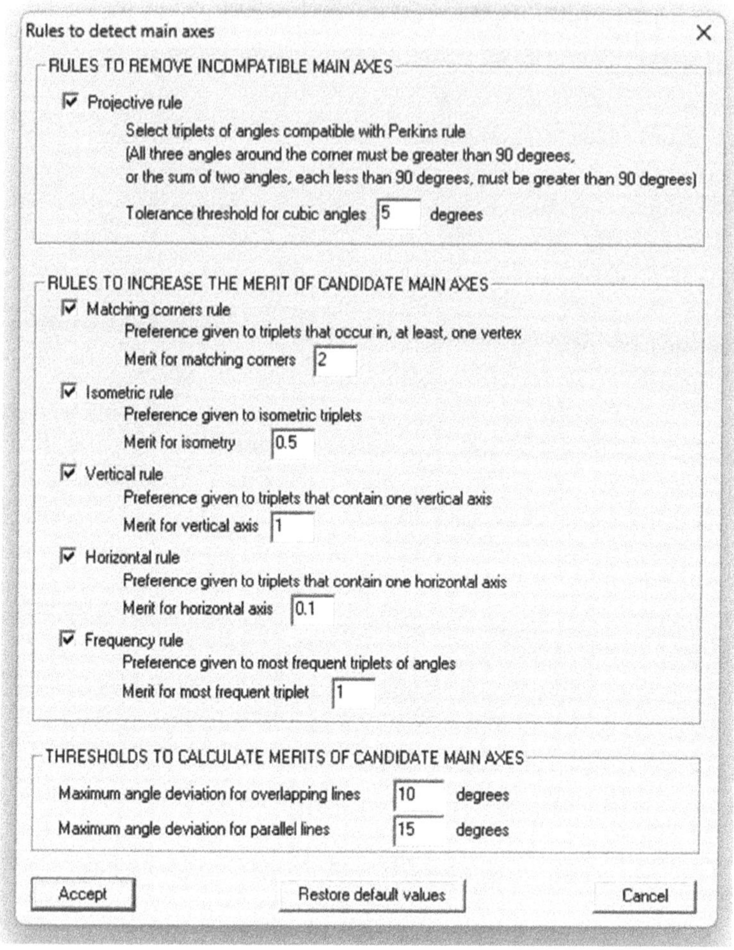

Figure 3.94. *Main Axes Rules* Dialog.

which can be accessed by clicking the *Select* button, allows the user to interactively select three edges from the offline vectorized line-drawing via a dedicated selection dialog, as shown in Figure 3.95.

Note that vectorization must be performed beforehand in order for the dialog to display the edge structure. 2D refinement is not strictly required. Also, the three selected edges do not need to meet at a common vertex, as only their orientations are used to compute the principal axes. An example is shown in Figure 3.96.

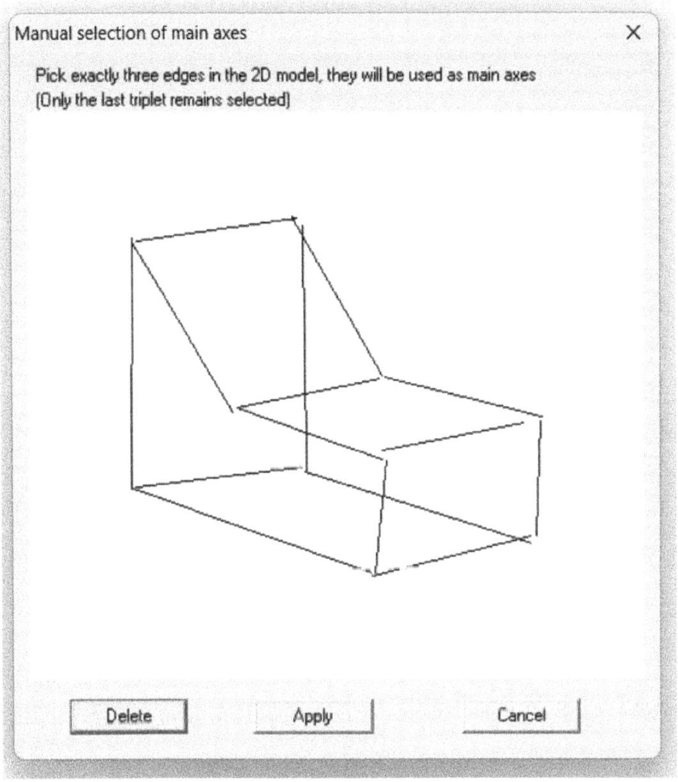

Figure 3.95. Dedicated selection dialog box for user-selected edges.

3.7.5.13 *Cue faces*

The identification of faces is a critical step in the interpretation and reconstruction of polyhedral objects from 2D representations (Heath *et al.*, 1993). Faces provide essential structural information, enabling the inference of spatial relationships, surface connectivity, and ultimately, depth. However, detecting faces in 2D line drawings is inherently ambiguous and does not always lead to a unique or reconstructible solution. Various methods have been proposed in the literature to address this challenge, and REFER implements several of them, which can be accessed via the *Cue Faces* dialog.

An effective strategy for extracting faces from natural line drawings was proposed by Roberts (1963). This method begins face construction by

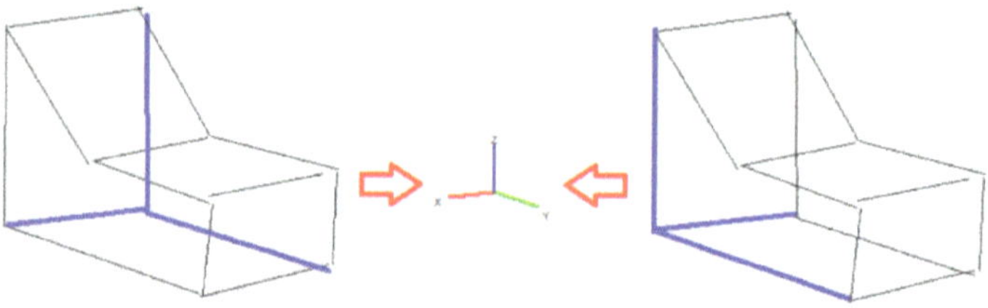

Figure 3.96. Only the orientations of the selected edges are used to compute the principal axes.

selecting a random starting edge. The next edge in the face is identified as the one that (1) connects to the terminal vertex of the current edge and (2) forms the smallest angle according to a predefined rotational direction (clockwise or counterclockwise). This process is repeated iteratively, moving from edge to edge, until the starting vertex is reached again, forming a closed loop that defines a face.

To distinguish between the outer (infinite) faces and internal faces, the traversal direction is inverted: the contour is traced in one direction, and all interior faces in the opposite. Completion is achieved when every internal edge (i.e., edges not on the contour) belongs to exactly two faces—illustrated in Figure 3.97 with edges 1-2, 1-4, and 1-6.

More general approaches for detecting faces interpret the line drawing as a graph. In one category of methods, purely topological techniques from graph theory are employed to identify faces by embedding the graph in a plane. These approaches are constrained to structures that can be represented by planar graphs, limiting their applicability. A second category of methods involves identifying a representative set of cycles in the graph that are hypothesized to correspond to faces (Paton, 1969). These cycles are then filtered based on geometric, perceptual, or heuristic criteria. While more flexible, such methods often incur high computational costs, making them impractical for interactive applications. Moreover, the heuristics used in these methods are typically ad hoc and lack systematic evaluation.

Varley and Company's approach (2010) is arguably the most advanced and systematic technique currently available for face detection. This method calculates faces from a graph-like representation of a polyhedral shape. Input data is partly topological (an undirected graph \(G (V, E)\)) and partly geometric ((x, y) coordinates for each vertex of the graph). The

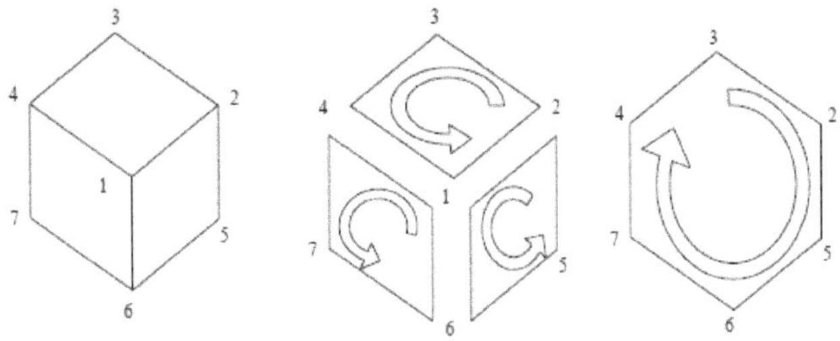

Figure 3.97. Extraction of faces from natural line drawings.

graph must satisfy two constraints: (i) no self-loop edges are permitted, and (ii) no two vertices are connected by more than one edge.

The algorithm is formulated in terms of half-edges–directed versions of each edge in the graph. Formally, each undirected edge {a, b} in (G) is represented in a derived directed graph (G′(V, A)) by two half-edges: one from a to b, and one from b to a. A cycle, in this context, is defined as a closed sequence of such half-edges. The primary goal is to construct a complete set of cycles such that every half-edge belongs to exactly one cycle. For the solution to be valid, these cycles must also be geometrically realizable, i.e., it must be possible to assign z-coordinates to the vertices such that every cycle lies approximately in a plane (with some tolerance for planarity).

In cases where only one valid set of cycles exists, the problem is reduced to finding that set. However, if multiple valid solutions exist, the algorithm seeks the interpretation that best aligns with what a human observer would consider the most plausible structure. It is assumed that at least one such valid set of cycles exists; the related but more difficult problem of determining whether any valid configuration exists is not addressed by this algorithm.

- *Input*: A 2D graph (G (V, E)) and coordinates for each vertex.
- *Constraints*: No self-loops; no duplicate edges.
- *Representation*: Convert to directed graph (G′(V, A)) using half-edges.
- *Definition*: A cycle is a closed sequence of half-edges.
- *Objective*: Identify a set of cycles covering all half-edges exactly once.
- *Geometric condition*: Each cycle must be approximately planar.
- *Goal*: Find the most perceptually plausible cycle configuration, if multiple exist.

Although powerful, Varley's approach has notable limitations, illustrated through the following representative examples:

- *Input restrictions*: The method only works with wireframe-style drawings (see Figure 3.98) and cannot be applied to natural sketches.
- *Face definitions*: A face is defined as a non-repeating cycle of alternating vertices and edges. As a result, even coplanar regions such as the one shown in Figure 3.99 are interpreted as two separate faces.
- *Internal face exclusion*: The algorithm is effective in detecting and excluding inappropriate internal faces (see Figure 3.100).
- *Incorrect loop elimination*: The algorithm also excludes many incorrect loops that are deceptively valid in 2D projection—an especially challenging task. An example is shown in Figure 3.101.

Successful detection of faces requires tolerance settings to handle sketch imperfections such as noise or slight geometric inconsistencies. Within the *Cue Faces* dialog, three different face detection approaches

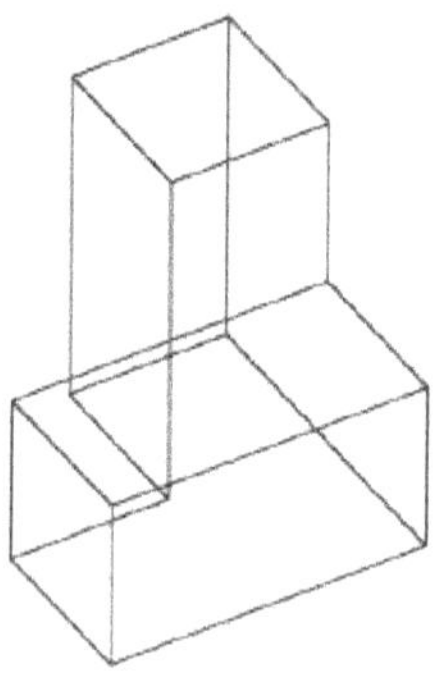

Figure 3.98. Wireframe-style drawing.

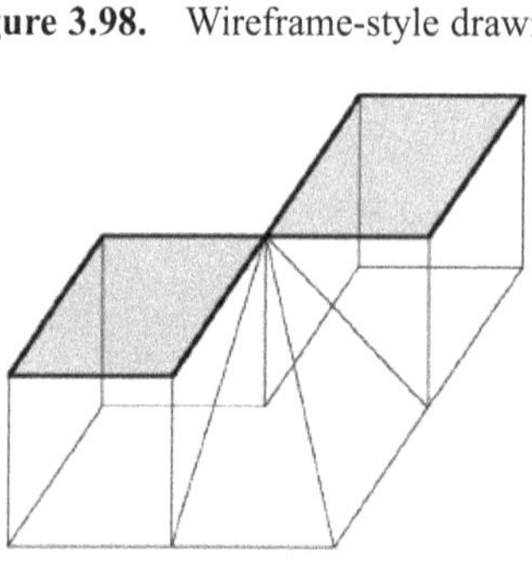

Figure 3.99. Coplanar regions.

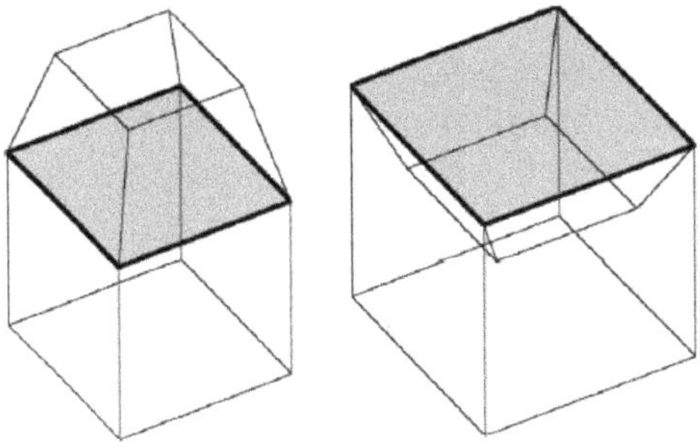

Figure 3.100. Exclusion of inappropriate internal faces.

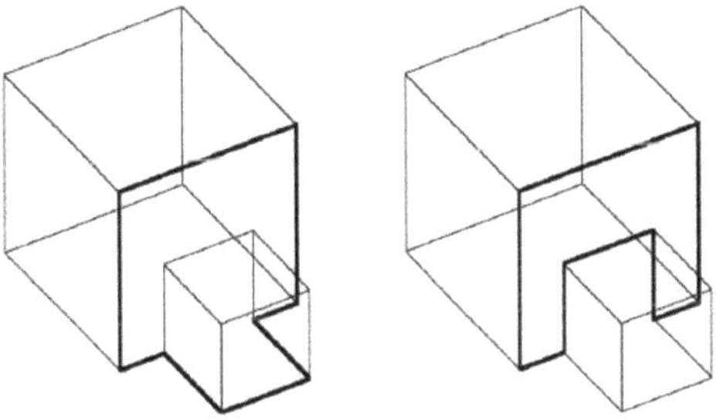

Figure 3.101. Incorrect loop elimination.

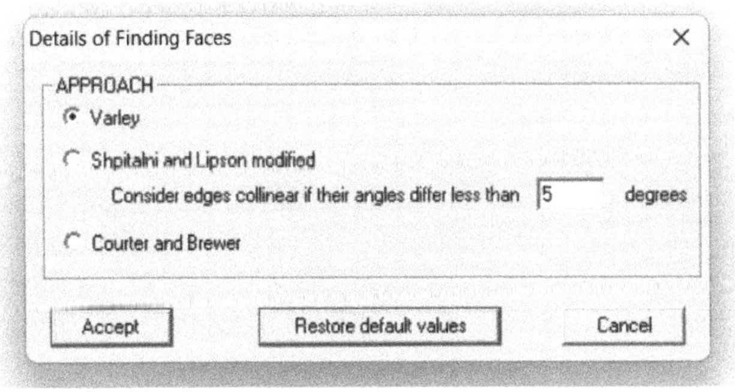

Figure 3.102. *Cue Faces* dialog box.

can be selected, as shown in Figure 3.102. The dialog is available via the corresponding *Details* button in the *Cue Detection Configuration* dialog box.

- Varley's Approach (Varley and Company, 2010) is currently the most complete and systematically developed method implemented in REFER. It combines both topological and geometric constraints to derive a set of valid, planar cycles (i.e., faces) from a graph-based representation of a wireframe model. It aims to match human perceptual expectations where multiple valid interpretations exist.

- Modified Shpitalni and Lipson (1996) is based on the algorithm originally proposed by Shpitalni and Lipson but has been adapted for integration in REFER. The method constructs a pool of candidate edge circuits—closed loops in the graph representation of the sketch—which are then evaluated using a variety of heuristics to determine which are likely to correspond to actual object faces. While flexible and perceptually motivated, this approach can be computationally expensive, especially for complex drawings, and may not be suitable for real-time or interactive use.
- Courter and Brewer's approach (1986): This method relies purely on topological reasoning, operating without direct geometric constraints. It begins by generating a spanning tree of the line-drawing's graph structure, from which the fundamental cycles (i.e., the minimal set of independent loops) are identified. Candidate faces are then formed by combining these fundamental cycles using the symmetric difference operation—the union of two cycles minus their overlapping edges. The technique is grounded in two observations from graph-theory: (1) A complete set of base cycles exists such that no edge appears in more than two of them, and (2) the correct face configuration minimizes the total number of edges used across all cycles.

Despite its theoretical elegance, this approach exhibits significant limitations. More specifically, it tends to fail in drawings with moderate to high face complexity or in cases involving two-connected graphs, i.e., graphs where the removal of any two edges results in disconnected subgraphs. In such scenarios, the method cannot distinguish between "true" (on the same side of the object) and "false" (on opposite sides) bifurcations, leading to misidentification of face boundaries. An example failure is illustrated in Figure 3.103.

As a general observation, many failures in face detection algorithms can be traced back to the presence of two-connected graphs. In such cases, ambiguity arises in determining which subgraph connections correspond to coherent 3D surface structures. Consequently, exploring algorithms that assess and leverage graph connectivity (Schmidt, 2013) may offer promising directions for enhancing the robustness and accuracy of face detection in such cases.

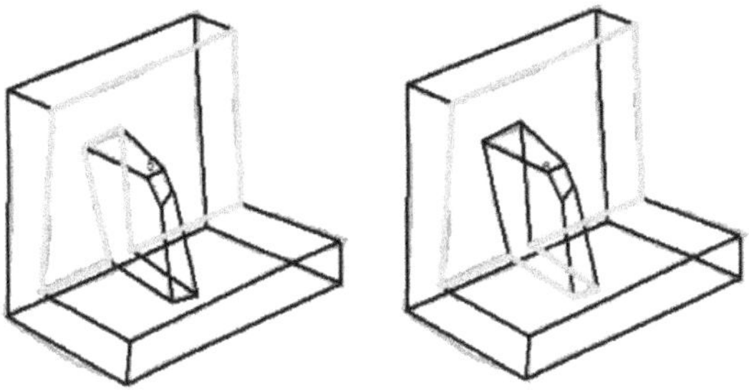

Figure 3.103. Face identification failure in a bracket with a divider.

3.7.5.14 *Cue typology*

This cue is based on the principle of "divide and conquer," the more accurately the typology of the object depicted in a 2D sketch can be determined, the better suited the selected reconstruction strategy will be. It is widely acknowledged that no single reconstruction strategy is universally applicable. Instead, the most effective strategy depends on the specific type of object being reconstructed. De Bonet (1995) proposed that reconstruction should be guided by the "grammar" of the object. Building on this idea, REFER implements a set of rules tailored to different polyhedral typologies.

The grammar focuses on distinguishing among various types of polyhedral shapes. However, traditional classifications of polyhedral, such as those based on symmetry or face regularity, are not helpful for deciding which inflation or reconstruction criteria should be applied. As such, a custom classification system is formulated that focuses solely on automating the selection of the most appropriate reconstruction method. A key challenge in this process is that typological classification must be performed prior to reconstruction. Therefore, all inferences must rely exclusively on the 2D sketch, without any 3D geometric context.

Despite the vast number of possible polyhedral configurations, REFER identifies three main typologies that can be reconstructed once detected in the 2D sketch (Conesa, 2001).

- Normalons are polyhedra in which all edges meeting at a corner form 90-degree angles, and all adjacent faces define 90-degree dihedrals. Quasi-normalons are shapes that, while not strictly normalons, can be reduced to one by removing a limited set of edges—without eliminating any vertices or disconnecting any part of the graph. These forms commonly appear in mechanical part designs and rough architectural sketches. A drawing is heuristically classified as a normalon if its edges predominantly align with three, and only three, principal directions. Small deviations are tolerated to account for sketch imperfections. If a figure fails to meet the strict criteria for normalons, but removal of misaligned edges does not compromise connectivity, it is classified as a quasi-normalon.

- Prismatic polyhedra are characterized by a prismatic surface bounded by two typically parallel planes, forming bases connected by parallelogram faces. If face information is available, the classification involves a topological check: the shape must have two non-adjacent faces with the same number of sides, while all other faces should be quadrilaterals. A geometric check then confirms whether the lateral edges (linking the bases) are parallel and whether both bases have similar areas. When faces are not known, an indirect approach is used. The drawing is considered prismatic if one direction clearly dominates among the links. Specifically, if the number of parallel edges in a given direction is at least 50% of the total number of vertices (or "nodes") in the image, the shape is classified as a prism. This decision is based on the expectation that the two bases will account for all vertices—each containing half—and the connecting edges will follow the dominant direction. For lower-quality sketches, the threshold may be relaxed to 40% to compensate for inaccuracies in drawing.
- Pyramidal polyhedra are defined by having a single base (typically polygonal), with all remaining faces being triangular and converging at a single apex. If the faces are known, classification is straightforward: the presence of exactly one non-triangular face (the base) suffices. When faces are unknown, the following heuristic conditions are applied:
  - One vertex is connected to all others (the apex).
  - All other vertices are connected to exactly three vertices (defining the triangular faces).

Shapes that do not conform to any of the three typologies above are classified as having an undefined typology. Two subtypes are distinguished:

- *Undefined with a dominant direction*: where most edges are oriented along a single direction.
- *Undefined in general*: where no dominant feature allows for classification.

To apply the classification criteria outlined above, certain cues must first be computed, including the detection of main axes and the valences of vertices, both of which provide critical structural information.

The implemented detector has a clear limitation: it only analyzes polyhedral shapes. However, it serves as a demonstrator for several important conceptual points. First, as noted earlier, only information available in the 2D drawing is used, since the objective is to infer the model type prior to reconstruction. Second, the detector yields estimates rather than certainties, which encourages the consideration of multiple candidate typologies before selecting the most plausible one. A purely sequential search strategy can lead to false positives. For instance, a sketch may (albeit with low probability) resemble a quasi-normalon, even though it more likely represents a pyramid. Third, the approach assumes that topological information—vertices, edges, and faces—is reliable, making the method less sensitive to the kinds of imperfections typically found in hand-drawn sketches. In contrast, geometric information is treated as approximate and must be evaluated in terms of how well it supports different hypotheses. For example, a shape is classified as pyramidal if all vertices except one have valence three, while the remaining vertex (the candidate apex) has valence equal to the total number of vertices minus one. This topological criterion provides a high degree of confidence, assuming correct vertex detection and that the object is indeed polyhedral. In contrast, a body is considered a normalon if its edges are approximately aligned with the three principal axes. This classification depends on a geometric threshold used to accommodate sketching inaccuracies. Varying this threshold allows the definition of a figure of merit, a quantitative measure of the likelihood that the drawing represents a normalon.

Cues primarily provide local information, whereas certain situations require a more global perceptual understanding. A clear example is illustrated in Figure 3.104(a): two edges that appear nearly parallel to the vertical main axis should not both be labeled as such, since they are consecutive edges within the same triangular face. This inconsistency can only be resolved if face-level (i.e., global) information has already been identified by the algorithm. By contrast, the two consecutive edges at the base are correctly labeled as parallel to the horizontal main axis. Although they also form an obtuse angle, their configuration suggests near alignment rather than the near overlap observed in the triangular face.

A more complex example is shown in Figure 3.104(b), which is incorrectly classified as quasi-normalon, mainly because edges that seem

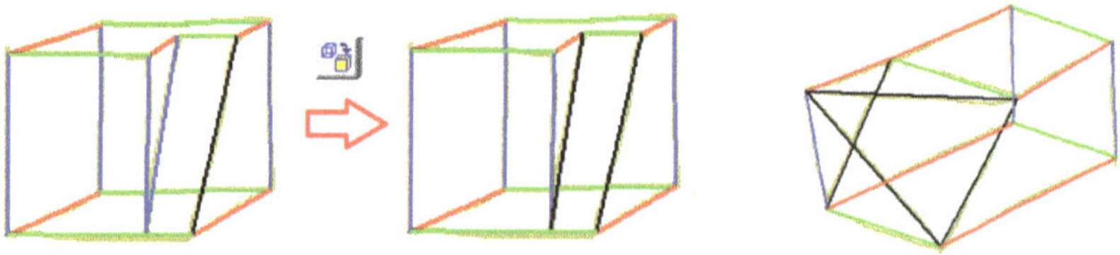

(a) (b)

Figure 3.104. A shape in which two edges that appear nearly parallel to the vertical main axis should not both be labeled as such (a) and a shape incorrectly classified as quasi-normalon (b) due to edges that seem "nearly" collinear in 2D cannot be collinear in 3D (they belong to faces that are not coplanar).

"nearly" collinear in 2D—like those marked in red and shared by the quadrilateral and the triangular upper faces and the quadrilateral and the triangular side faces—cannot be collinear in 3D, as they belong to faces that are not coplanar (otherwise they would be the same face, with none internal separation like the green line connecting upper faces and the black line connecting side lines).

The use of typology detection to select reconstruction strategies was eventually discontinued. The reasons include the overwhelming complexity of classifying increasingly diverse and intricate objects, and the fact that whole-object classification often provides limited benefit unless it is directly tied to an effective reconstruction method, such as the axonometric inflation applied to normalons.

As a result, REFER currently detects only a limited set of typologies, and the practical value of this classification is likewise constrained. Nevertheless, the original concept of using a "grammar" has evolved. Rather than classifying the entire object, the focus has shifted toward analyzing smaller subcomponents, which are then classified, reconstructed, and reassembled. This shift underpins the new feature-based methodology and the CSG modeling strategy. Consequently, the "divide and conquer" principle has evolved from object-level classification to identifying and categorizing individual features and structural components, enabling more targeted cue detection and more effective reconstruction.

The results of typology detection are not explicitly displayed, but the current classification is shown in the status bar of the interface, as illustrated in Figure 3.105. For shapes identified as normalons or quasi-normalons, the normalon display option becomes available for visualization.

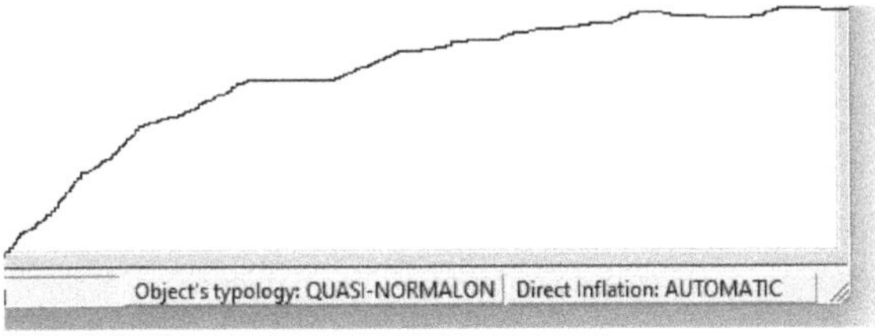

Figure 3.105. Status bar in REFER showing the current classification.

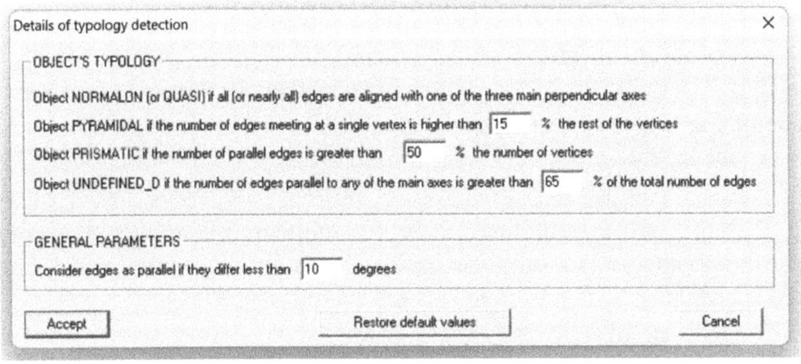

Figure 3.106. Parameters in the *Cue Typology Dialog* box.

The parameters related to the cue typology can be configured in the *Cue Typology* dialog box, which can be accessed via the corresponding *Details* button in the *Cue Detection Configuration* dialog box, as shown in Figure 3.106.

The tolerance parameters vary depending on the type of shape being detected:

- Detecting NORMALONS requires a general threshold that groups together lines that are “nearly” parallel to account for imperfections in hand-drawn sketches.
- Detecting PYRAMIDS is based on the assumption that a single vertex connects to “most” of the edges in the figure.
- Detecting PRISMS relies on the assumption that a “significant number” of edges are parallel to one another.
- Undefined shapes are classified as type *D* (dominant) if a “significant number” of their edges are parallel to the main axes.

3.7.5.15 *Cue hidden edges*

This cue aims to distinguish between lines in a wireframe that represent visible edges and those representing hidden ones. We distinguish three types of drawings: natural line drawings (which show only the lines visible from a specific viewpoint), wireframe drawings (where visible and hidden lines are shown indistinctly), and hidden-line drawings (where different line styles differentiate between visible and hidden lines). This algorithm applies to the second case, where the set of edges in a graph-like representation of a wireframe drawing of a polyhedral shape can be divided into two subsets: visible and hidden edges.

The problem differs from predicting the hidden parts of shapes depicted in natural drawings, which was studied by Varley (2003) and recently revisited by Tanaka *et al.* (2022). A key distinction in this problem is that hidden edges are identified before any 3D model is obtained. Therefore, conventional approaches to detecting hidden lines in 3D polyhedral shapes are not applicable. To the authors' knowledge, the only algorithmic attempt to produce hidden-line drawings from wireframe input (see Figure 3.107) is contributed by Conesa (2001).

The labeling of edges as visible or hidden provides valuable perceptual cues for SBM. This classification is incomplete when edges are partially visible and partially hidden. However, by recursively subdividing these hybrid edges, all resulting segments can eventually be labeled as either visible or hidden.

REFER can also leverage information from dashed lines provided by grouping non-solid strokes to infer hidden edges. This capability requires enabling the *Group non-solid (dashed) strokes* option in the *2D Vectorization Configuration* dialog box and activating the *Use dashed strokes*" option in the cue hidden edges dialog.

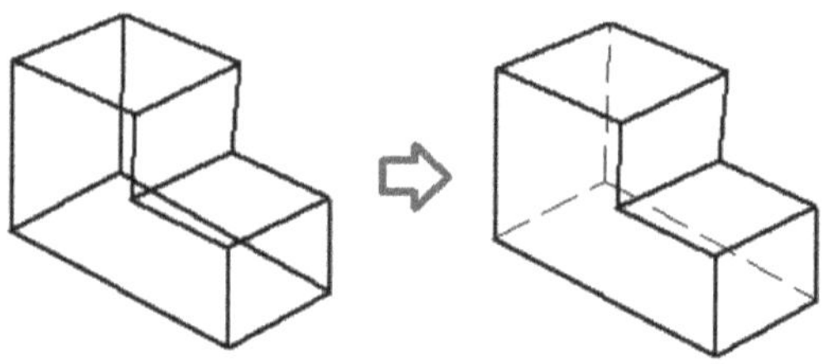

Figure 3.107. Production of a hidden-line drawing from wireframe input.

The list of hidden edges inferred from dashed strokes can be refined by enabling *Refine dashed strokes by re-searching hidden lines* in the *Cue hidden edges* dialog. This refinement uses face information, which must be computed in advance. Alternatively, hidden edges can be detected from scratch by selecting *Detect hidden edges* in the *Cue configuration* toolbar.

The exhaustive method implemented in REFER computes hidden edges from a graph-like representation of a wireframe polyhedral shape. The input (the model stored in the database) must contain vertices and edges. The output updates the database by marking edges as visible or hidden. As a derived or second-level cue, this method depends on prior cues like faces and/or perimeters. Two approaches are available:

- Exhaustive search over all possible combinations of visible/hidden faces. This approach requires both faces and perimeters to be pre-computed.
- Exhaustive search over all possible combinations of visible/hidden edges. This approach requires only faces to be pre-computed.

The two approaches differ in their heuristics and the order in which invalid permutations are discarded. In the faces approach, each candidate solution is called a *variation* and is accepted only if it satisfies a set of heuristic rules. For each variation, edges occluded by visible faces are labeled as hidden:

- Declaring a face as visible implies all its edges are visible. Otherwise, they are hidden.
- Variations in which a face is hidden while all surrounding faces are visible are redundant, as their edges are ultimately visible.

A variation is considered invalid if it violates any of the following rules:

- Perimeter edges cannot be hidden (unless they intersect another perimeter edge).
- Visible edges cannot intersect other visible edges.
- Vertices inside visible faces must be hidden (they cannot connect to visible edges).

- If hidden edges do not intersect any visible edge, then at least one of their vertices must be hidden (and all connected edges must also be hidden).

Typically, more than one valid solution is found. One solution is then selected either automatically or interactively. In the edges approach, candidate solutions are again referred to as *variations* and validated through heuristics. First, the search space is reduced by applying standard rules from descriptive geometry:

- Perimeter edges are marked as visible.
- Edges on the front faces are marked as visible (thus selecting one of the two Necker reversal interpretations).
- All edges connected to an internal vertex must share the same visibility. If one such edge is visible, all connected edges are also marked as visible.

Next, the remaining edges are incrementally labeled as visible or hidden, and each resulting variation is tested. A variation is marked as invalid if it violates any of the following rules:

- Visible edges cannot intersect other visible edges.
- If hidden edges do not intersect visible ones, then at least one of their vertices must be hidden (along with all connected edges).
- Vertices located inside a visible face must be hidden and not connect to visible edges.

A final valid solution is selected either automatically or interactively, depending on the interaction mode.

Both the selection of the approach (faces vs. edges) and the interaction mode (automatic vs. interactive) can be configured in the *Hidden Edges* dialog. Valid solutions are ranked from fewest to most hidden edges. In automatic mode, the solution with the fewest hidden edges is selected. Otherwise, the user is prompted to choose.

Identifying hidden edges in wireframes, prior to obtaining any 3D information, remains an open problem. The two approaches implemented in REFER perform well in simple examples but still face significant limitations:

- They are not fully automatic. The final selection relies on user input to choose the perceptually correct result.

- They are computationally expensive. Despite search space reduction techniques, the process remains nearly exhaustive and scales poorly with increasing edge or face count.
- The overall success rate is still insufficient for general application.
- The approaches rely on prior detection of faces and contours, which are themselves unsolved challenges.

As an alternative to exhaustive search, a more efficient deterministic strategy may be adopted, beginning from a face fixed as visible. Fixing one face as visible introduces no loss of generality, as it merely corresponds to selecting one of the two Necker reversal interpretations. Since these interpretations represent perceptual variants of the same solution (like bipolar pairs), grouping them as one would reduce the number of candidate solutions without loss. This method is particularly helpful when only two solutions exist, making automatic selection feasible. These bipolar pairs are easy to detect: edges on the contour remain visible in both, while the remaining edges have opposite visibility in the two interpretations.

Fixing the upper face as visible is a promising strategy, and it has already been adopted. However, the heuristic proposed by Varley (2003) to detect the upper face leads to errors. A possible improvement would be selecting the face that contains the uppermost vertex and has the highest minimum y-coordinate among all faces sharing that vertex. This criterion may prevent selecting solutions that correspond to unnatural viewpoints, such as views from below.

The parameters used to configure the cue hidden edges can be accessed via the corresponding *Details* button in the *Cue Detection Configuration* Dialog box. One of the options enables the use of vectorized dashed lines as indicators of hidden geometry. When activated, this feature interprets dashed strokes as hidden lines in the model. If no dashed strokes are found in the input drawing, or if the user has enabled the option to refine dashed strokes, the system proceeds to apply the primary algorithmic approach for hidden edge detection.

Hidden edges can be computed using one of two algorithmic strategies, as explained earlier. If multiple solutions are found, the system can either select one automatically or present them interactively, based on the current settings. Valid solutions are sorted in ascending order by the number of hidden edges. If automatic selection is enabled, the first solution in the list (i.e., the one with the fewest hidden edges) is selected as the

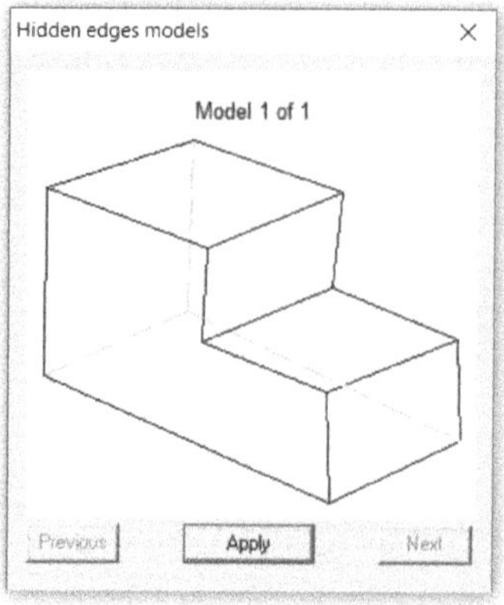

Figure 3.108. Dialog window with the possible solutions calculated during the cue detection process.

output. If interactive mode is enabled, a dialog window appears during the cue detection process, presenting the user with each possible solution in sequence, and allowing them to manually choose the most perceptually accurate interpretation, as shown in Figure 3.108.

3.7.5.16 *Cue levels of vertices*

The implemented algorithm determines the levels of vertices within polyhedral shapes. The input is a list of vertices and edges; the output is a list of labels. The main characteristics of the algorithm are:

- Hidden edges (if available) are used to calculate levels of the vertices.
- Labels are assigned to all the vertices.
- Levels depend on the configuration of the vertices. The configuration is the angular distribution of the edges connected to the vertices.
- Valid levels range from L0 to L17.

The approach estimates vertex depth by determining their levels within a graph-based representation of the wireframe structure of a polyhedral shape. It implements a perceptually motivated, direct (non-iterative) form of inflation, where the z-coordinates of vertices are inferred from the configuration of the junctions to which they belong. The method relies on the assumption that there is a strong correlation between the angular arrangement of edges at a vertex (i.e., the junction shape) and that vertex's relative depth in 3D space. This assumption holds reasonably well for simple, convex, and symmetric polyhedral forms, particularly when the drawing reflects a typical viewpoint—such as observing the object from a slightly elevated position (as if held in hand and tilted forward or viewed from above at a slight pitch).

The algorithm implemented in REFER leverages this perceptual assumption to assign distinct depth levels to each vertex. The algorithm operates as follows:

- *Input*: A set of vertices and edges representing a wireframe model.
- *Output*: A label (or level) assigned to each vertex, estimating its relative depth.
- Key features include:

(1) If available, hidden-edge information is used to improve level estimation.
(2) All vertices in the wireframe are assigned a discrete level label.
(3) Levels are determined based on the local geometric configuration of each vertex—that is, the angular distribution of connected edges.
(4) The output levels range from L0 to L17, each corresponding to a specific junction type and visibility context.

Vertices are first categorized by their 2D junction configuration, as follows:

Level 0 = inverted Y
Level 1 = inverted W
Level 2 = lateral W
Level 3 = lateral inverted W
Level 4 = W
Level 5 = Y

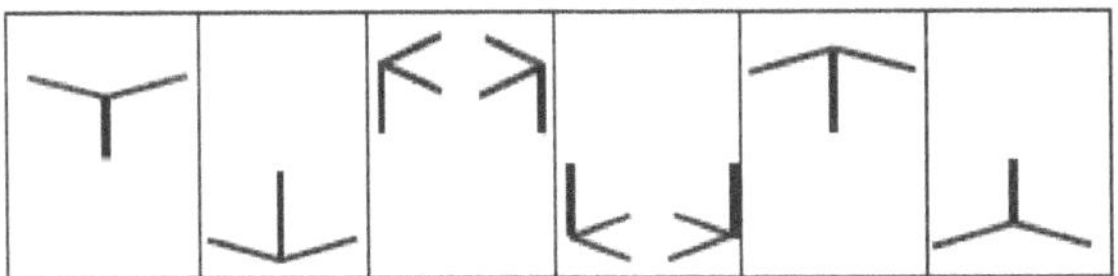

Level 5	Level 4	Level 3	Level 2	Level 1	Level 0
Y	W	Lateral inverted W	Lateral W	inverted W	inverted Y

If hidden/visible edge information is available—either from the initial input or from the cue hidden edges—it is used to further refine level assignments. Vertex levels are divided into three main categories based on edge visibility:

Levels 0–5: vertices with all connected edges hidden
Levels 6–11: vertices with a mix of hidden and visible edges
Levels 12–17: vertices with all connected edges visible

3.7.5.17 *Cue bilateral symmetry*

This approach estimates planes of bilateral symmetry from a graph-based representation of a polyhedral shape. The underlying algorithm and methodology have been described in prior studies (Piquer *et al.*, 2003, 2004; Plumed *et al.*, 2016).

The algorithm identifies all potential bilateral symmetry planes that are present in a line drawing of a polyhedral object. It operates through a two-stage process:

(1) Detection of skewed facial symmetry axes. In the first stage, the algorithm analyzes each face of the polyhedron, previously extracted in 2D using the *Cue Faces* procedure, to identify its axis of skewed symmetry. These are axes that represent approximate facial symmetry based on the drawing's geometry. An example is shown in Figure 3.109.
(2) Identification of candidate symmetry circuits. In the second stage, the algorithm identifies closed chains (or circuits) of facial symmetry axes that indicate a global plane of bilateral symmetry.

This circuit is interpreted as the 2D projection of a symmetry plane intersecting the 3D polyhedral structure represented in the drawing, as illustrated in the examples in Figure 3.110.

Because the process is applied directly to the 2D line drawing—prior to any 3D reconstruction—the identification of valid symmetry circuits relies on a heuristic strategy. This method involves several tunable parameters that govern the selection criteria which can be configured via the

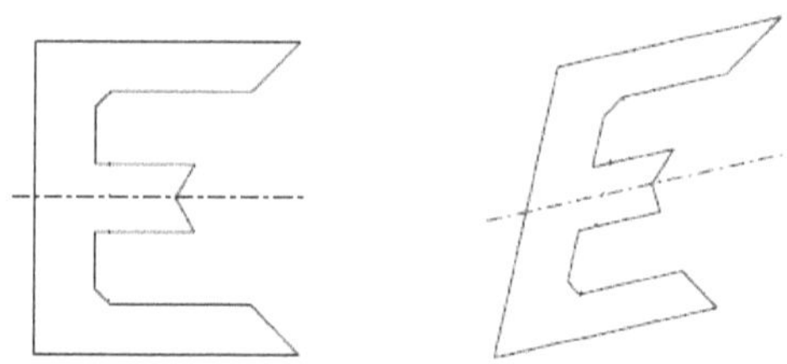

Figure 3.109. Detection of skewed facial symmetry axes.

Bilateral Symmetry dialog. Bilateral symmetry detection has been explored through alternative techniques in more recent work by Hahnlein *et al.* (2022), who provided a complementary perspective and further validation of symmetry cues in 2D-to-3D inference tasks.

The parameters to configure the cue bilateral symmetry can be accessed via the corresponding *Details* button in the *Cue Detection Configuration* dialog box, as shown in Figure 3.111.

Bilateral symmetry can be computed using three different approaches, which rely on a set of structural rules designed to reduce the number of potential symmetry planes (Piquer *et al.*, 2003). The reduction process

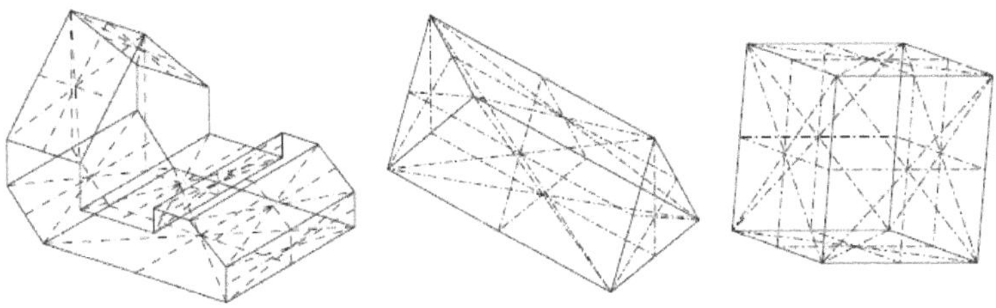

Figure 3.110. Identification of candidate symmetry circuits.

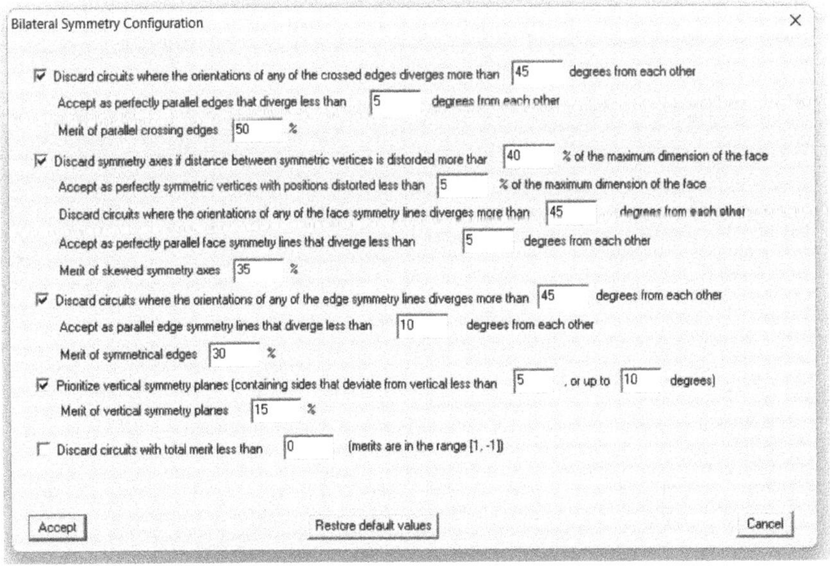

Figure 3.111. *Bilateral symmetry configuration* dialog box.

ensures that only geometrically consistent symmetry axes are considered as valid candidates. The primary filtering rules are:

(1) *Edge midpoint continuity*: If a symmetry axis across a face terminates at the midpoint of an edge, then the adjacent face on the opposite side of that edge must also include a symmetry axis that ends at the same midpoint. This rule ensures geometric alignment across neighboring faces.

(2) *Vertex continuation constraint*: When a symmetry axis of a face intersects a vertex, it may extend in two possible directions: (1) along another symmetry axis that passes through the same vertex but belongs to a different face, or (2) along an edge that is not part of the original face. In the latter case, a new symmetry axis must meet the opposite endpoint of the edge, and this new axis must belong to a third face (one that is distinct from both the original face and any other face sharing the edge).

(3) *Face closure rule*: Suppose two consecutive candidate lines *l1* and *l2* share a vertex *v*, forming segments between vertices *v1*-*v* and *v*-*v2*. If *v1* and *v2* belong to the same face, then for *l1* and *l2* to define part of a symmetry plane, the face must include a line connecting *v1* and *v2*, thereby closing the loop.

    Any axis or line that violates one or more of the above conditions is eliminated from the set of candidates. Often, eliminating a single inconsistent axis enables the pruning of entire chains of connected axes, which significantly reduces the search space for valid symmetry planes. Nevertheless, even after applying these constraints, it is common for the system to identify an excessive number of candidates.

    To further refine the results, an additional heuristic filter is employed—based on empirical testing across numerous sketches—which has proven effective in practice:

(4) *Single-face axis exclusion*: Candidate axes that connect two vertices entirely within a single face are discarded. This rule applies only to axes crossing the interior of faces and not to the actual edges of those faces.

The combination of structural constraints and heuristics helps focus the detection process on the most plausible planes of bilateral symmetry, improving both computational efficiency and the quality of the results.

3.7.6 *Features*

3.7.6.1 *Introduction*

Design features can be defined as sketched configurations that, in the designer's mind, serve a specific purpose related to the functional intent of the object (Plumed *et al.*, 2013b). These are conceptually distinct from machining features, which are geometric shapes typically created through subtractive manufacturing processes such as drilling, milling, or cutting. The broader concept of feature is explored in more depth in the following section.

Feature recognition from 3D input is a well-established research domain—though still an open problem in many respects—with numerous methods and applications having been proposed over the years (Babic *et al.*, 2008; Verma and Rajotia, 2010; Sunil *et al.*, 2010). In contrast, the recognition of features from 2D input, such as line drawings or sketches, has received comparatively little attention (Company and Varley, 2010; Company *et al.*, 2012). The concepts are illustrated in Figure 3.112.

Early work by Tanaka *et al.* (2022) focused on recognizing form features, which are geometric elements that allow solids to be decomposed into simpler volumetric components. However, this early approach did not address the underlying design intent associated with those shapes. More recent work by the same group has expanded to include specific types of design features (Tanaka *et al.*, 2024).

REFER implements the design feature recognition method proposed by Plumed *et al.* (2014, 2022), which specifically targets features in the

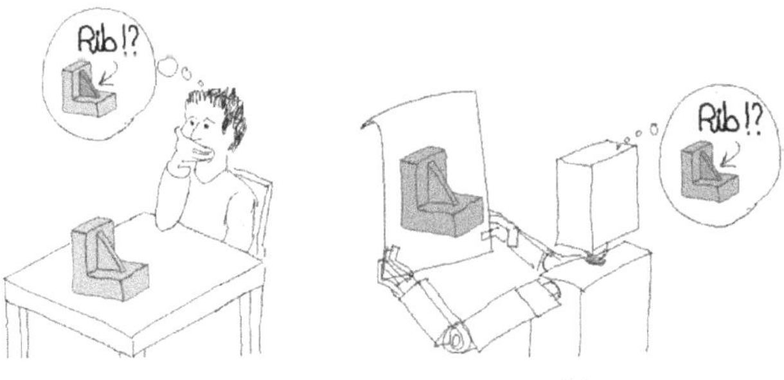

Figure 3.112. Feature recognition from 3D input (a) vs. feature recognition from 2D input (b).

context of sketch-based modeling. To manage features in REFER, the following key components must be considered:

(1) Feature computation is configured through the *Feature Configuration* toolbar.
(2) Detected features are visualized within the *2D Sketch* window interface.
(3) Tools for displaying and interacting with features are available under Display Features toolbar.

The feature-based reconstruction workflow follows a different logic and processing sequence than optimization-based reconstruction. This alternative pipeline has been described in detail in prior studies (Plumed *et al.*, 2014, 2022) and is also outlined in the *Reconstruction Flow* section.

3.7.6.2 *Definition*

The concept of feature remains subject to debate (Sanfilippo and Borgo, 2016). In a critique that promoted designing primarily in terms of manufacturing features, Vandenbrande and Requicha (1993) proposed a distinction between two key types of features. They defined machining features as geometric forms produced by specific manufacturing processes–typically subtractive in nature; and design features as configurations sketched by the designer with a particular functional intent in mind. While manufacturing methods may evolve multiple times over the lifecycle of a product, the object's intended function generally remains constant. Consequently, the authors argued that design should be driven by functional requirements rather than the constraints of manufacturing technology.

To this discussion, Han (1996, 1997) added a third category: form features. These are geometric descriptors of shape that do not inherently convey any functional purpose or manufacturing implication. Han advocated for the use of form features during the design phase, with the conversion to machining features occurring only during later stages of product development. His work includes documented procedures for translating form features found in CAD models into machining features specific to various manufacturing technologies. Additional feature categories such as analysis features, material features, and tolerance features have been identified in other contexts, but they are unrelated to the problem of interpreting pictorial sketches and are therefore not considered here. In summary, three main types of features are relevant for our purposes:

- Design features. Conceptual elements that exist in the designer's mind, driven by the object's function.
- Form features. Explicit geometric configurations that are present in sketches or CAD models.
- Machining features. Shapes interpreted in terms of manufacturability, associated with specific production processes.

The mappings among these three types are not one-to-one. A single design feature may be represented by various form features, and a given form feature can correspond to different machining features depending on the manufacturing context. Since the publication of Vandenbrande and Requicha's work (1993), significant progress has been made in converting form features into machining features. The field of feature-based design has also matured, evolving from early efforts focused on feature recognition for manufacturing into a broader paradigm that supports interactive, feature-driven design workflows (Shahin, 2008). Despite these advancements, relatively little work has addressed the automatic inference of design features, particularly from pictorial or sketch-based inputs. Recent developments in the use of Artificial Intelligence for machining feature recognition in 3D models (Zhang *et al.*, 2018; Rai *et al.*, 2021) show promise. These advances suggest a potential path forward for similar AI-driven approaches aimed at detecting design features in 2D sketches (Miles *et al.*, 2023).

In REFER, once a sketch has been vectorized and refined, a feature detection stage is launched. This step attempts to identify as many features as possible before initiating the 3D reconstruction process. The features are illustrated in Figure 3.113.

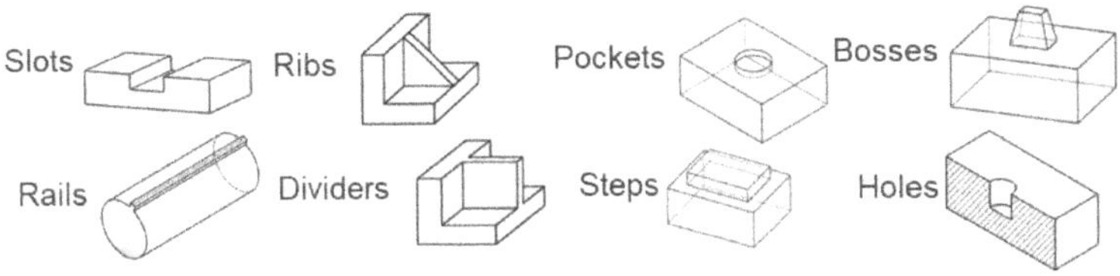

Figure 3.113. Supported features.

It is important to note that feature names are not standardized or universally applied. A common example is the distinction between *rib* and *gusset*. Both refer to thin structural elements that reinforce a part. In welded components and metal structures, these features are typically referred to as ribs and often have triangular profiles, as they commonly reinforce stepped regions. In contrast, within injection molding, the term gusset is used for similar features, while rib is reserved for reinforcements that strengthen the cavities of hollow, shell-like components. An example is shown in Figure 3.114.

This distinction is important because it influences both the type and the relative importance of geometric cues that, in 2D, indicate the presence of a corresponding 3D feature. However, since feature detection remains an evolving capability, REFER does not yet subdivide main feature categories into more specific subtypes. As a result, the terminology used may not always align with the expectations of certain users—for example, those familiar with sheet metal ribs. Nonetheless, this aspect does not undermine the broader premise that any feature can, in principle, be identified in 2D, provided that the most relevant revealing cues are properly defined and systematically detected.

3.7.6.3 *Feature configuration toolbar*

The *Feature Configuration* toolbar is located over the parent window and is shown in Figure 3.115.

The toolbar contains the following commands:

- Detect ribs and dividers.
- Detect rails and slots

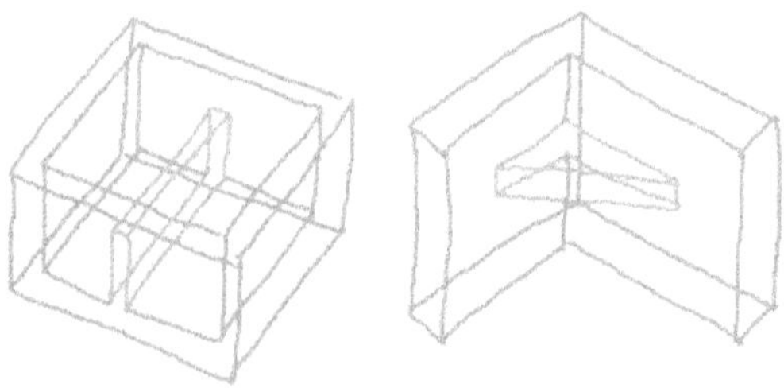

(a) (b)

Figure 3.114. Injection molded rib (a) and injection molded gusset (b).

Figure 3.115. *Features Configuration* toolbar.

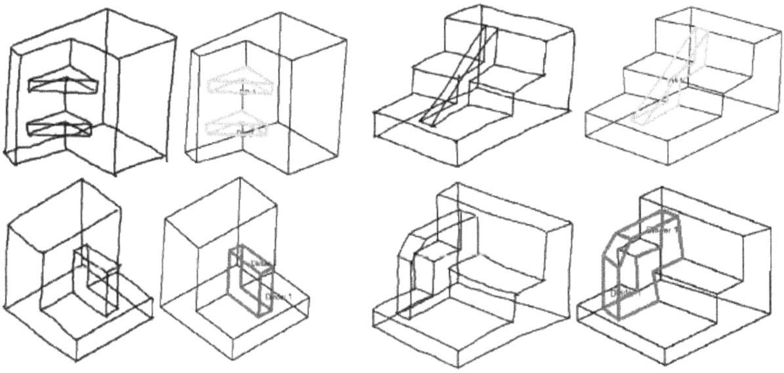

Figure 3.116. Detection of ribs and dividers.

- Detect steps and pockets.
- Detect chamfers and rounds (not yet implemented).
- Configure detection of features.

3.7.6.4 *Ribs and dividers*

This command detects ribs and dividers (see Figure 3.116) using a set of heuristic rules, as detailed by Company *et al.* (2012).

The key geometric characteristic of ribs is that they are thin walls. The primary cue for identifying such a wall is the presence of narrow, rectangular border faces. In an axonometric 2D view, a 3D rectangle often appears as a parallelogram; hence, a 2D parallelogram is taken as an indirect cue for a rectangular face in 3D.

Since ribs are attached to the body of the object, most of their border faces are embedded within it (for webs, all border faces are embedded). However, a rib usually has at least one visible, non-embedded border face. The first step in detection, therefore, is identifying thin border faces—typically quadrilateral faces with two short and two long opposing edges.

Second, as ribs represent "walls," they can be described as extruded shapes. The border face should appear as a parallelogram, and the two faces connected to its longer edges should be geometrically identical. In wireframe inputs, both adjacent faces are known, which helps disambiguate rib candidates.

Third, ribs often "interrupt" the surrounding faces. For polyhedral objects, this means the external edges (i.e., edges connected to the lateral faces of the rib but not part of the rib itself) are typically collinear with corresponding external edges on the opposite side. While this collinearity is not a strict requirement, it acts as a reinforcing cue. When present, the algorithm increases the figure of merit that determines the likelihood of a candidate being classified as a rib.

Finally, since the algorithm detects thin walls, it can also identify dividers, which are thin walls used to partition space or restrict movement between regions. While ribs serve a structural or design function, dividers are geometric in nature. For example, dividers often have multiple border faces and quadrilateral lateral faces. These differences are leveraged by the algorithm: candidates with multiple border faces or parallelogram lateral faces are more likely to be *dividers* and are reclassified accordingly.

The example shown in Figure 3.117 highlights a key limitation of the current method. Since the input is a 2D boundary representation (B-Rep), the algorithm cannot distinguish between true ribs and negative ribs (i.e., hollow features that appear similar but are not protrusions).

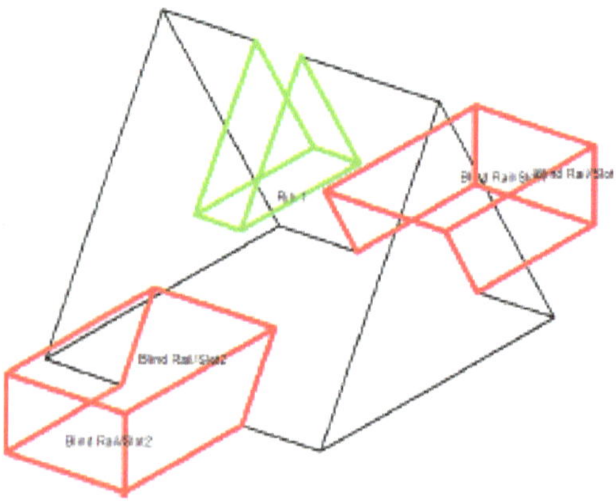

Figure 3.117. Failure to distinguish between true ribs and negative ribs.

3.7.6.5 *Rails and slots*

This command detects rails and slots (see Figure 3.118) using a set of heuristic rules, as detailed by Company *et al.* (2012).

Slots are long, narrow openings typically intended to accommodate inserted parts, possibly allowing sliding motion. While some linear ribs serve structural reinforcement purposes, others act in conjunction with slots in adjacent components to enable sliding joints. These specialized ribs are referred to as rails. The shape of a rail complements the shape of the slot in its counterpart part. An example is shown in Figure 3.119. As such, the procedures for detecting rails and slots are very similar.

The main cues used for detection are as follows:

- Slot geometry. A machined slot typically appears as a long, narrow cavity on a flat surface, resembling an incomplete prism. The bottom and lateral faces of such slots are usually quadrilaterals.
- Homothetic lateral faces: The lateral faces should have the same shape (i.e., they should be homothetic), reinforcing the symmetry of the slot structure.
- Collinear external edges. Since rails and slots interrupt the faces in which they are embedded, collinearity between external edges on opposite sides of a candidate slot or rail increases the likelihood of a correct match. This condition boosts the associated figure of merit.

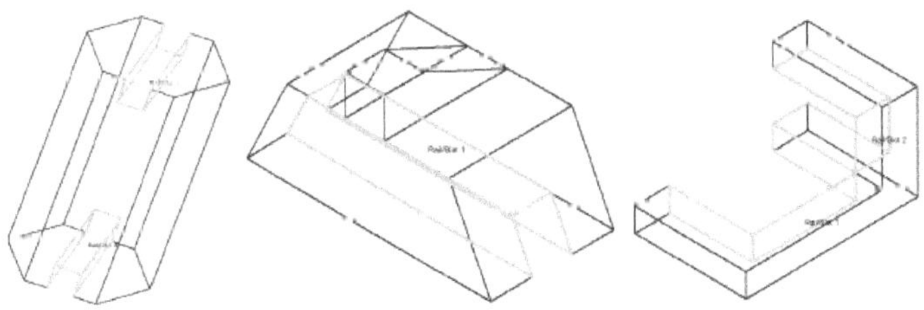

Figure 3.118. Detection of rails and slots.

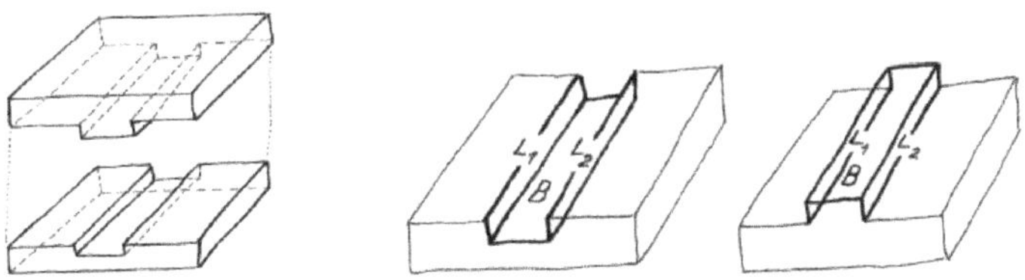

Figure 3.119. Rails and slots.

- Slenderness: Because most common rails and slots are wider than they are deep, narrow quadrilateral lateral faces contribute positively to the figure of merit.

To avoid incorrect classification, a heuristic rule is applied to prevent the largest face of the model from being considered part of a rail, and thus, mistakenly identifying the outer contour as a rail.

A known limitation of the method is its inability to detect rails that do not have rectangular cross-sections. For example, triangular-section rails such as the one shown in Figure 3.120 require different heuristic strategies for detection.

Finally, although the heuristic rules perform well for isolated rails and slots, they may produce false positives when the rail or slot interacts with surrounding geometry in a more complex way—even when such complexity is not visually apparent, as illustrated in Figure 3.121.

3.7.6.6 *Steps and pockets*

This command detects steps and pockets using heuristic rules detailed by Plumed *et al.* (2013b). A pocket is an extruded subtractive volume that does not meet the criteria to be classified as a hole or a slot. It is not deep enough to be considered a hole (see Figures 3.122(a) and (b)), or long enough to be a slot (see Figures 3.122(c) and (d)). Specifically:

- A feature with a uniform profile along its depth but a low aspect ratio is considered a pocket, not a hole (Figure 3.122(a)).

Figure 3.120. Failure to detect a slot with a non-rectangular cross-section.

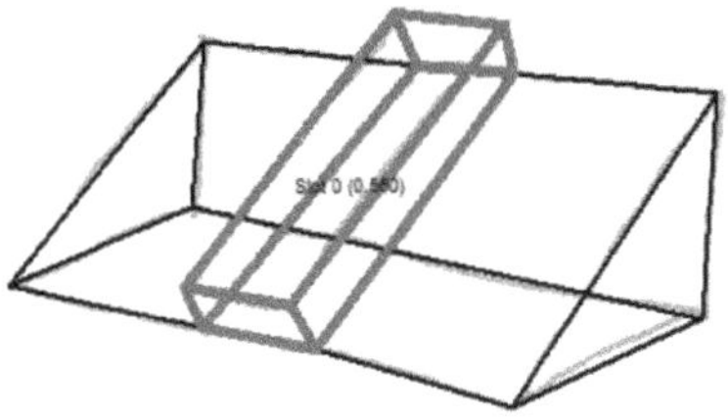

Figure 3.121. False positive detected as a rail in REFER.

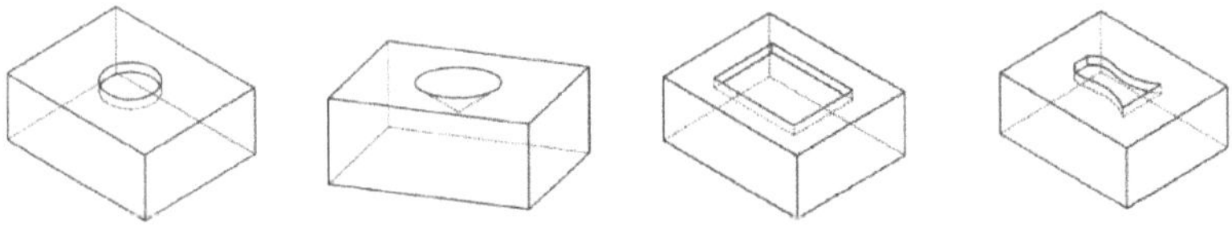

(a) (b) (c) (d)

Figure 3.122. Examples of pockets.

- A feature with a high-aspect ratio but a non-uniform profile is also considered a pocket (Figure 3.122(b)).
- A uniformly extruded subtractive feature over a short length is considered a pocket rather than a slot (Figure 3.122(c)).
- Similarly, a long feature with a non-uniform profile along its length is also classified as a pocket (Figures 3.122(d)).

The following geometric cues are considered when determining whether a wireframe drawing depicts a pocket:

- Does the candidate meet the topological criteria for a slot?
- Does the bottom face have a larger area than the side faces? This cue is the key geometric criterion for distinguishing pockets from slots.
- Are there smaller bosses embedded within the bottom face? If so, the main feature cannot be a slot, but it may still be a pocket.

Pockets can have complex borders, either regular or irregular (Figures 3.123(a) and 3.123(b)), and do not require a flat base (Figure 3.123(c)). When pockets exhibit multiple complex elements such as irregular profiles and contoured bases, they may resemble sunk reliefs (Figure 3.123(d)). However, because reliefs are decorative rather than functional, they are not considered here.

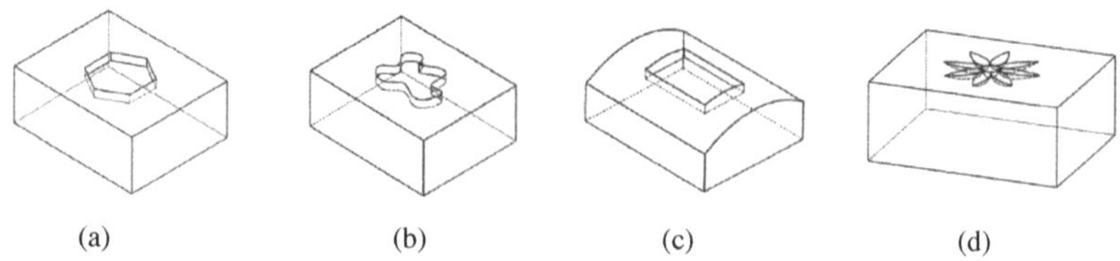

Figure 3.123. Examples of complex pockets.

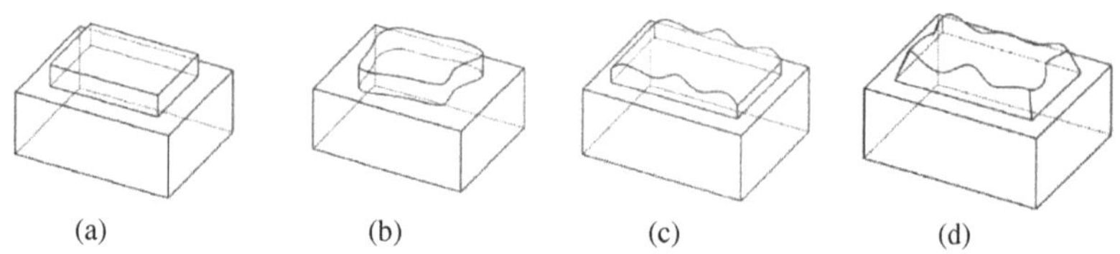

Figure 3.124. Examples of steps.

A step is an extruded additive volume with a low aspect ratio (wider than it is tall) and may have a regular (Figure 3.124(a)) or irregular (Figure 3.124(b)) contour. Steps can take various extruded additive profiles (Figure 3.124(c)) and may be uniform or non-uniform in any direction (Figure 3.124(d)). Like pockets, steps may also feature flat or contoured surfaces. When a step combines many of these complexities—such as an irregular and non-uniform cross-section with a contoured top—it is better classified as a bas-relief.

The geometric cues to consider when evaluating whether a wireframe drawing depicts a step are the following:

- Does the candidate meet the topological criteria for a rail?
- Does the bottom face have a larger area than the side faces? This geometric criterion distinguishes steps from rails.
- Are there smaller bosses embedded within the upper face? If so, the main feature cannot be a rail but may still be a step.

The example of a *contiguous* step and pocket in Figure 3.125 highlights both a current limitation and a major challenge of this approach. The limitation is that the algorithm only accepts inner subgraphs as candidates for defining steps and pockets. The more significant problem is that the drawing omits the edge where the step ends and the pocket begins, since these features share a common face. In other words, there is no explicit "end of step" or "start of pocket" edge. For this reason, the current algorithm cannot use this shared border to identify either feature.

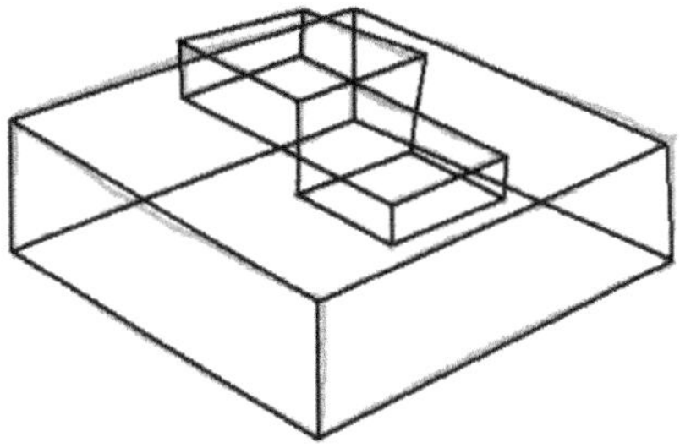

Figure 3.125. Example of a contiguous step and pocket.

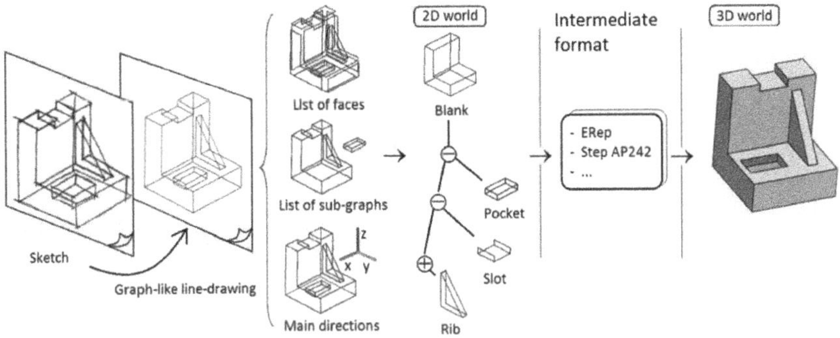

Figure 3.126. CSG reconstruction process from a sketch in REFER.

3.8 CSG Modeling

3.8.1 *Introduction*

REFER's goal is to develop an approach for capturing the design intent embedded in a 2D sketch and automatically generating a 3D model that is ready for manufacturing. To this end, we aim to construct a Constructive Solid Geometry (CSG) feature tree that combines all the design and manufacturing features inferred from the sketch in a meaningful manner. The detection of design and manufacturing features is a key topic in automated feature recognition, a research area that typically begins with a 3D model as input. REFER's objective is to detect features prior to producing any 3D models. The idea is illustrated in Figure 3.126.

Some existing approaches that directly address 2D drawings include Meeran and Pratt's (1993), who designed an experimental system for automated process planning based on 2D orthographic views. The authors developed general rules for recognizing common machining features in prismatic parts and certain interacting features. Meeran and Taib (1999) later introduced a two-stage, hint-based feature recognition system also based on orthographic views. Their method identified profiles in 2D drawings and extruded them to form feature volumes. While both approaches apply general rules for detecting features, they are limited in the number of feature types they can handle, and all features must be aligned with either the x- or y-axis.

The FlexiCAD system proposed by Tyan and Devarajan (1998) takes a 2D orthographic drawing (possibly derived from a 3D model) as input and identifies features through geometric patterns, classifying them into a hierarchical structure based on characteristic attributes. FlexiCAD is constrained to features that (1) do not interact, (2) have uniform thickness, and (3) are included in a predefined pattern library.

Approaches originally developed to reconstruct 3D CAM models from legacy 2D blueprints typically rely on multiple orthographic views. These methods assume that all necessary information is present, though distributed across different views. In contrast, REFER's input is a single pictorial view, which inherently contains incomplete information. Therefore, detecting cues or hints that may indicate the presence of specific design or manufacturing features becomes a critical step in REFER's approach.

REFER adopts a two-stage approach: (1) Reverse detection of features in 2D, and (2) Definition of their mutual relationships through datum extraction. The general procedures for configuring and controlling the CSG modeling process are available via the *CSG Modeling* toolbar (see Figure 3.127).

Three commands are available for CSG modeling: one initiates the process, another resets the current CSG model, and the third opens the configuration settings:

- Build CSG from line-drawing. When toggled on, CSG modeling will run automatically whenever the reconstruction process is launched via the *Reconstruction* toolbar.

Figure 3.127. *CSG Modeling* toolbar.

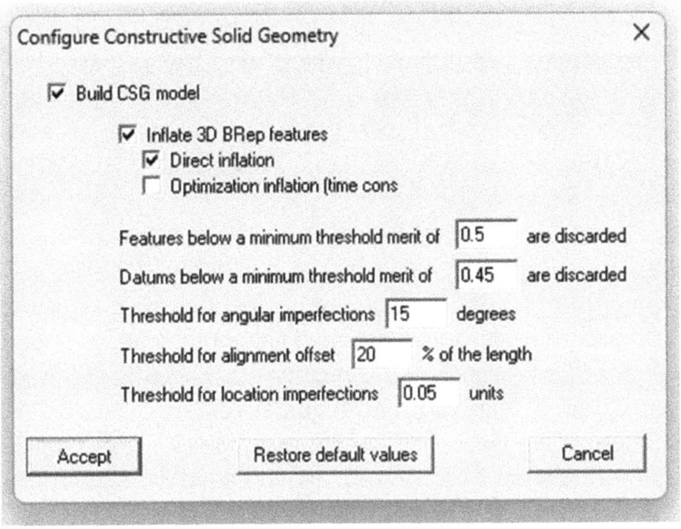

Figure 3.128. *CSG Modeling Configuration* dialog.

- Delete current CSG model.
- Configure CSG. Opens the *CSG Modeling Configuration* dialog, which allows the user to customize the reconstruction process. The dialog is shown in Figure 3.128.

The user can manage the activation of each stage in the sequence independently: (1) CSG deconstruction, and (2) CSG tree extraction. The user is also given the option to enable or disable the automatic inflation of features detected during the deconstruction process. Inflated features can be visualized in the CSG model window by activating the *Display CSG Features* command. The option to disable this functionality is provided, as inflation increases computational cost and may introduce reconstruction errors. This limitation arises from the fact that axonometric inflation is only reliable for normalon or quasi-normalon geometric configurations.

Additionally, since a merit value is automatically computed for each detected feature, the dialog provides a parameter to set the minimum merit threshold. This parameter ensures that only features with sufficient relevance are retained in the final model. For the same reason, the datum threshold is used to discard datums detected with low credibility. Two additional thresholds allow the user to adjust the tolerances used to account for angular and positional inaccuracies.

3.8.2 CSG deconstruction in 2D

REFER implements and approach where the CSG feature tree is constructed in reverse, starting from child features and progressing toward parent features (Plumed *et al.*, 2013c, 2014). This reverse construction method has been extensively explored by Tanaka *et al.* (2015, 2017, 2020).

In this method, the CSG tree is built in a bottom-up fashion. Initially, features that are more likely to be children are detected, added to the tree, and removed from the sketch. Their parent features are then searched for recursively until the blank is reached. This entire process is performed directly on the 2D drawing using features that are identified before inflating the sketch into 3D, as shown in Figure 3.129.

Constructing a reverse design history involves identifying features, determining which are child features, and linking them to their respective parents, as illustrated in Figure 3.130.

REFER's method recognizes features in the sketch and ranks them by significance. Child features are considered less important, while parent features are deemed more significant. Heuristic rules help assess this hierarchy and feature types are used to define the Boolean operations connecting them. Once a child is processed, it is removed from the drawing to facilitate identification of its parent. The process is repeated recursively until the blank is reached. The process is illustrated in Figure 3.131.

To guide feature detection, the following hierarchical assumptions are applied:

- Rounds and chamfers are typically final design refinements and thus the least significant.
- Ribs are usually added later in the process to reinforce existing geometry.
- Other features can be introduced at any design stage.

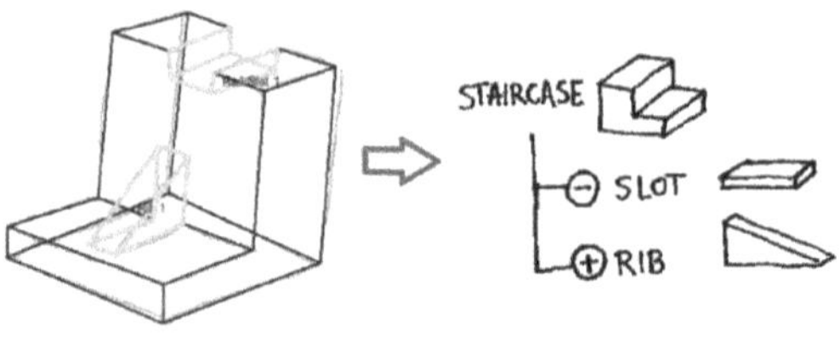

Figure 3.129. CSG deconstruction in 2D.

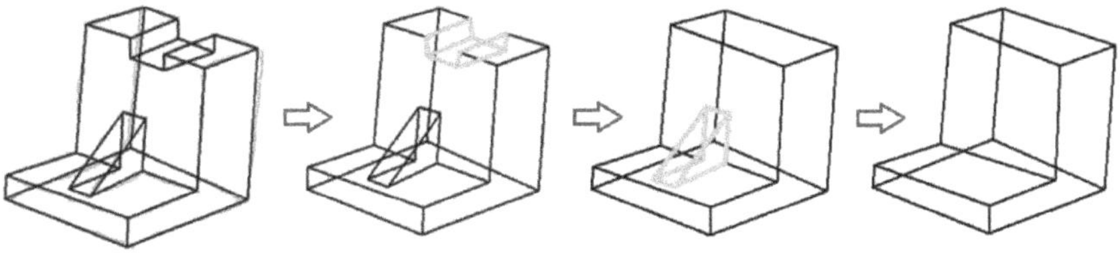

Figure 3.130. Construction of a reverse design history.

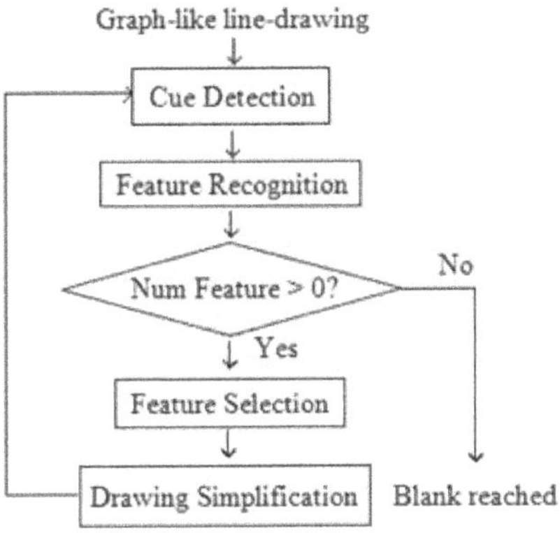

Figure 3.131. Feature recognition process in 2D.

When features are of equal priority, a size-based strategy is used. Designers generally progress from larger features, which may act as containers, to smaller ones embedded within them. Therefore, features with smaller bounding boxes are considered first.

The algorithm also models common behavior observed in designers using 2D CAD applications:

- Designers tend to begin with additive features (e.g., rails, steps, bosses, etc.) before moving to subtractive features (e.g., slots, pockets, holes, etc.).
- Features on faces that are parallel to coordinate planes are typically drawn before features on slanted faces.

Once a feature is identified, its parent is sought using direct and indirect cues. In most cases, a feature includes a contained face that rests on a face from a potential parent (the container feature). When this relationship is found, the container is removed in the next iteration, allowing the algorithm to trace branches in the CSG tree until the last container belongs to the blank. If additional features remain, a new branch is initiated from the next leaf.

Although the algorithm can handle interacting child-parent pairs, it currently supports only independent branches. Intersections between branches are not yet supported.

Feature removal simplifies the sketch by deleting the corresponding edges and vertices, which is straightforward if the feature forms an isolated subgraph. Otherwise, additional cleanup may be required to maintain a valid drawing structure.

3.8.3 *CSG datums in 2D*

Two related methods for deriving a CSG feature tree from a 2D sketch have been proposed by Suh (2007) and Plumed *et al.* (2022). Suh's approach implicitly constructs the procedural structure of the model during the creation process, while Plumed's method explicitly leverages design features and organizes them according to parent-child relationships.

Figure 3.132 illustrates the strategy adapted from Plumed *et al.* (2022) that REFER uses to construct the CSG feature tree. To complete the tree, additional information is required, including the spatial position of each feature within the model and its associated extrusion parameters. On the left, the figure shows the simplification process, which follows a "reverse design order" from child to parent; at each step, the system produces a list of candidate features, each assigned a figure of merit in the range [0, 1]. Concurrently, the right side of the figure addresses the definition of

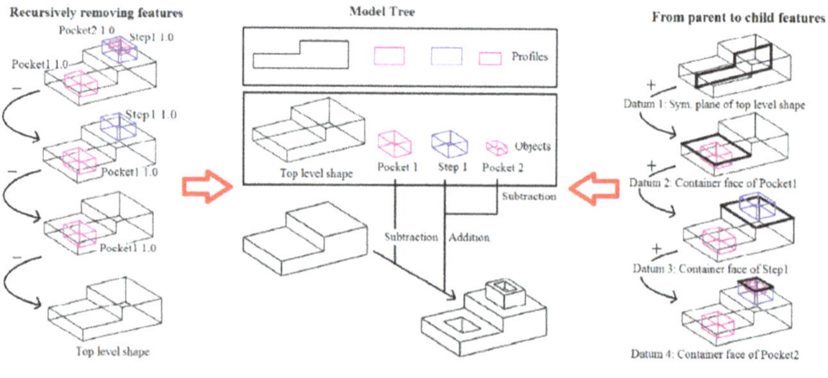

Figure 3.132. Strategy to construct the CSG feature tree (Plumed *et al.*, 2022).

datums, which are used to position features, while the center section captures the geometric extrusion profiles necessary to finalize the feature tree.

According to Plumed *et al.* (2022), two key principles guide the definition of reference systems for CSG reconstruction: (a) each feature should have its own explicit reference system, ideally anchored to the global reference system of the top-level shape; and (b) reference systems must operate in 3D, as reconstruction inherently involves spatial modeling.

The process begins with a *Drawing Graph*, where nodes represent sketch vertices and edges correspond to sketch lines. To construct the CSG feature tree, a recursive strategy is applied in reverse design order, from child to parent. At each step, a feature is identified, and its associated

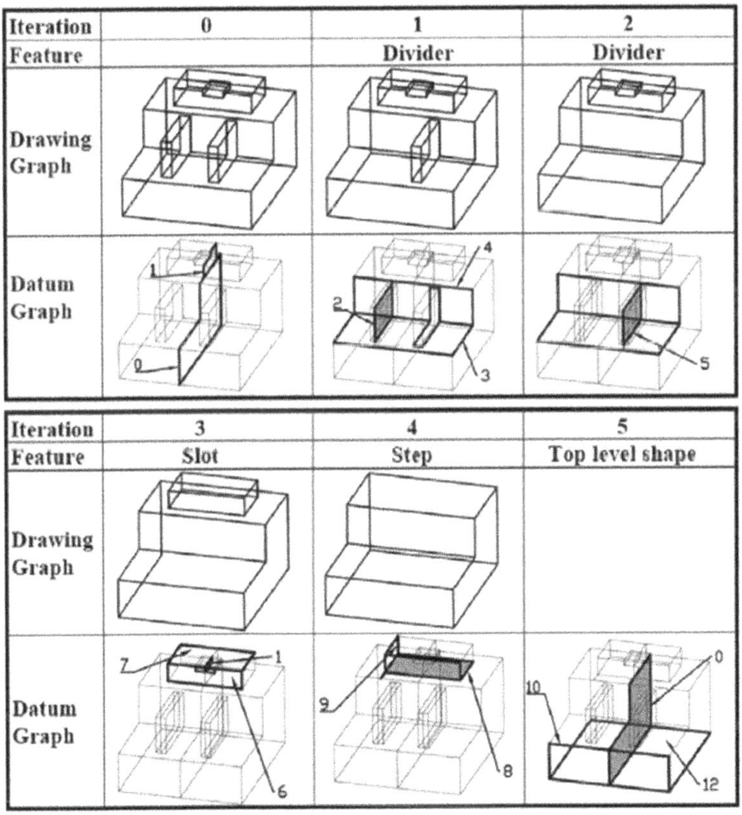

Figure 3.133. Drawing and Datum graphs.

lines and nodes are removed from the Drawing Graph. Simultaneously, a copy of the original graph, referred to as the *Datum Graph*, is progressively enriched with datum information. Datum planes, essential for locating features in 3D space, are defined by identifying closed circuits within the Datum Graph. The process is illustrated in Figure 3.133.

The algorithm searches the Datum Graph for axis-aligned planes. Unlike other methods such as (Varley *et al.*, 2005), which work after inflation to 2-1/2D, our method is based on a path search algorithm where the inputs are two nodes of a graph representing a line drawing that only contains 2D information about a 3D model. The algorithm determines locations, measuring distances between contiguous elements and obtaining their absolute positions by combining relative locations. This simple algorithm works correctly as long as all the lines of the graph are connected. In the future, we will reduce errors by adding redundant distance calculations to reduce the impact of errors on the "beautification" of the vector drawing.

3.9 B-Rep Model Inflation

3.9.1 *Introduction*

One method for recovering the depth information lost when a B-Rep 3D model is projected into a flat pictorial view is known as *inflation*. Specifically, inflation refers to the process of assigning a third coordinate (Z) to the most critical points in a 2D drawing, thereby converting it into a 3D model.

Inflation can be viewed as the inverse of an orthogonal projection, as illustrated in Figure 3.134. To simplify the mapping between 2D junctions in the drawing and 3D vertices in the model, a Cartesian coordinate system is defined such that the XY plane coincides with the drawing plane. Under this setup, inflation ensures that each junction retains its (X, Y) coordinates from the 2D drawing, while the Z coordinate is computed. The resulting 3D model is thus characterized by a set of depth values $z = (z_1, z_2, ..., z_n)$, where n is the number of vertices in the sketch.

Inflating is calculating the (X_I, Y_I, Z_I) coordinates of the vertices of a 3D model from the (X_D, Y_D) coordinates of the input 2D drawing. Therefore, the approach relates the so-called *drawing system* to the *inflation system*. A third model coordinate system (X_M, Y_M, Z_M) is also typically used.

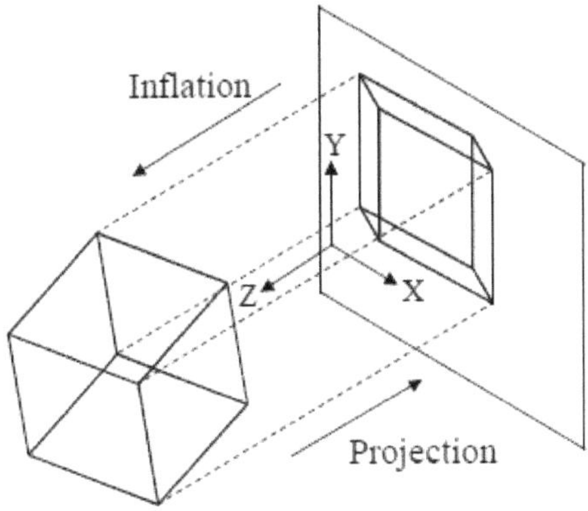

Figure 3.134. Relationship between inflation and projection processes.

The main limitation of this approach is the assumption that the input is an orthogonal projection. As a result, inflating drawings from oblique views typically yields distorted models. Nonetheless, well-constructed orthogonal axonometric projections of certain object types can still be reconstructed successfully. Depending on the nature of the object and the types of curves present, two inflation strategies are commonly used:

(1) Inflating nodes and control points of freeform shapes.
(2) Inflating vertices of analytic shapes.

Takeo Igarashi is one of the most influential figures in the development of inflation for freeform shapes. His pioneering software TEDDY (Igarashi *et al.*, 1999) gained widespread attention and inspired numerous successors. CrossSketch (Alexis *et al.*, 2007), for example, applies Perkins's cubic corners method, originally devised for analytic shapes and Liu and Lee (2010) introduced minor improvements to this technique.

A notable overview of sketch-based input for engineering objects with functional curves was provided by Wang *et al.* (2009), while Roth-Koch and Westkaemper (2010) analyzed their relevance in engineering design. Xu *et al.* (2014) extended inflation by interpreting sketches as semantically meaningful representations, arguing that "designers leverage descriptive curves to effectively convey 3D information in 2D drawings." Their system parses sketches to extract and prioritize this rich information during inflation. This strategy aligns with recent CSG modeling approaches that detect and analyze features within 2D sketches. Other recent contributions depart from purely geometric methods and incorporate perceptual rules for inferring depth (Gryaditskaya *et al.*, 2020).

Since REFER is intended for generating CAD/CAM/CAE models, its inflation strategies focus on analytic shapes. When the inflation stage is enabled in the reconstruction flow, it is executed automatically by the reconstruction module. The process is managed via the *Reconstruction* toolbar, which includes commands for launching the reconstruction and resetting the resulting model. Inflation-specific controls are provided via the *Inflation* toolbar and the corresponding *Inflation Configuration* dialog. Two main types of inflation are supported:

(1) Direct inflation. This approach uses heuristic rules to derive a 3D model from the available 2D information. A classic case is axonometric inflation, where edges assumed to align with the 3D coordinate axes are traced. The corresponding coordinate values are incremented along each path, leading to 3D coordinate estimates for all vertices. A more recent variant predicts depth (Z) values by sampling along rays defined by each 2D point, while recovering (X, Y) positions from a density map (Gao *et al.*, 2024).
(2) Optimization-based inflation. This method iteratively adjusts the 2D drawing until a 3D model is formed that best approximates the object implied in the drawing. Shape changes are driven by an optimization process that refines the Z coordinates of the vertices. Similarity between the resulting 3D model and the intended form is evaluated using mathematical measures based on compliance with visual regularities or depth cues inferred from the 2D drawing.

3.9.2 *B-Rep inflation coordinates*

To improve efficiency, the inflation process typically preserves the X and Y coordinates of the vertices from the 2D line drawing and assigns only their Z coordinates. This strategy transforms drawing vertices into 3D model vertices while reducing the number of variables to be determined to one per vertex, rather than three–thereby streamlining computation during iterative inflation processes, as illustrated in Figure 3.135.

Four coordinate systems are either implicitly or explicitly involved in the inflation process:

- *Drawing coordinates*: The 2D line drawing is placed on a virtual sheet of paper, with its elements referenced in a 2D coordinate system

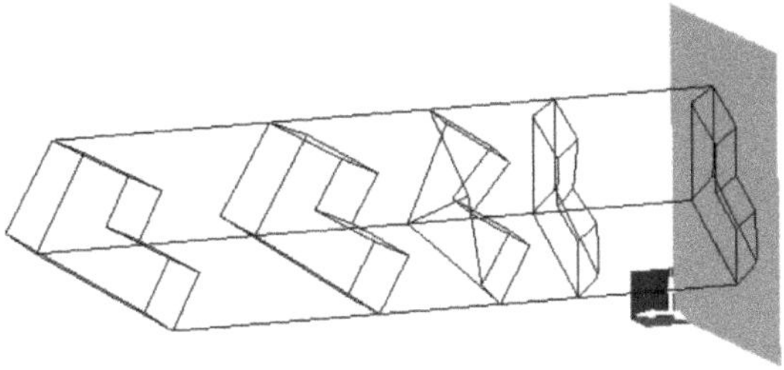

Figure 3.135. Inflation as an iterative process.

(X^D, Y^D) corresponding to the paper plane. It is the reference system used to input the strokes of the sketch, and to vectorize.

- *Axonometric coordinates*: The drawing's lines follow axonometric projection rules, defined by three axes (X^A, Y^A, Z^A) lying on the same plane. Although these axes are coplanar, human perception interprets them as forming a three-dimensional orthogonal system. It is the (implicit) reference system in the designer's mind while selecting the orientations of the lines to draw a pictorial view.
- *Inflation coordinates*: These are the coordinates of the 3D model obtained through the inflation process (X^I, Y^I, Z^I). By design, the inflation assumes $X^I = X^D$ and $Y^I = Y^D$, simplifying the transformation without loss of generality. It is the most convenient system for iterative inflations. The problem is that the final model is rotated relative to the directions of inflation. However, fixing the X and Y coordinates is not convenient in other inflation methods.
- *Main coordinates*: These coordinates refer to the principal directions of the final inflated 3D model (X^M, Y^M, Z^M), which, under certain conditions, can be inferred prior to inflation. It is the best system that helps understand and convey the design intent of the model, although it is not always easy to calculate before obtaining the model.

The four coordinate systems are illustrated in Figure 3.136.

The inflation coordinate system works effectively because in pictorial projections vertices typically do not overlap—unlike in orthographic views, where overlapping is common. This limitation renders inflation unsuitable for orthographic representations.

Conversely, most 3D models have a natural main coordinate system that can often be detected through the main axes cue. Therefore, it is useful to allow the user to select which reference system—either the inflation system or the main coordinate system—is used to represent the model.

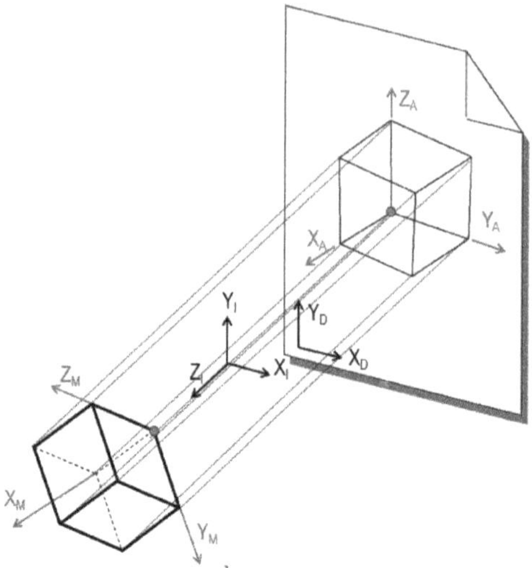

Figure 3.136. Relationships between the four coordinate systems involved in the inflation process in REFER.

To support this option, REFER includes a tool for computing the transformation between these two systems, provided the main axes have been identified. The transformation is computed using a matrix derived from the direction cosines of the model's principal axes with respect to the inflation axes. By default, B-Rep 3D models are displayed relative to their inflation coordinate system. Users can interactively switch to the main coordinate system as needed.

There is a size difference between the projected lines and the model lines. In the case of axonometric projections, this size difference is expressed through the so-called axonometric coefficients (e_x, e_y, e_z). There is a well-known dependence that links the axonometric coefficients and the angles between every pair of the reference axis projections ($X\hat{O}Y$, $X\hat{O}Z$, $Y\hat{O}Z$). It is described in nearly every standard Descriptive or Constructive Geometry text (Hohenberg, 1956), most of which go back to Polkhe's Theorem. The dependence is established between the angles defined by the orthogonal projections of three concurrent orthogonal segments and the angles that each of those segments determine with its own projection. In classical references, the problem was always solved graphically. We, instead, leverage Perkins (1971) and Attneave and Frost (1969) to find its analytical formulation and apply it to visual perception problems.

Perkins (1971) studied the bearing of projective geometry on the perceptual processes by which pictures are "read" for spatial information. To be noticed that Perkins emphasized the logical ambiguity of line drawings (their lack of distinct three-dimensional information), along with the active role of the visual system in making assumptions to resolve this ambiguity. Attneave and Frost (1969) used that formulation to explore the differences between perceived and geometrical depth and orientation.

Computing the direction cosines is not trivial. In our implementation, we leverage previous formulations to determine the angle (for instance, 90°–γ) between every axis (OZ^M) and its own projection (OZ^A) when three orthogonal axes are connected to the same central vertex (O). It is a function of the angles (XÔZ and YÔZ) between the orthogonal projection of OZ^M (OZ^A) and the orthogonal projections of the other two orthogonal axes (OX^A and OY^A). Therefore, we can calculate the "pitch" of each of the axonometric axes from its 2D representation to its corresponding axis of the 3D model, as shown in Figure 3.137.

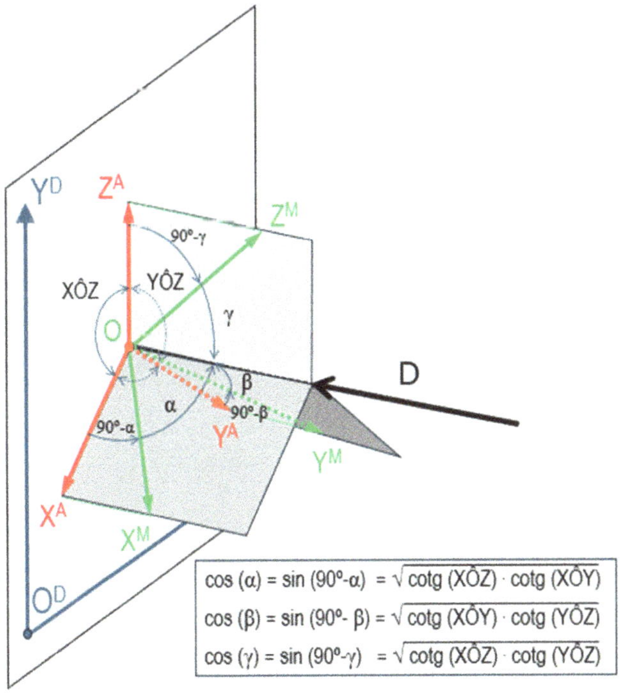

Figure 3.137. Calculation of the axonometric axes.

The only input to calculate the pitch angles are the relative angles between adjacent axonometric axes in 2D, provided that the projection is orthogonal. It is important to note that the axonometric coefficient that relates the length in the projection to the length in the 3D model is $e_Z = OZ^A/OZ^M = \cos(90° - \gamma)$.

In turn, the angles that define the orientation of the orthogonal projection *D* relative to the model's reference system can be related to the direction cosines, as explained in Figure 3.138.

The spherical angles are helpful to define two consecutive rotations to convert *Main Axes* into *Drawing Axes*. We first apply a rotation equal to 90°+φ around the Z^M axis, resulting in a rotated system (O, $X^M{}_1$, $Y^M{}_1$, $Z^M{}_1$) whose X axis is contained within the drawing plane, and parallel to the X^D axis. The process is illustrated in Figure 3.139.

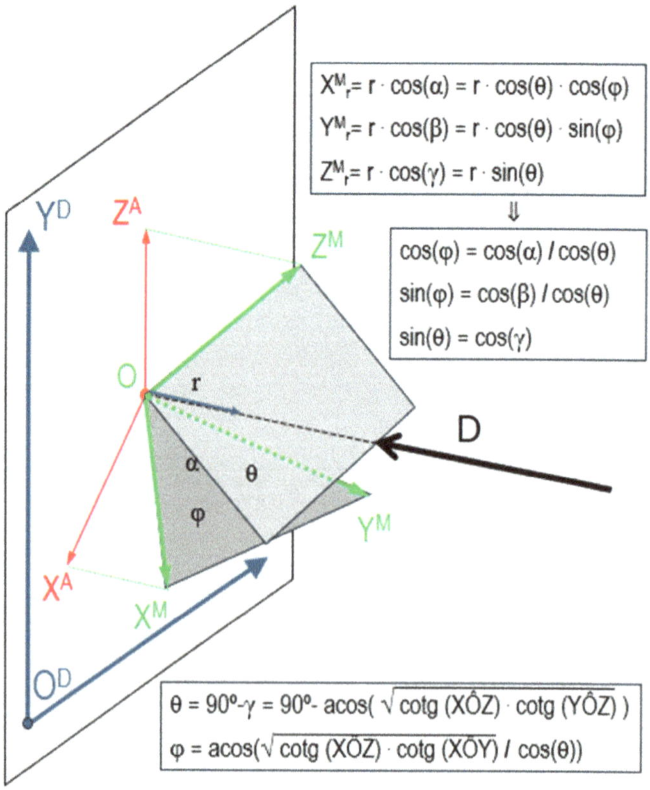

Figure 3.138. Orientation of the orthogonal projection *D* relative to the model's reference system.

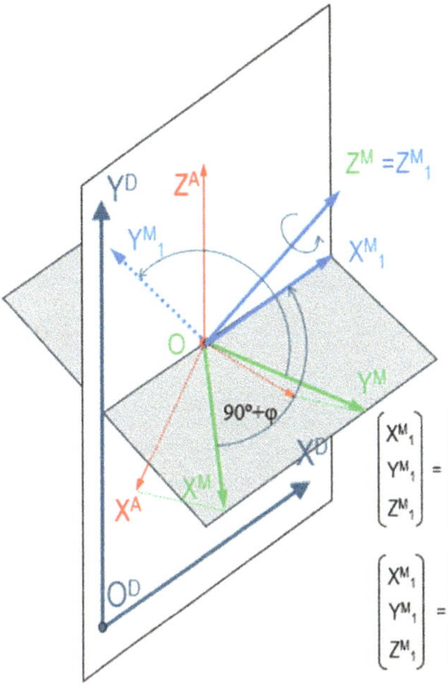

$$\begin{pmatrix} X^M_1 \\ Y^M_1 \\ Z^M_1 \end{pmatrix} = \begin{pmatrix} \cos(90^\circ+\varphi) & -\sin(90^\circ+\varphi) & 0 \\ \sin(90^\circ+\varphi) & \cos(90^\circ+\varphi) & 0 \\ 0 & 0 & 1 \end{pmatrix} \cdot \begin{pmatrix} X^M \\ Y^M \\ Z^M \end{pmatrix}$$

$$\begin{pmatrix} X^M_1 \\ Y^M_1 \\ Z^M_1 \end{pmatrix} = \begin{pmatrix} \sin(\varphi) & -\cos(\varphi) & 0 \\ \cos(\varphi) & \sin(\varphi) & 0 \\ 0 & 0 & 1 \end{pmatrix} \cdot \begin{pmatrix} X^M \\ Y^M \\ Z^M \end{pmatrix}$$

Figure 3.139. First rotation of 90° + φ around the Z^M axis.

We then apply a second rotation equal to angle θ around the $X^M{}_1$ axis, thus obtaining a rotated system (O, $X^M{}_2$, $Y^M{}_2$, $Z^M{}_2$) whose Y axis is contained within the drawing plane (and parallel to the Y^D axis), while its Z axis is perpendicular to the drawing plane. The process is shown in Figure 3.140.

The transformation matrix that converts the main reference system to the inflation system (or to drawing system, if we make $Z^D = 0$) can be obtained by simply combining the two previous rotations, as shown in Figure 3.141.

Although the illustrations show the drawing coordinate system offset from the inflation system, the problem is simplified, without loss of generality, by aligning their origins. As a result, only a rotation is required, with no translation needed.

The above matrix includes both rotation between systems and the shortening of lengths due to projection (axonometric coefficients). Alternatively, we can obtain the transformation matrix that accounts for rotation while ignoring scaling.

A key limitation of the axonometric inflation approach is its reliance on the assumption that the input is an orthogonal projection (Company *et al.*,

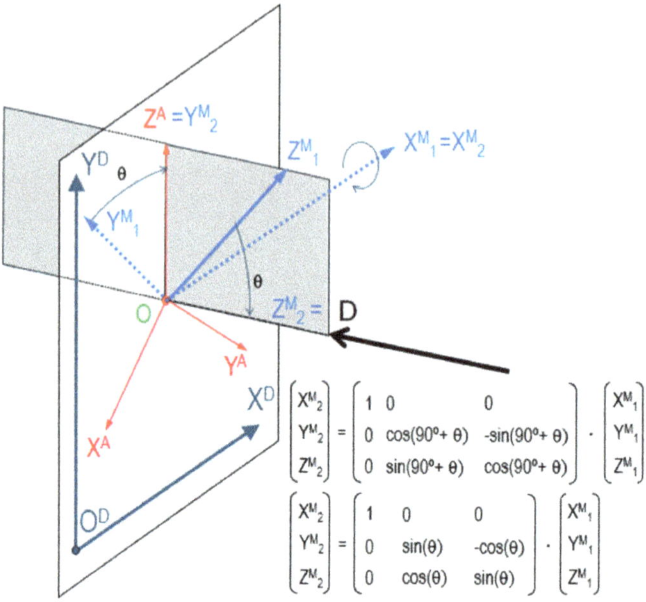

Figure 3.140. Second rotation of θ around the $X^M{}_1$ axis.

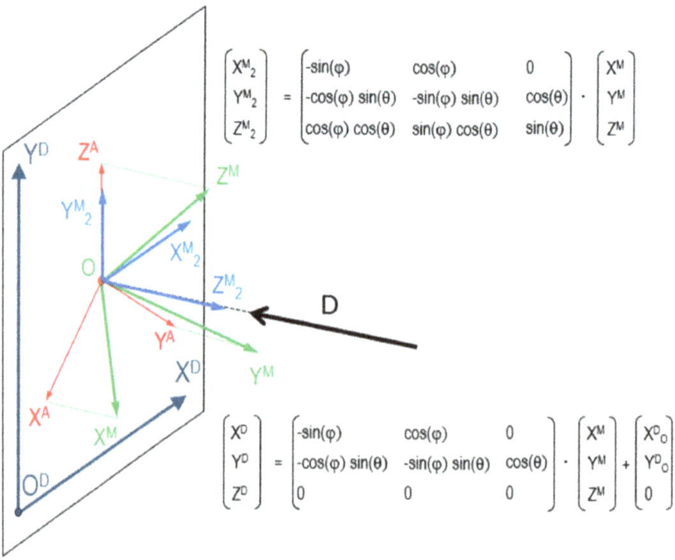

Figure 3.141. Transformation matrix.

2001, 2004). Consequently, inflating drawings based on oblique views often results in distorted models—or in some cases, computational failures (e.g., NaNs resulting from trigonometric calculations on singular angles like 90 degrees). Fortunately, the axonometric inflation method can operate independently of the inflation coordinate system. This is the strategy used by the *2011* method, which supports the inflation of normalon or quasi-normalon objects even when represented using oblique cabinet or military axonometric projections. Nevertheless, users of SBM applications may wish to switch between the model and drawing coordinate systems. For this purpose, the following transformation is provided (see Figure 3.142).

In the image, the X-axis is foreshortened, which implies that the angle YÔZ is 90°, and the scale factors for the Y and Z axes are both equal to 1 ($e_Y = e_Z = 1$). The angle XÔY and the scale factor e_X are determined as shown in the figure.

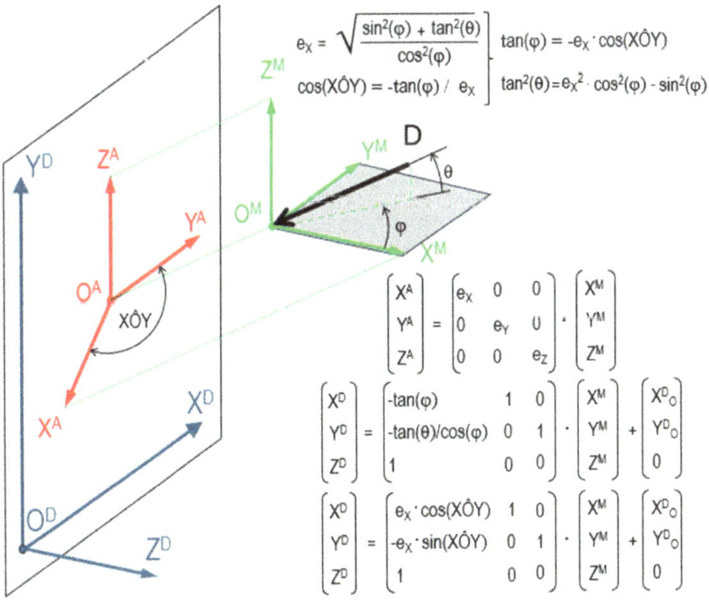

Figure 3.142. Transformation matrix to switch between the model and drawing coordinate systems.

Two important considerations arise at this stage:

(1) Since designers typically do not account for the oblique direction of projection (denoted as D) when sketching, the angle XÔY is measured directly from the sketch, while the scale factor e_X is assigned an arbitrary value.
(2) Although the specific calculations vary slightly, a similar approach applies when any of the other two coordinate planes in the model system are parallel to the drawing plane.

In particular, the transformation matrix for oblique projections with foreshortening along the x-axis is derived from the composition of two successive rotations, as illustrated in Figure 3.143.

Likewise, when the y-axis is foreshortened, the transformation is calculated as illustrated in Figure 3.144.

Finally, when the z-axis is foreshortened—the XY plane is parallel to the drawing plane—the required transformation is calculated as illustrated in Figure 3.145.

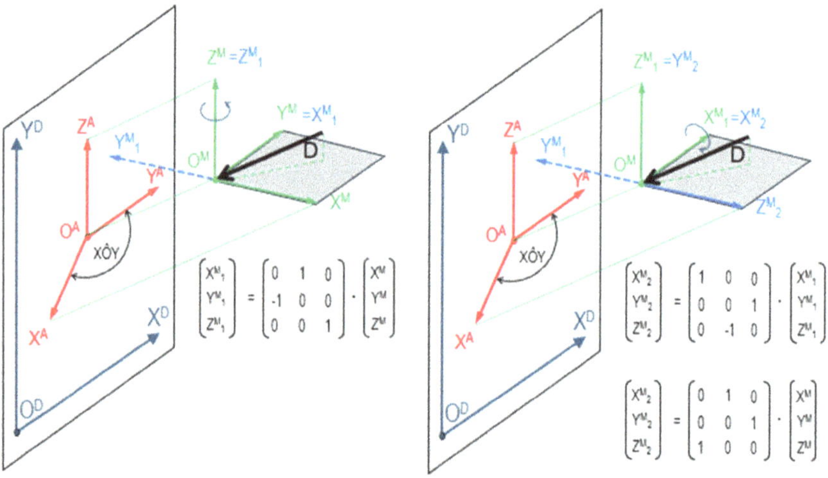

Figure 3.143. Derivation of the transformation matrix for projections with foreshortening along the x-axis.

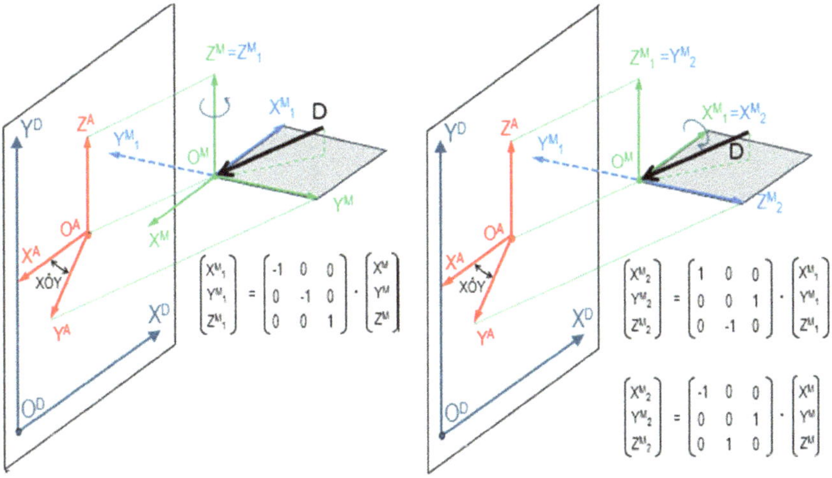

Figure 3.144. Derivation of the transformation matrix for projections with foreshortening along the y-axis.

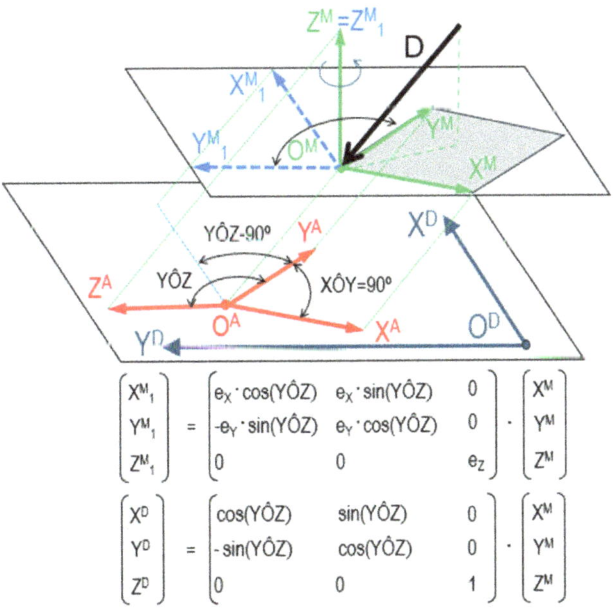

Figure 3.145. Derivation of the transformation matrix for projections with foreshortening along the z-axis.

3.9.3 *B-Rep inflation toolbar and configuration*

The *Inflation* toolbar (shown in Figure 3.146) is located over the parent window.

The toolbar contains the following commands:

- Inflate (from a drawing).
- Inflate from previous 3D model.
- Delete previous inflation.
- Inflation configuration.

The inflation mode can be selected via the *Inflation Configuration* command. Two different strategies that can be activated independently or executed in sequence:

(1) *Direct inflation*: A fast method suited to a limited range of model types. However, it is sensitive to imperfections in the input sketch.
(2) *Optimization-based inflation*: A more general method that does not require prior knowledge about the shape to be inflated. However, it often converges to incorrect models due to local minima in the optimization process.

The Configure Inflation dialog is shown in Figure 3.147.

Certain geometric regularities such as edge parallelism and face flatness are trivially satisfied in a 2D sketch. As a result, running optimization-based inflation directly on the flat sketch may lead the algorithm to converge to a local minimum that simply reproduces the 2D figure.

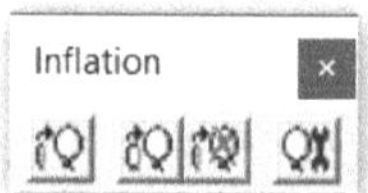

Figure 3.146. Inflation toolbar.

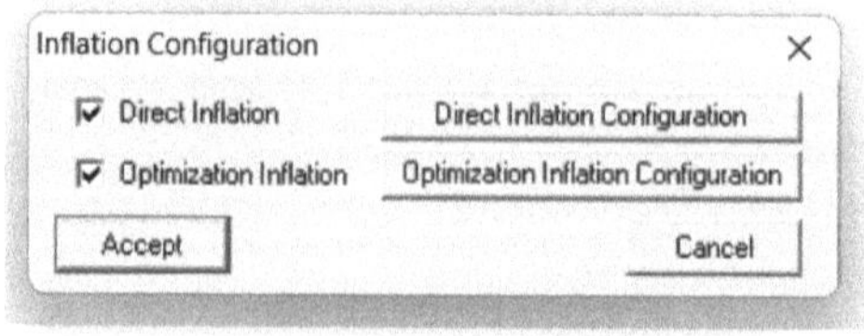

Figure 3.147. Inflation Configuration dialog.

Conversely, even a simple direct method such as random inflation can produce a rough 3D model that helps guide the optimization toward a better global minimum, using regularities derived from the intermediate geometry. The *Inflation Configuration* dialog also provides access to configuration options for both inflation modes: direct inflation and optimization-based inflation.

3.9.4 *B-Rep direct inflation*

3.9.4.1 *Direct inflation*

Direct inflation generates 3D models from 2D line drawings by computing the third coordinate (depth) of each vertex in a single step. Its earliest precedents include the methods proposed by Kanatani (1986) and by Lamb and Bandopadhay (1990). Direct inflation methods were initially introduced to produce "tentative models," preliminary 3D models that serve as valid, non-trivial starting points for optimization. Their purpose is to prevent the optimization process from converging prematurely to local minima.

The simplest strategy in the direct inflation category is random inflation, which assigns arbitrary depth values to the vertices of the input drawing. A more structured approach based on geometric labeling is the level-inflation method, implemented in REFER.

REFER also includes a method specifically designed to detect and reconstruct normalons and quasi-normalons. Normalons are polyhedra where all edges at a corner meet at 90 degrees and all intersecting faces form right dihedral angles. Quasi-normalons are those whose wireframe structure can be reduced to a normalon by removing certain edges without eliminating any vertices. Many mechanical part skeletons and simplified architectural sketches fall within these categories. The reconstruction method relies on the relationship between the angles formed by orthogonal projections of three concurrent orthogonal segments, and the angles that each segment forms with its projection. As a result, exact 3D models are recovered when the input 2D drawings are precise axonometric projections of normalons or quasi-normalons, while useful tentative models can still be produced in less ideal cases.

In the approach by Lamb and Bandopadhay (1990) any face parallel to one coordinate plane was considered as "main" face, while the rest were deemed "oblique." In the core phase, the coordinates of any point in the object could be obtained by calculating the distance, along the x, y and z axes, between this point and the origin. Likewise, if the coordinates of one point were known, the coordinates of adjacent points could be derived relative to the known point. The system leveraged this concept to determine all the coordinates of an object relative to the junction selected as the origin. Therefore, parallelism and proportionality relations in axonometric views are the bases of this approach.

The reconstruction method proposed by Lamb and Bandopadhay requires users to interactively determine the projections of the reference axis. REFER's implementation of this method is known as *Axonometric Inflation* (Company *et al.*, 2004). The axonometric approach computes depth values sequentially using a spanning tree. In one version of the method, the process starts by identifying the dominant angles from the angular distribution of lines to determine the principal directions. The model is then initialized by assigning a depth coordinate $z_i = 0$ to an arbitrary node (this step fixes the model's position relative to the image plane but does not affect reconstruction accuracy). The algorithm then propagates depth assignments to adjacent vertices based on their connectivity. If the sketch contains three dominant directions, the method can typically infer depth values for all vertices, yielding a consistent 3D model. Subsequent improvements expanded the algorithm's applicability to a broader range of model types and enhanced its robustness.

All direct inflation methods are heuristic in nature, relying on general perceptual assumptions. As such, they can produce reasonably accurate models for many polyhedral shapes but may struggle with complex geometries or sketches that contain noise or inaccuracies.

3.9.4.2 *Random inflation*

In the direct generation category, the simplest strategy for obtaining a tentative 3D model is *random inflation*, which assigns an arbitrary depth value to each vertex in the 2D drawing. This approach was first proposed by Brown and Wang (1996). In REFER's implementation, the depth values *(z_i)* are selected within a normalized interval defined by *max (Delta x, Delta y)*, based on the assumption that the input model is relatively regular.

While random inflation performs poorly when used to generate final models, it is a viable method for producing tentative models-preliminary

3D reconstructions used as starting points for optimization. These models help avoid trivial or undesired local minima in the solution space. Surprisingly, experiments indicate that, as long as the shapes are not overly complex, optimization processes initialized with random tentative models tend to yield topologically valid solutions. Indeed, a low failure rate was observed (approximately one in fifteen attempts). In cases where optimization fails (e.g., when the final objective function values remain significantly above zero), subsequent random initializations often succeed in producing the correct topology. In this regard, random inflation can serve as a practical fallback strategy in scenarios where other tentative modeling techniques are unsuccessful.

3.9.4.3 *Levels-inflation*

The reconstruction by levels-inflation of vertices (Company *et al.*, 2004) is a simple, heuristic method of direct inflation (non-iterative), in which a set of *z* coordinates is assigned based on the "typologies" of the vertices. Vertex typologies have previously been used in labeling approaches to validate sketches for reconstruction. However, typologies are not the central element of this method. Instead, the approach is based on the strong empirical correlation between a vertex's typology in the sketch and its z-coordinate in the reconstructed model.

The method assumes that vertex typologies can be arranged into sequential "levels," with each increasing level corresponding to a higher z-coordinate. In other words, rather than performing a full labeling check or validating graph consistency, the algorithm performs a basic classification of each vertex, assigns a level based on its typology, and then uses heuristically defined z-coordinates for each level to generate a tentative 3D model.

Drawing from aeronautical terminology, the perspective of the model is described using the *yaw*, *roll*, and *pitch* parameters, as illustrated in Figure 3.148. The yaw angle is defined as the angle between the Z-axis and the horizontal projection of line BD. When this angle deviates significantly from zero, it causes symmetrical vertices (such as C and E) to be assigned different z-coordinates.

Psychologically, the assumption is that an observer favors minimal pitch and yaw to maintain symmetry while achieving a general viewpoint. Meanwhile, the roll is typically set to zero to preserve verticality in

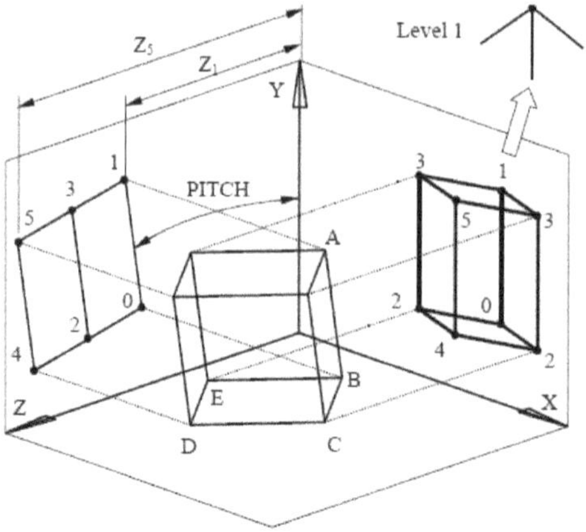

Figure 3.148. Model's perspective described using yaw, roll, and pitch parameters.

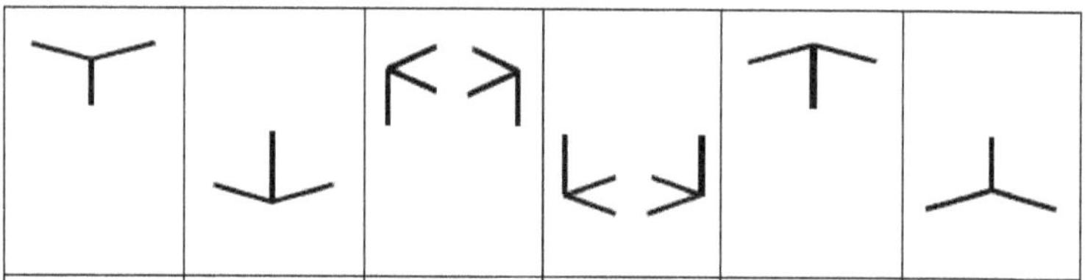

Level 5	Level 4	Level 3	Level 2	Level 1	Level 0
"Y"	"W"	"Lateral inverted W"	"Lateral W"	"Inverted W"	"Inverted Y"

Figure 3.149. Vertex typologies.

the resulting model. Based on these assumptions, six distinct vertex typologies are defined and ordered by increasing z-value, as shown in Figure 3.149.

In REFER, each vertex is assigned a z-value according to its typology, and the overall depth range is controlled by a scaling proportion, as illustrated in Figure 3.150.

3.9.4.4 *Axonometric inflation*

REFER implements a fast axonometric inflation algorithm inspired by the work of Sugihara (1984), Kanatani (1986), and Lamb and Bandopadhay (1990). The method applies to normalon and quasi-normalon objects. A normalon is a rectangular trihedral polyhedron, with corners formed by

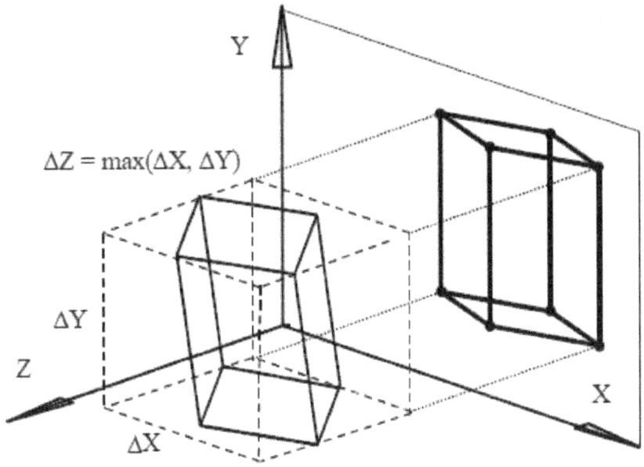

Figure 3.150. Scaling proportion to control the overall depth range.

three mutually perpendicular edges-an extension of Dori's definition of a normalon (1995). The concept closely aligns with the modern notions of "orthogonal polyhedral" or "Manhattan world" objects.

As explained in the B-Rep Inflation Coordinates section, assuming an orthogonal axonometric view, the 3D orientation of an arbitrary corner (consisting of one central and three lateral vertices) can be determined analytically. First, the orientation of the 3D reference frame is computed from its projection (Gomis and Company, 1995) by measuring sketch axes angles (α', β', γ') which allow the calculation of angles φ_A, φ_B and φ_C, as illustrated in Figure 3.151.

Second, for every cubic corner, the variation in z coordinates between the central (D) and the lateral vertices (A, B, C) depend on the lengths of the lateral edges and their respective angles φ_A, φ_B and φ_C. The lengths of the lateral edges are calculated as shown in Figure 3.152.

Two Z coordinates are valid in the analysis of vertex C depending on the orientation considered for the axonometric angle φ_C, as illustrated in Figure 3.153.

$$Z_{C1} = Z_D - C'D' \tan(\varphi_C)$$

$$Z_{C2} = Z_D - C'D' \tan(\varphi_C)$$

We arbitrarily inflate the first edge of the initial triplet in one of two possible directions, thereby selecting one of the two Necker reversals. From there, we exploit a key property of regular polyhedra—the orthogonality of all corner angles—to determine the appropriate orientation of subsequent angles as the inflation propagates. The step is illustrated in Figure 3.154.

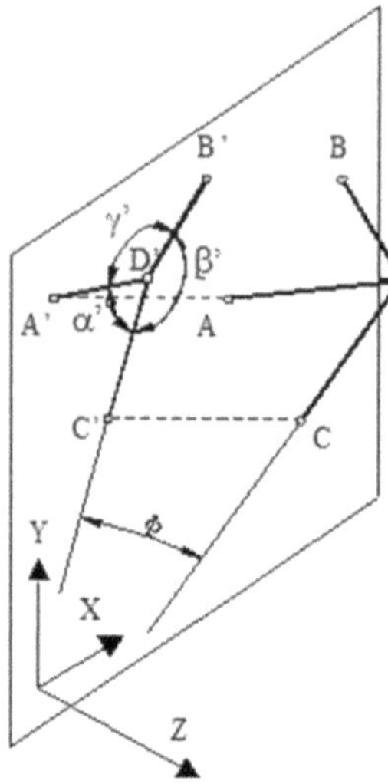

Figure 3.151. Computation of the orientation of the 3D reference frame from its projection.

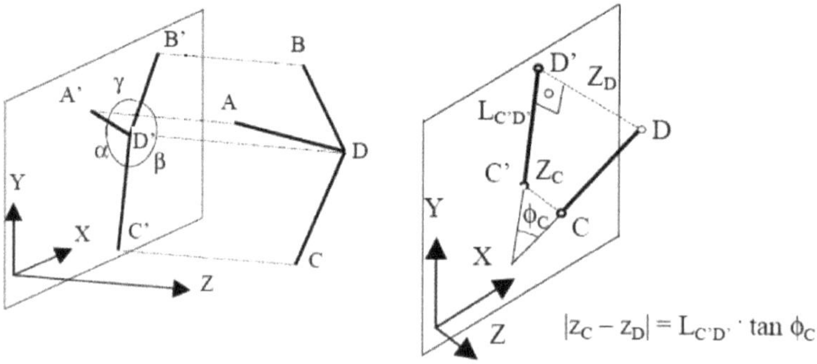

Figure 3.152. Calculation of the length of the lateral edges.

Instead of arbitrarily selecting the initial corner, the algorithm identifies the most "central" vertex according to the following hierarchy of criteria:

- Must connect at least three distinct normalon edges.
- Must contain the longest normalon edge among those satisfying the first condition.
- Must be closest to the center of mass among those satisfying the previous criteria.

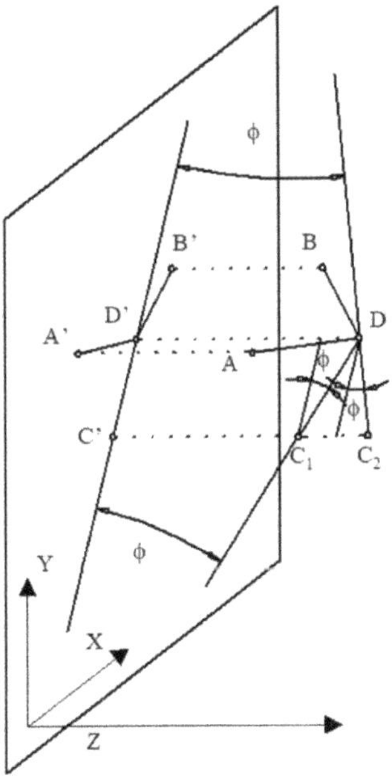

Figure 3.153. Valid Z coordinates for vertex C based on the orientation considered for the axonometric angle φ_C.

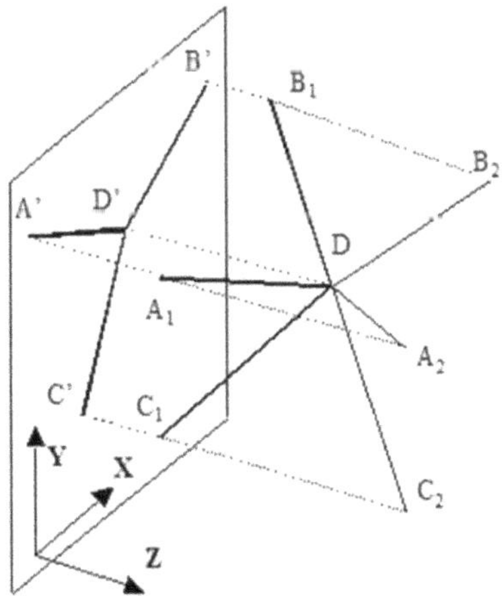

Figure 3.154. Arbitrary inflation of the first edge of the initial triplet in one of two possible directions.

From this central vertex, neighboring vertices are visited in order of decreasing edge length. The lateral vertex connected by the longest edge is then treated as the new central vertex in the next iteration. This ordering assumes that longer edges are less susceptible to drawing or discretization errors.

Once vertex D is selected to receive an arbitrary Z-coordinate, and vertex C is resolved (with either C_1 or C_2 selected), the edge CD (either C_1D or C_2D) becomes the reference for determining the orientation of edges AD and BD. These orientations allow the computation of the Z-coordinates of vertices A and B. The inflation of edges AD and BD is constrained such that the resulting angles at vertex D are orthogonal (90-degrees). This orthogonality condition is then propagated: edges AD and BD, now resolved, serve as new reference edges to determine the orientation of all other edges converging at vertices A and B, respectively.

REFER determines the orientation of the initial reference edge based on the following criteria (see Figure 3.155):

- If $Y_{D'} < Y_{C'}$ then: $Z_C = Z_D - C'D' \tan(\varphi)$
- If $Y_{D'} > Y_{C'}$ then: $Z_C = Z_D + C'D' \tan(\varphi)$
- If $Y_{D'} = Y_{C'}$ then:
    - If $X_{D'} < X_{C'}$ then: $Z_C = Z_D - C'D' \tan(\varphi)$
    - If $X_{D'} > X_{C'}$ then: $Z_C = Z_D + C'D' \tan(\varphi)$

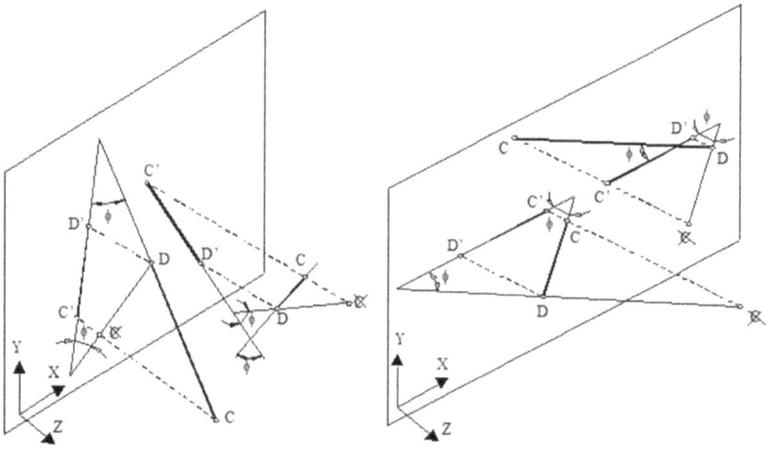

Figure 3.155. Determining the orientation of the initial reference edge in REFER.

The algorithm then propagates through a spanning tree, converting previously determined lateral vertices into central ones of neighboring corners, which guarantees topological consistency. A modified version of Kruskal's algorithm is used to build a spanning tree of edges connecting these successive central vertices, as illustrated in Figure 3.156.

The process continues until all vertices are defined. If any lateral vertex cannot be promoted to central status without forming a cycle (i.e., all neighboring vertices have already been visited), the branch is abandoned, and another begins from the next longest unexplored edge. Numerical inconsistencies such as those caused by trigonometric operations must also be managed during this process.

This method requires three orthogonal edges to meet at each central vertex, but no information about faces is necessary. The edge valence of lateral vertices is irrelevant. Thus, the method works as long as a spanning tree can be formed in which all central vertices have valence three and lateral vertices are accessible accordingly. Only edges in the tree must align with the three principal directions, which allows extension to quasi-normalon shapes-those that become normalons upon removing non-aligned edges, without deleting any vertices.

This approach is not directly applicable to non-orthogonal projections such as cabinet, cavalier, or military perspectives. However, the algorithm accommodates these cases approximately by arbitrarily adjusting angles that are incompatible with orthogonal projection (e.g., 90 or 135 degrees), thereby preventing division-by-zero errors. Additionally, the lengths of receding lines are scaled according to a user-defined factor—set to 1/2 by default—to simulate depth compression consistent with these projection types.

For *Class 1 quasi-normalon* models (where simplification yields a connected graph), axonometric inflation is feasible if a spanning tree can

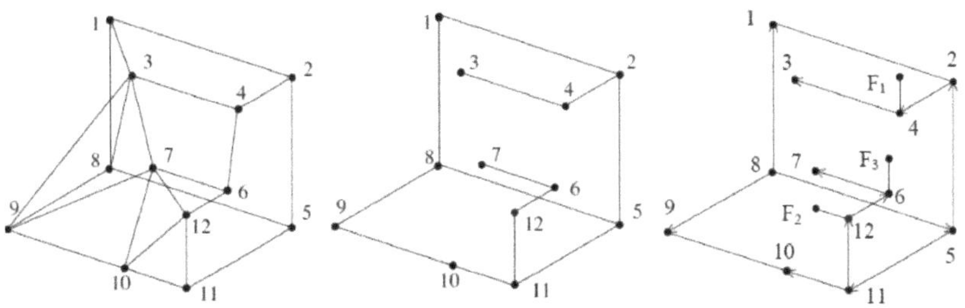

Quasi-normalon model | Equivalent normalon | Transformed model

Figure 3.156. Construction of spanning tree of edges.

be formed in the resulting normalon structure, which requires that the removal of non-aligned edges does not eliminate any vertices and keeps the graph connected.

Occasionally, even with a valid equivalent normalon, axonometric inflation cannot proceed if certain vertices are unreachable through a valid spanning tree (e.g., when junctions are only connected to vertices of valence less than three). In these cases, fictitious line segments (with unit length and aligned with missing main directions) can be added. This solution is consistent with the normalon assumption and solves the connectivity issue.

When vertices of valence greater than three occur, the algorithm still succeeds if a valid propagation tree avoids making such vertices central. If it is not possible, the algorithm defaults to choosing three of the incident edges at random to continue the process and reconstruct a potential model.

Class 2 quasi-normalons are cases where simplification results in a disconnected graph. When this situation occurs and the user has not manually connected components, REFER automatically generates auxiliary lines to rejoin them (see Figure 3.157). This operation must satisfy the following criteria:

(1) Connecting edges must be parallel to the XYZ axes to preserve axonometric inflation propagation.
(2) Connecting edges and associated vertices must be coplanar, which requires face detection in the 2D sketch (a challenging task).

REFER also includes an unpublished enhanced algorithm, internally referred to as "2011," with a recursive structure for compact implementation.

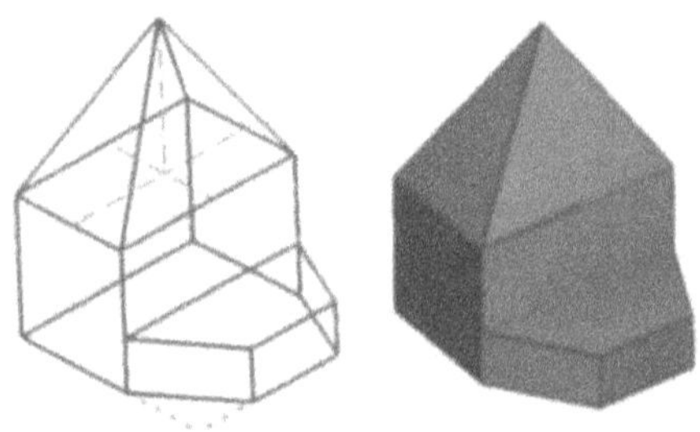

Figure 3.157. Generation of auxiliary lines in REFER to connect components.

The primary distinction of this variant is that inflation is carried out within the Main coordinate system. Once the model has been inflated, the transformation matrices detailed in the B-Rep Inflation Coordinates section are applied to map the model back to the inflation coordinate system.

This version supports two spanning tree traversal strategies: depth-first and breadth-first. Users can select between these options via the *Direct Inflation Configuration* dialog. While both are functional, breadth-first traversal is recommended, as depth-first tends to be more susceptible to early errors propagating through the structure. Nonetheless, determining which method is more resilient to sketching errors remains an open research question.

This improved version also supports cabinet (cavalier or military) projections, through their own transformation matrices. The main strengths of this method are its speed and its ability to inflate normalon objects even in the absence of detected faces. Axonometric inflation is robust enough to reconstruct normalon models from natural drawings—i.e., sketches without hidden lines—such as those found in the Ribald example folder. However, the algorithm does not attempt to reconstruct missing or occluded lines. This limitation becomes evident when the inflated model is rotated, as shown in Figure 3.158, a slight leftward rotation exposes discontinuities where lines, previously hidden in the 2D sketch, were not extended into the 3D reconstruction. The main weakness of this approach lies in its sensitivity to alignment errors—an inherent characteristic of hand-drawn sketches—which tend to propagate along the spanning tree, often resulting in polyhedral models with distorted or tangled faces.

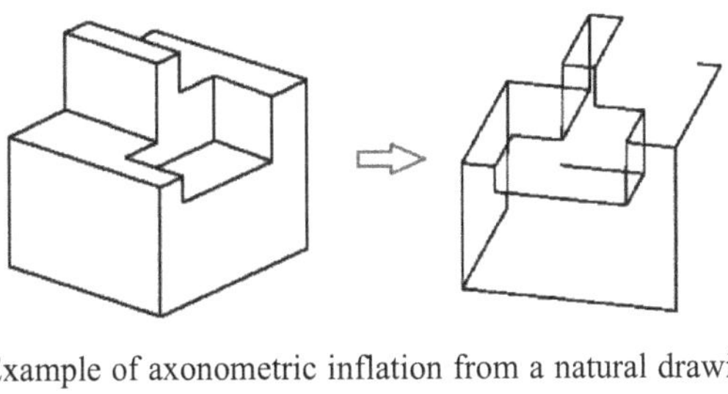

Figure 3.158. Example of axonometric inflation from a natural drawing. The figure on the right shows that a slight leftward rotation reveals discontinuities in the lines that were previously occluded in the natural drawing and do not extend into the 3D model.

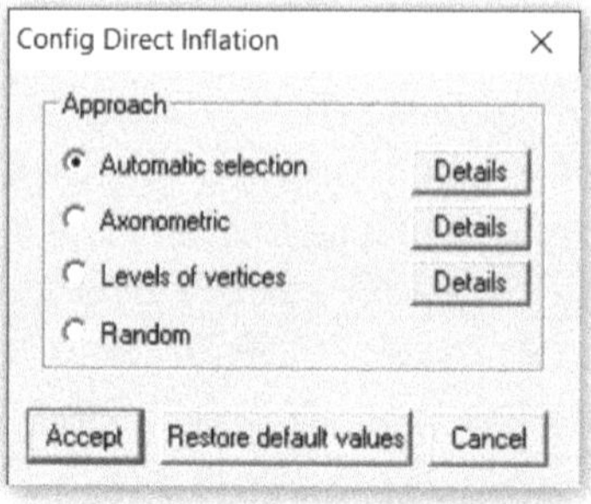

Figure 3.159. *Direct Inflation Configuration* dialog.

3.9.4.5 *Direct inflation configuration*

The parameters of the direct inflation process can be configured via the *Direct Inflation Configuration* button in the *B-Rep Inflation Configuration* dialog (see Figure 3.159).

The dialog enables users to select from four available direct inflation methods:

(1) *Automatic Inflation*: This mode automatically selects one of the three available inflation methods based on the object's previously identified cue typology. The selection criteria are detailed in the dialog shown in Figure 3.160.
(2) *Axonometric Inflation*: This option enforces the use of the inflation strategy described in the *Axonometric Inflation* section. Two variants are available via the *Details* button, as shown in Figure 3.161.

    (a) For the *2001* approach, cabinet views may cause inflation to fail. Two parameters are provided to adjust problematic angles to nearby safe values.
    (b) For the *2011* approach, the breadth-first traversal option is recommended, as depth-first is more sensitive to inaccuracies in the initial vertex positions (errors in early z-coordinates tend to propagate). The 2011 algorithm also detects cabinet projections by identifying approximately 90-degree angles between main axes (within a tolerance) and calculates the appropriate axonometric coefficients.

(3) *Levels-Inflation*: This option enforces the inflation strategy described in the *Levels-Inflation* section. Two variants are available via the

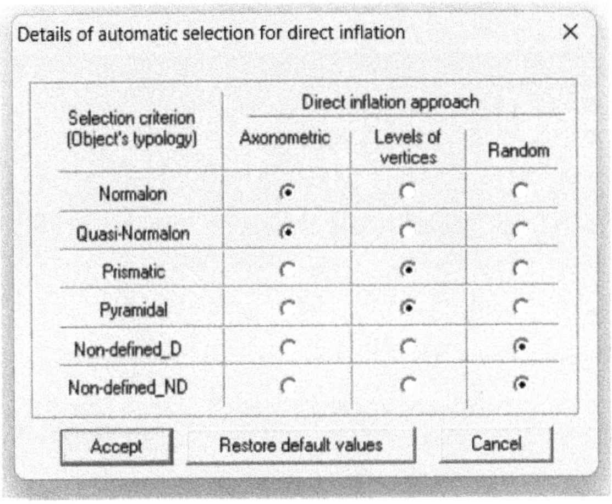

Figure 3.160. *Automatic inflation* dialog.

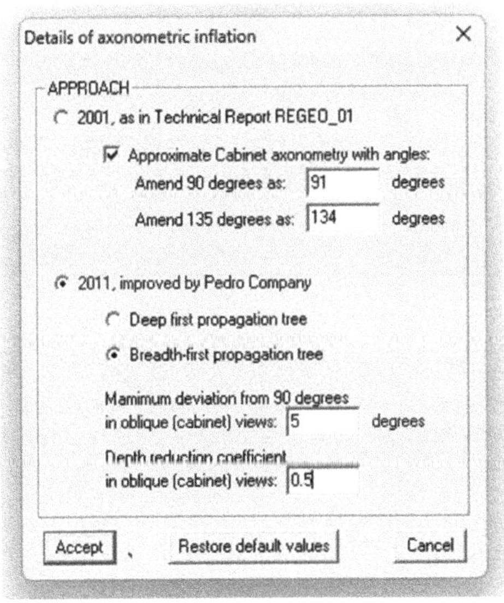

Figure 3.161. *Axonometric inflation* dialog.

Details button, along with the corresponding tuning parameters, as shown in Figure 3.162.

If *Cue Hidden Edges* is enabled, hidden lines will be included in the inflation process. Details are shown in Figure 3.163.

(4) *Random Inflation*: This option enforces the inflation strategy described in the *Random Inflation* section.

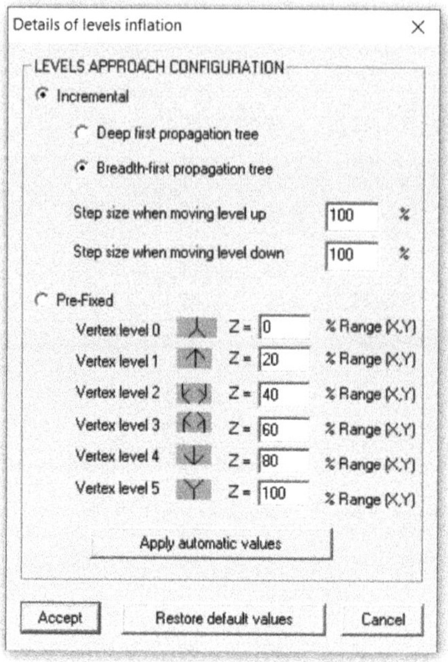

Figure 3.162. *Levels-inflation* dialog.

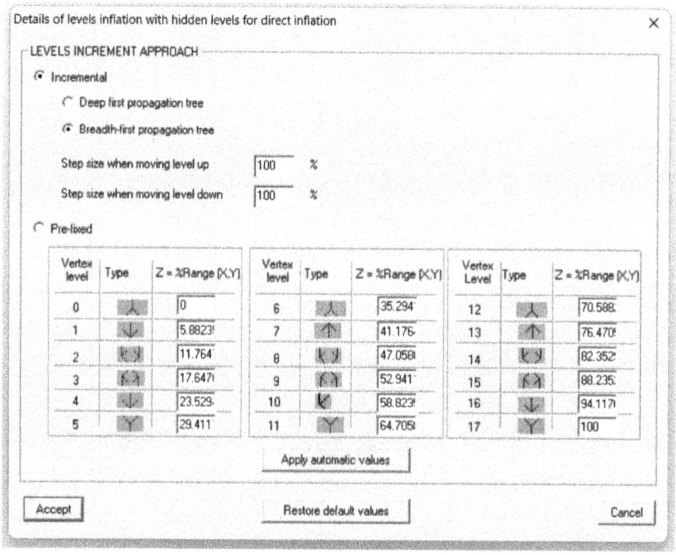

Figure 3.163. *Levels-inflation with hidden edges* dialog.

3.9.5 *B-Rep optimization-based inflation*

3.9.5.1 *Optimization-based inflation*

Optimization-based inflation reconstructs 3D models from 2D line drawings by framing the task as an optimization problem (Wang and Grinstein, 1993). This process enables the generation of geometrically unrealizable (but perceptually plausible) 3D models from a single projection. Psychologists have long observed that humans can infer 3D structures from 2D images by relying on a small set of perceptual heuristics (Hoffman, 2000). Notable contributions in this field include the works of Marill (1991), Leclerc and Fischler (1992), and Lipson and Shpitalni (1996).

In the context of inflation, it is assumed that the x and y coordinates of the 3D model vertices match those in the 2D drawing. The z coordinates are then determined through a mathematical optimization process, where they serve as the optimization variables. The algorithm minimizes an objective function that penalizes tentative 3D models failing to comply with selected geometric regularities.

The optimization algorithms implemented in REFER (Company *et al.*, 2004) address limitations of earlier optimization-based reconstruction methods. Enhancements include algorithmic adjustments, simplified model representations, and a categorized set of regularities to streamline the objective function. These improvements help reduce the risk of local minima and significantly increase computational efficiency.

Optimization-based inflation requires fine-tuning a large number of parameters. Users can access the configuration interface (see Figure 3.164) via the *Optimization-Based Inflation Configuration* button.

Through this dialog, users can configure how cues, or regularities, influence the optimization process. A key configuration option includes setting the threshold used to determine whether two angles or directions in the 2D drawing should be treated as nearly equal. For example, lines diverging by less than this threshold are considered nearly parallel and are included in the calculation of the parallelism regularity cost. Similarly, the threshold applies when identifying candidate lines for perpendicularity, collinearity, and orthogonal corners in the reconstructed 3D model. The dialog also includes options for customizing the weights and cost functions of the objective function, particularly when these are user-defined.

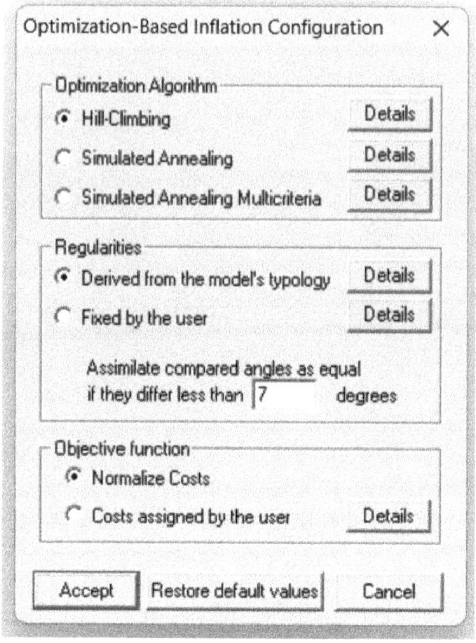

Figure 3.164. *Optimization-Based Inflation Configuration* dialog.

The default parameters in the dialogs have been extensively tested to achieve a balanced compromise between computational efficiency and numerical stability. These defaults can be restored at any time using the *Restore default values* button. Because the optimization process is computationally intensive, REFER displays a progress dialog while the reconstruction is running. This dialog provides real-time progress feedback and can be canceled at any time, if desired.

3.9.5.2 *Optimization-based inflation algorithms*

There is no universal method for optimization. Instead, optimization methods are broadly classified into discrete and continuous approaches, with several variants within each category. For the task of calculating the optimal set of z-coordinates from 2D sketches, continuous methods are most appropriate.

Optimization methods can also be divided into deterministic and stochastic categories. Deterministic methods tend to be slower and are more prone to getting trapped in local minima. Stochastic methods are generally more complex, requiring careful parameter tuning, but offer better global search capabilities, though they may occasionally fail or become unstable.

REFER implements one deterministic method (Hill Climbing) and two stochastic variants of Simulated Annealing. Since all optimization approaches are inherently complex, iterative, and highly sensitive to parameter settings, the user can configure the relevant parameters in the *Optimization-Based Inflation Configuration* dialog. The three optimization algorithms are described as follows:

(1) Hill Climbing is a gradient descent algorithm (Curry, 1944), grounded in a method proposed by Augustin-Louis Cauchy in 1847. It begins with an arbitrary solution and iteratively attempts to improve it by incrementally adjusting a single variable. If the adjustment yields a better solution, it is accepted, and the process continues until no further improvement is possible. As a greedy local search algorithm, it is prone to becoming trapped in local minima. The configuration dialog is shown in Figure 3.165.

Recommended parameter configurations for balancing speed, precision, and robustness are shown in a separate dialog (see Figure 3.166) available via the *Recommended parameters* button.

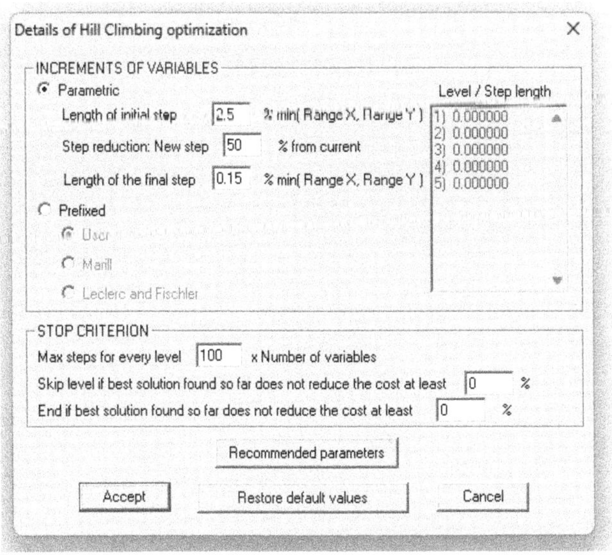

Figure 3.165. *Hill Climbing Optimization* dialog.

Pre-defined configurations for adjusting the Hill Climbing algorithm

PARAMETERS	FUNCTION	RANGE				VALUES RECOMMENDED FOR		
		Minimum		Maximum				
		Value	Remarks	Value	Remarks	Speed	Accuracy	Functional
Initial step length	Escape from local optima	1	Slow Convergence (Advances some steps for the same variable)	25 (10)	Scapes from local optima (too big steps). More dangerous for complex shapes	10	2	2.5
Step reduction	Complements to the parameter "step length"	10	The increments decrease too soon. Convergence results decelerated	50	Abort if it does not become close enough to optimum (oscillate around the optimum)	20	40	50
Final step length	Speed/precision	2	Small parts are not reconstructed (unless when the parameter "step length" goes below its minimum)	10	The variation of the increment of the variables may not be perceived. Increasing parameter "steps s by level" is better choice	0.5	0.06	0.15
Steps by level	Complements to the parameter "number of steps"	50	Abort before ending linnear search (may disminish when the parameter "step length" increases up to 25 and the parameter "steps by level" disminishes dawn to 20)	500	It takes too long when the las increments are too small (parameter "step reduction" is small and parameter "number of steps" is large)	100	500	100
Cost decrease		0	Influences the linear search	1	Improvments for every step are always too small. Yo can never ask for big improvments in one single step	0	0	0

Apply

Figure 3.166. Predefined setting for the Hill Climbing Optimization algorithm.

To reduce sensitivity to input dimensions, we normalized the initial step size *s_1* relative to the maximum sketch dimensions:

s_l=p · max(Delta x, Delta y), where p ∈ (0.1, 0.3). This strategy implicitly encodes the psychological assumption of proportionality. A loop count between five (fast, lower precision) and ten (slower, higher precision) offers good performance. Avoiding large step variations between iterations reduces instability. Empirical testing has shown that exponentially decreasing step sizes yield better objective function behavior (for example, $s_{i+1} = d \cdot s_i$, with d ∈ (0.3, 0.4). However, a slower rate of decrement (e.g., d ∈ (0.6, 0.8)) is more effective at escaping local minima, which is crucial when the model is highly entangled. The minimum s_i value should be selected in accordance with the maximum deviation allowed for fulfilling the regularities.

(2) Simulated Annealing (Kirkpatrick *et al.*, 1982) is a more computationally demanding algorithm but excels at exploring large, complex search spaces. At each iteration, the algorithm evaluates neighboring states and makes a probabilistic decision to either transition to a new state or stay in the current one. By occasionally accepting worse solutions, the algorithm can escape local minima and seek global optima. This method simulates physical annealing-a process of heating and gradually cooling a material to reach a low-energy state. The algorithm's temperature parameter controls this probabilistic behavior. While convergence is

theoretically guaranteed with probability close to 1, in practice, convergence may take exponentially long in some cases (Lundy and Mees, 1986). Configuration parameters are shown in Figure 3.167.

Multi-Criteria Simulated Annealing (MCSA) extends the basic Simulated Annealing framework (Bohachevsky *et al.*, 1986; Dekkers and Aarts, 1991) to address problems involving multiple conflicting objectives. This algorithm introduces additional complexity in comparing solutions, defining acceptance criteria, and managing a set of non-dominated (Pareto-optimal) solutions. Key differences from traditional Simulated Annealing include:

(a) Number of objectives: Single vs. multiple.
(b) Acceptance criterion: Scalar-based vs. vector-based with dominance relationships.
(c) Cooling strategy: Fixed vs. potentially adaptive.
(d) Results: A single optimal solution vs. a Pareto front.
(e) Complexity: Simpler vs. greater due to multi-objective considerations.

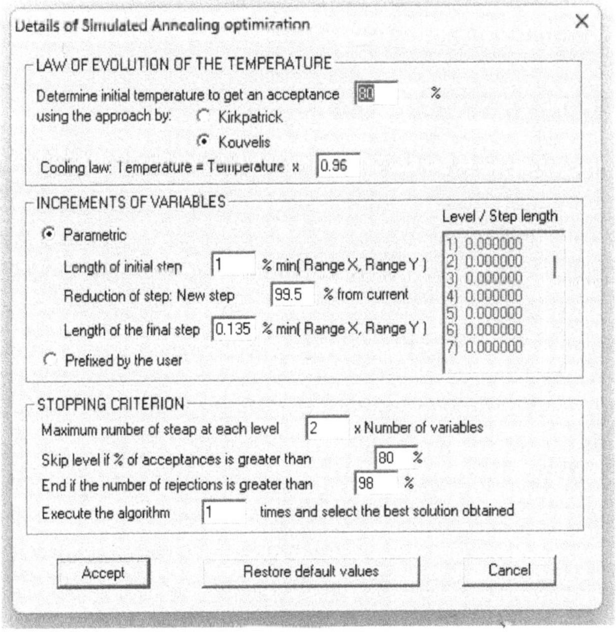

Figure 3.167. *Simulated Annealing Optimization* dialog.

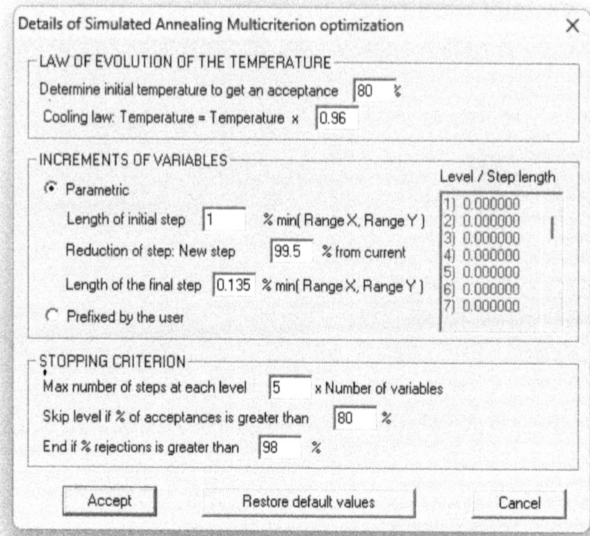

Figure 3.168. *Multi-Criteria Simulated Annealing (MCSA) Optimization* dialog.

The configuration parameters for MCSA are summarized in the dialog shown in Figure 3.168.

3.9.5.3 *Optimization-based inflation regularities*

The *Optimization-Based Inflation Configuration* dialog enables users to specify which regularities should guide the optimization process. For simplicity, we refer to *regularities* as a combination of cues and regularities. Distinguishing between the two is challenging, as many authors use the terms interchangeably. Historically, geometric regularities were introduced to manage the cost of inflation optimization. An objective function is defined which reaches its minimum when certain geometric regularities are satisfied in the inflated model. In contrast, cues are perceptual hints present in the 2D drawing that suggest certain characteristics that the corresponding 3D model must exhibit. Regularities based on detectable cues in the 2D drawing are sometimes referred to as *true regularities*, while those based on assumptions about the model's nature are called *false regularities*. Another key distinction is that true cues or regularities reflect specific features of the depicted model, whereas false regularities are grounded in general assumptions.

Determining whether a regularity is present in the drawing, and whether it should be applied in optimization, cannot be reduced to a binary yes/no criterion. 2D sketches are inherently imprecise, so regularities are detected probabilistically. Lipson and Shpitalni (1996) were the first to propose a mathematical formulation to incorporate probability into regularity detection. They introduced a continuous compliance factor μ(x), which evaluates to 1.0 when x = a and tapers to 0.0 following a standard distribution curve with σ = b as x moves away from a.

$$\mu_{a,b}(x) = \max(0, 1.1e^{-\left(\frac{x-a}{b}\right)^2} - 0.11) \tag{1.1}$$

REFER uses this compliance factor—at least for the initial set of regularities—as well as other strategies for additional cues and regularities.

In the context of inflation optimization, REFER treats cues and regularities interchangeably, as long as a cost function can be mathematically defined to quantify the extent to which each is satisfied in the various tentative models explored during optimization. Nonetheless, distinguishing between true and false regularities remains important: false regularities help guide the optimization toward the global optimum, whereas true regularities assist in initiating inflation and escaping local minima.

Some regularities are calculated automatically, either during the generation of the sketch or as the inflation optimization is applied. Other regularities will only be available during the optimization process if the user activates the corresponding cue commands to ensure they are calculated prior to inflation. Furthermore, some regularities are used at other stages of the reconstruction and are not available for controlling inflation through optimization. The regularities that may contribute to calculating the cost of the objective function that governs inflation optimization include:

- MSDA computes a cost function that reaches zero when all angles between edges converging at a vertex are equal, and increases as the angles deviate from this condition. This heuristic was introduced by Marill (1991), who observed that nearly regular and convex polyhedra often conform to this geometric property.
- Leclerc and Fischler (1992) proposed enforcing planarity of faces. Naturally, faces must be detected in the 2D line-drawing prior to inflation.

- The work of Perkins (1971) led to the development of the cubic corners regularity, which assigns a value of zero when the three edges of each corner form right angles. The value increases when corners that are perceived as orthogonal in the drawing are not reconstructed with perpendicular edges.
- Lipson and Shpitalni (1996) improved the state-of-the-art by adding an extended set of regularities:
    - Line parallelism assumes that parallelism between two 2D lines is likely to reflect parallelism in their 3D edge counterparts.
    - Line collinearity builds on the fact that lines that are collinear in the sketch plane are also collinear in 3D space.
    - Line verticality assumes that lines marked as vertical in the sketch must correspond to edges parallel to the vertical axis in 3D (which requires prior detection of the main axes).

REFER uses these three regularities to optimize the coordinates of the 2D drawing, adjusting the vertex positions so that "nearly" parallel lines become truly parallel, among other refinements. As a result, these regularities are trivially satisfied in the refined 2D line drawing and should not be used to initiate the inflation process.

- *Isometry*: Lengths of entities in the 3D model are uniformly proportional to their lengths on the sketch plane.
- *Corner orthogonality*: A junction of three lines that mathematically qualifies as a projection of a 3D orthogonal corner is orthogonal in space.
- *Skewed facial orthogonality*: A face contour that shows skewed orthogonality is likely orthogonal in 3D space.
- *Skewed facial symmetry*: A face showing skewed symmetry in 2D denotes a truly symmetrical face in 3D.

According to Lipson and Sphitalni (1996), certain "false regularities" can be used to initially inflate the flat projection into a rough convex 3D shape, providing a better starting point for the subsequent optimization process:

- *Line orthogonality*: All line pairs in a junction, except those that are collinear, are perpendicular in 3D.
- *Face perpendicularity*: All adjacent faces must be perpendicular.

Company *et al.* (2004) reorganized and classified the existing regularities to clearly distinguish true regularities from false ones, and prioritize those that best contribute to optimization:

- Minimum standard deviation of angles (MSDA) was successfully used by Marill (1991) to introduce an optimization approach, but it is a false regularity unless applied to regular polyhedra.
- Face planarity is a true regularity for polyhedral shapes and is arguably the most important. Its main limitation is that it only becomes directly applicable after the faces have been identified in the line drawing. A second drawback is that this regularity is trivially satisfied in the 2D line drawing.
- Line parallelism is a true and powerful regularity. Its main disadvantage is that parallelism is invariant only in pictorial (non-perspective) views. Therefore, it is a direct cue in pictorial views and an indirect one in perspective views (where *Cue Vanishing Points* must be applied instead). It is also trivially fulfilled in the 2D line-drawing.
- Line collinearity can be seen as a specific case of line parallelism, except that it is invariant under perspective projection. It serves as a cue for a true and direct constraint of the model: alignment. However, like other regularities, it is trivially satisfied in the 2D line drawing. In REFER's implementation, collinearity is defined in the strict sense (applying only to consecutive lines that share a vertex). As discussed in the *2D Refinement* section, this regularity is preserved primarily for historical reasons. In practice, it has limited value, as the option to merge collinear edges often makes it redundant.
- Line verticality is a true regularity, but it primarily relates to the model's orientation rather than its shape. Like other geometric regularities, it is trivially satisfied in the 2D line drawing. As discussed in the *2D Refinement* section, once the three main directions are detected, any edges parallel to those directions can be refined accordingly. We have not found any particular advantage in handling vertical lines separately at an earlier stage. Nonetheless, this regularity has been updated so that its cost is computed based on the angle between each candidate edge and the Z axis of the main coordinate system, rather than the Z axis of the inflation coordinate system.
- Line orthogonality is not a projective invariant, as originally proposed. However, it can be reformulated to relate indirectly to a parallelism cue: two edges may be considered orthogonal if their projections are

respectively parallel to two directions perceived as orthogonal. This reformulation ties it to the detection of the main orthogonal directions, introducing similar limitations under perspective projection as those affecting line parallelism. Only pairs of lines that share a common vertex and are nearly parallel to one of the three main directions are considered. In any case, line orthogonality remains a false regularity, as its application is not based on any perceptual cue present in the 2D image. Moreover, it generalizes corner orthogonality and overlaps with another false regularity: minimizing the standard deviation of angles. As such, its use is strongly discouraged, since it does not correspond to actual features detected in the sketch and may not reflect the sketcher's intent.

- Isometry is a variant of the regularity that minimizes the standard deviation of segment lengths (Brown and Wang, 1996). While it may represent a true regularity—since a human observer often assumes the object depicted in the line drawing has similar dimensions along different directions—it cannot usually be verified directly in the sketch due to the effects of projective distortion. Therefore, it is considered a false, or at best, an indirect cue. Moreover, while designers' sketches often preserve proportions, they rarely reflect precise measurements, further limiting the reliability of this regularity.
- Corner orthogonality is a specific case of the minimum standard deviation of angles (MSDA), where all angles formed by connected edges are assumed to be 90 degrees. As such, it is considered a false regularity unless specifically applied to normalons. To address this situation, we have reformulated it as a cubic corner cue by restricting its application to vertices of valence 3. In this form, it serves as a kind of post-refinement step, useful after the direct inflation of normalons.
- Face perpendicularity (in manifolds) is a specific case of corner orthogonality. Its relevance is limited to normalons, where it represents a true regularity. Therefore, it may be effectively used as a post-refinement step following the direct inflation of normalons.
- Skewed facial orthogonality is a specific case of line orthogonality, extended to all edges of a given face. It holds true for the faces of normalons (rectangular polyhedra).
- Skewed facial symmetry is a true and noteworthy regularity that was extensively studied by Varley and Martin (2000c). However, we consider facial symmetry to be better understood as an indirect cue indicating a more general model-level symmetry.

Despite the significant contributions outlined above, the selection, formulation, and application of regularities for optimization remains an open problem. Users can explore these issues in REFER by configuring different cost functions to assess their effect on the inflation process. Two options for selecting cues or regularities are available:

(1) Automatic selection of regularities based on the model's typology. The default recommended regularities for each type of B-Rep model are shown in Figure 3.169.
(2) Manual selection of regularities. This option allows users to explore the impact of each regularity on B-Rep model reconstruction. The default recommended regularities are highlighted in Figure 3.170.

3.9.5.4 *Optimization-based inflation objective function*

The *Optimization Based Inflation Configuration* dialog allows the user to configure the details of the objective function costs. In principle, the costs of all regularities contribute equally to the global cost, which is calculated by adding all the partial costs. However, in some cases, it may be useful

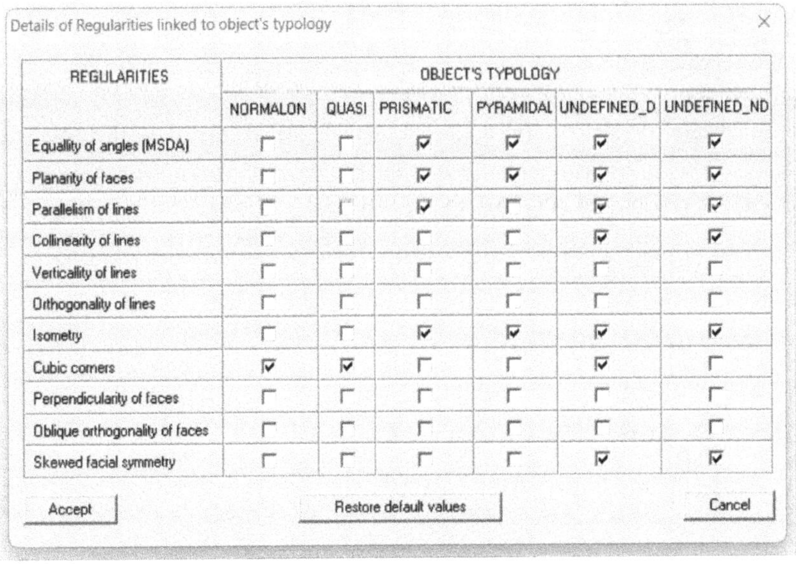
Details of Regularities linked to object's typology

REGULARITIES	OBJECT'S TYPOLOGY					
	NORMALON	QUASI	PRISMATIC	PYRAMIDAL	UNDEFINED_D	UNDEFINED_ND
Equality of angles (MSDA)	☐	☐	☑	☑	☑	☑
Planarity of faces	☐	☐	☑	☑	☑	☑
Parallelism of lines	☐	☐	☑	☐	☑	☑
Collinearity of lines	☐	☐	☐	☐	☑	☑
Verticality of lines	☐	☐	☐	☐	☐	☐
Orthogonality of lines	☐	☐	☐	☐	☐	☐
Isometry	☐	☐	☑	☑	☑	☑
Cubic corners	☑	☑	☐	☐	☑	☐
Perpendicularity of faces	☐	☐	☐	☐	☐	☐
Oblique orthogonality of faces	☐	☐	☐	☐	☐	☐
Skewed facial symmetry	☐	☐	☐	☐	☑	☑

Accept — Restore default values — Cancel

Figure 3.169. Default recommended regularities for each object's typology.

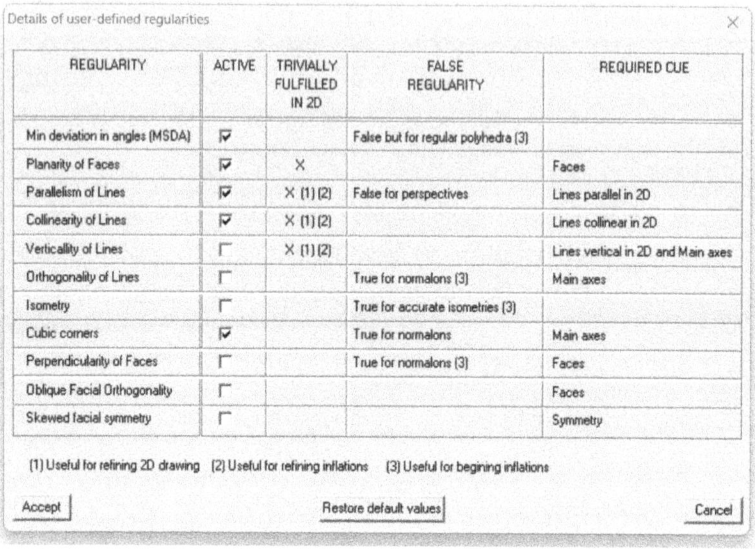

REGULARITY	ACTIVE	TRIVIALLY FULFILLED IN 2D	FALSE REGULARITY	REQUIRED CUE
Min deviation in angles (MSDA)	☑		False but for regular polyhedra (3)	
Planarity of Faces	☑	X		Faces
Parallelism of Lines	☑	X (1) (2)	False for perspectives	Lines parallel in 2D
Collinearity of Lines	☑	X (1) (2)		Lines collinear in 2D
Verticallity of Lines	☐	X (1) (2)		Lines vertical in 2D and Main axes
Orthogonality of Lines	☐		True for normalons (3)	Main axes
Isometry	☐		True for accurate isometries (3)	
Cubic corners	☑		True for normalons	Main axes
Perpendicularity of Faces	☐		True for normalons (3)	Faces
Oblique Facial Orthogonality	☐			Faces
Skewed facial symmetry	☐			Symmetry

Figure 3.170. Default recommended user-defined regularities.

to prioritize certain regularities over others. This prioritization can be done by assigning *weights*, which are coefficients that multiply the costs of the various regularities before adding them to obtain the total cost.

Some regularities such as face parallelism are inherently satisfied in the 2D drawing, so applying them early in the optimization process can be counterproductive. However, introducing them later, as the solution approaches a global optimum, can be beneficial. To this end, the ability to define variable weights to each regularity has been included.

The parameters in the first column in the dialog shown in Figure 3.171 are the fixed weights assigned to different regularities (with a default weight of 1). In the second column, users can select and configure variable weights. The right side of the dialog allows users to define and adjust the parameters (A, B, ...) used for the variable weights.

3.10 3D Refinement

3.10.1 *Introduction*

Inflation does not correct the geometric imperfections inherent in the input sketches or line drawings (imperfections in the X and Y inflation coordinates). In fact, imperfect inflations may introduce undesired new

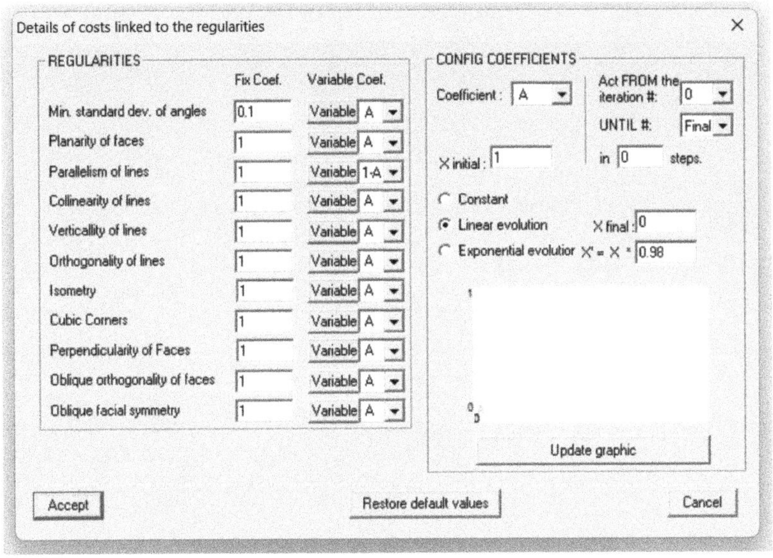

Figure 3.171. Details of costs linked to regularities.

imperfections in the Z coordinate. As a result, different *post-inflation* filters can help refine or "tidy up" the inflated 3D models.

The 3D refinement stage is somewhat controversial, as some authors argue that heuristic approaches may alter the original design intent embedded in the model obtained by inflating the input sketch or line drawing. However, REFER implements several 3D refinements that seem to be beneficial in producing 3D models that better align with the original design intent, at least for certain types of models. REFER users can activate or deactivate these filters based on their preferences. The filters can be managed through the *3D Refinement* toolbar, which controls the procedures for refining the 3D B-Rep model. The following filters can be applied:

(1) 3D refinement by reinforcing planarity of faces (Piquer, 2003; Company *et al.*, 2004).
(2) 3D refinement by reinforcing alignment of faces parallel to main reference planes.

3.10.2 *3D refinement toolbar and configuration dialog*

The *3D Refinement* toolbar (shown in Figure 3.172) is located over the parent window. Two commands control the refinement process (refine face planarity and refine face alignment) and another (*3D Refinement*

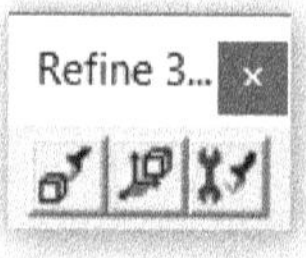

Figure 3.172. *3D Refinement* toolbar.

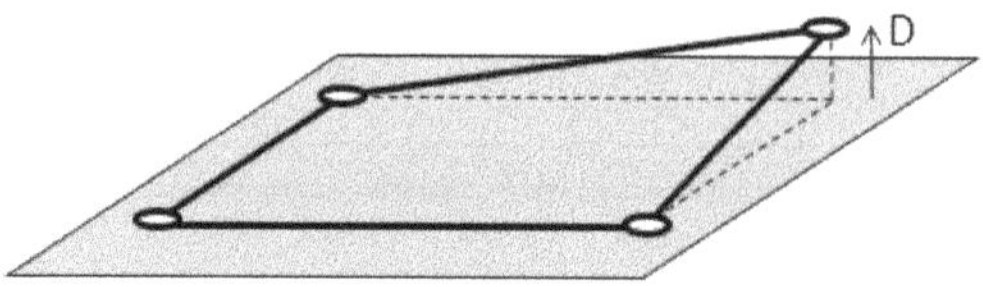

Figure 3.173. Face planarity.

Configuration) opens the configuration dialog, which allows users to configure the 3D face planarity and 3D face alignment refinement filters.

3.10.3 *Face planarity*

This tool enhances the 3D model by enforcing face planarity, a refinement step that can only be performed if the faces have already been computed. The method targets faces composed of polygons with more than three vertices, as triangular faces are inherently planar. It iteratively selects vertex triplets from each face to identify the plane that minimizes the distances from all remaining vertices to that plane. Each vertex triplet is evaluated to define a candidate plane, and the distances of the remaining vertices to this plane are computed. The plane that yields the minimum total distance is selected, and vertex coordinates are adjusted to align the face more closely with this optimal plane. The concept is illustrated in Figure 3.173.

The approach is *local* and *naive*, as it analyzes one face at a time, disregarding the effects of altering vertices shared with adjacent faces. Consequently, the sequence in which faces are processed can affect the result, because the coordinates changed to accommodate a later face may "unflatten" previously flattened adjacent faces (those sharing at least one vertex with the flattened face). To minimize this effect, faces with more vertices are processed first, as they are presumed to contain most of the imperfections.

The algorithm analyzes each 3D face of the B-Rep model to enforce planarity as follows:

(1) Identify the face with the greatest number of edges (not yet evaluated).
(2) Parse all possible planes defined on the face by sequentially using every vertex triplet.
(3) Select the optimal plane (that which minimizes the distances from the remaining vertices to its surface).
(4) Adjust the positions of the remaining vertices to align them with the selected plane.
(5) Repeat until all non-trivial faces have been processed (triangular faces are inherently planar).

Some notes on the method:

- To prevent redundant adjustments, each vertex is processed at most once.
- When a vertex has already been visited, it is treated as the optimal choice, and no further candidate triplets are evaluated.

3.10.4 *Face alignment*

This tool enhances the 3D model by improving the alignment of faces that are approximately parallel to the principal reference planes. This refinement is only possible if both the faces and the main axes have already been identified. The method works by detecting faces whose normals are "nearly" parallel to one of the three primary axes, indicating approximate parallelism with the corresponding reference plane.

In this approach, the B-Rep 3D model is represented relative to its main reference system (i.e., (X_D, Y_D, Z_D)), not the inflation system (i.e., (X_P, Y_P, Z_P)). Parallelism is reinforced by ensuring that all vertices are at the same distance from the reference plane. This outcome is achieved by calculating the average value of the corresponding coordinate (x, y, or z) and then applying this average value to all corresponding coordinates of the vertices of the face. This calculation is meaningful only for normalons and quasi-normalons.

3.11 3D Tessellation

3.11.1 *Introduction*

The inflation process generates a wireframe structure that, when combined with the faces identified from the input sketch or line drawing, result in a polygon mesh that can resemble a 3D boundary representation (B-Rep) model. However, a true B-Rep also requires face normals and, in certain cases, the triangulation of polygonal faces to be properly defined. In REFER, the 3D tessellation stage can be managed via the *3D Tessellation* toolbar. Activation of the tessellation stage requires face reorientation, to ensure that the vertex/edge sequence goes counter-clockwise (CCW) for normal pointing outward.

Not all B-Rep model formats support high-level semantic polygonal faces. To ensure compatibility, REFER includes an algorithm for triangulating polygonal faces. This process generates a tessellated B-Rep model suitable for such formats, though it reduces the semantic richness of the original polygonal faces by converting them into lower-level triangulated representations.

Triangulation could be applied during the initial face detection stage; however, since it is not immediately needed within REFER's workflow, it is implemented as a separate module. This module is executed at the end of the reconstruction process, immediately prior to exporting the explicit 3D model. In this way, the normals to the triangles may be calculated at the time.

An actual B-Rep also requires face normals, which cannot be computed when faces are first detected. At that point, faces are simply identified as candidate polygons in the 2D line drawing. Therefore, the only information available at this stage is a graph structure representing the 2D polygons. Only after the inflation process can we calculate the 3D normals for the faces, providing the necessary data to output a true B-Rep model.

3.11.2 *3D tessellation toolbar and configuration dialog*

The *3D Tessellation* toolbar (shown in Figure 3.174) is located over the parent window. Two commands control the 3D tessellation process (calculate face orientation, calculate face triangulation, and calculate face

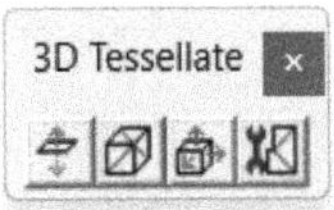

Figure 3.174. *3D Tessellation* toolbar.

normals), and another command opens the *3D Tessellatation Configuration* dialog, which allows users to configure the activation/deactivation of the 3D Tessellation mode, as well as select the filters to triangulate faces and calculate face normal.

3.11.3 *3D face orientation*

Candidate faces are initially detected from the original sketch or line drawing. Therefore, their orientation to ensure that their normals point outward from the object can only be determined after the model is inflated into 3D space.

The process begins by fixing the orientation of one face, referred to as the *seed*. This seed face is selected from the convex hull of the model, since all vertices of the object lie on one side of the plane defined by any convex hull face. Therefore, the outward direction of the normal points toward the empty half-space.

The vertices of the seed face are reordered so that a counter-clockwise (CCW) winding produces a normal pointing outward. Adjacent faces that share an edge with the seed are then reordered to ensure their normals also point outward. This outcome is achieved by reversing the direction of the shared edge in the adjacent face so that it is opposite to its orientation in the seed face. After processing all adjacent faces, the algorithm selects a newly ordered face as the next seed, continuing recursively until all faces have been correctly oriented. The process is illustrated in Figure 3.175.

3.11.4 *3D face triangulation*

This command triangulates the faces of the B-Rep model derived from an input sketch, which was previously vectorized as a 2D line drawing. The method tessellates polygonal faces, converting them into triangular facets. It returns triplets of vertex indices, each corresponding to a triangular facet. The algorithm assumes the faces are flat, although in practice, they

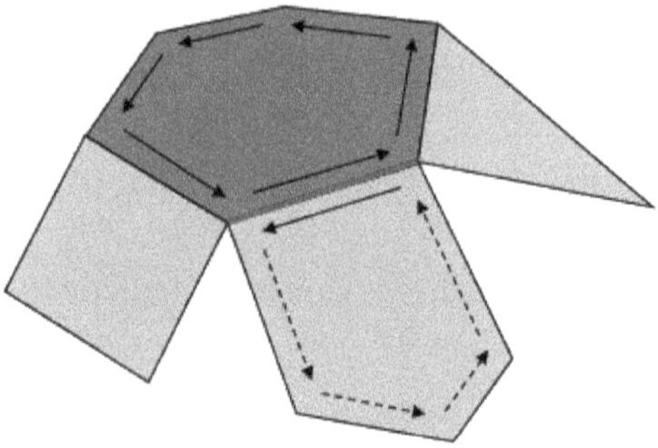

Figure 3.175. 3D face orientation.

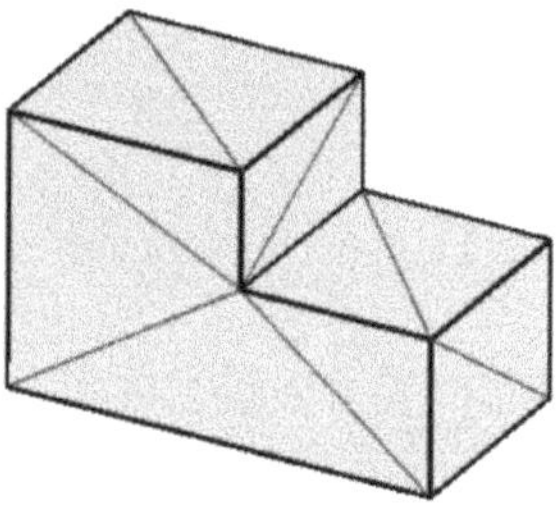

Figure 3.176. 3D model with triangulated faces.

are typically slightly slanted due to imperfections in the original sketches and inaccuracies introduced during the inflation process. An example of a model with triangulated faces is shown in Figure 3.176.

The approach triangulates simple polygons (either convex or concave) located in a two-dimensional plane, with vertices defined by 2D points. Triangulation is performed in the XY projection of the face, as it is simpler and generally projection-invariant, except in cases of degenerate projections. Such cases should not occur, as the faces were originally detected from a flat image in the XY plane.

The algorithm requires polygon vertices to be ordered counterclockwise (CCW). Therefore, all candidate faces are reviewed, and the sequence of their vertices or edges is reversed if necessary. The implementation of the algorithm is adapted from Moreno's code (1998c), which is based on the methods described in (1998a) and (1998b) which, in turn, build on Schildt's original approach (1997). Additional adaptations were made to fix bugs and integrate the method with REFER's data structures.

The algorithm identifies convex vertices in the input polygon. For each convex vertex found, it evaluates the triangle formed by:

- V_{prev}: the vertex preceding the convex vertex
- V_{curr}: the convex vertex itself
- V_{next}: the vertex following the convex vertex

If no other polygon vertex lies inside the triangle, it is guaranteed that the segment (V_{prev}, V_{next}) does not intersect any other edge of the polygon. The triangle is then output and V_{curr} is removed from the polygon.

If the triangle contains other vertices of the polygon, the algorithm attempts to find a vertex (V_s) that can be connected to V_{curr} without intersecting any edges. This process divides the polygon into two simple sub-polygons, which are then recursively triangulated. The recursion stops when a polygon has only three vertices, i.e., it is already a triangle.

Finding a vertex that can be joined to V_{curr} is non-trivial. A common but incorrect strategy is to select the vertex closest to V_{curr}, which can produce invalid triangles. Instead, the correct approach is to find the vertex P that lies farthest from the line defined by (V_{prev}, V_{next}), measured by perpendicular distance. As noted by Moreno (1998), "to obtain this point the algorithm does not directly compute a single distance (it would be too hard, since we are talking about perpendicular distance). Instead, the algorithm finds the point P for which the area of the triangle (V_{prev},V_{next}, P) is maximized."

3.11.5 *3D face normals*

REFER includes an approach to calculate the 3D normals (see Figure 3.177) of the B-Rep 3D model obtained through inflation.

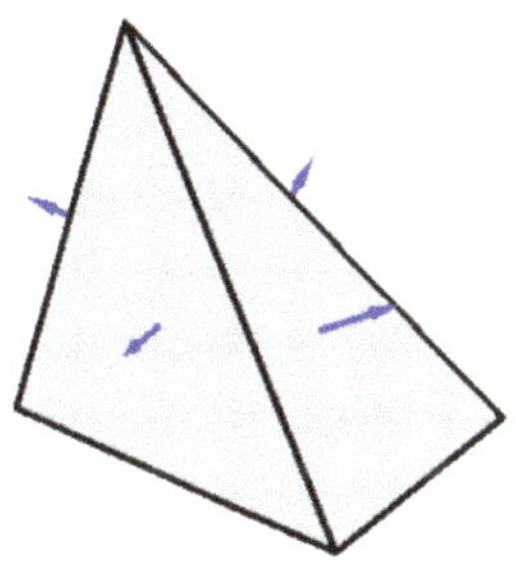

Figure 3.177. 3D normal (in blue) in a B-Rep 3D model in REFER.

The normal of each face is calculated using the scalar (cross) product of vectors formed by the first three vertices of its convex hull. The convex hull is used instead of the original polygon to avoid incorrect results in cases where the selected vertices belong to a concave region of the polygon. To ensure that normal point outward, faces must have been oriented beforehand.

If faces have been triangulated, the normals to the triangular facets are calculated, as well as the normal to the whole face—which is calculated as an average of the normals to the triangles.

Chapter 4

Display

4.1 Introduction

REFER is designed to assess the impact of various experimental Sketch-Based Modeling (SBM) approaches, rather than to produce a final output. With this goal in mind, the application can be configured to initiate different execution flows and apply distinct methods at each stage to test different behaviors. As a result, a wide range of display tools are provided for evaluating the full reconstruction process or specific stages.

In developing REFER, we have carefully selected the most effective visualizations and provided tools to configure and customize them. This chapter should not be viewed merely as a collection of tools for visualizing results, but rather as the outcome of applying our expertise to identify the graphical representations that are most helpful in detecting and correcting conceptual errors in SBM algorithms.

Beginning with the most fundamental options, the user can adjust the size and position of most images displayed in the windows using the standard zoom and pan controls. These controls also serve as search tools for identifying imperfections that may go unnoticed during sketching but can significantly affect the behavior of algorithms.

In addition, distinguishing between the depth perceived by users in pictorial views and the depth calculated by the algorithms requires continuous changes in viewpoint. Therefore, in windows displaying 3D information, users can rotate the three-dimensional shapes (i.e., change the point of view) using trackball control. To use the trackball, simply move the cursor while holding down the left mouse button to rotate the image.

More advanced and customizable display filters are provided, which can be accessed via the following toolbars:

- Display 2D Sketch toolbar
- Display Cues toolbar
- Display Features toolbar
- Display Direct Inflation toolbar
- Display Optimization Process toolbar
- Display B-Rep Model toolbar

4.2 Zooming and Panning

The size and position of most objects displayed in the various windows can be adjusted using standard panning and zooming commands. Zooming in or out alters the distance between the camera and the scene (i.e., the length of the line of sight), causing objects to appear larger or smaller on the screen. Similarly, panning changes the relative position of the image plane with respect to the scene, making the scene appear displaced within the screen. For panning, hold down the mouse wheel while moving the cursor. For zooming in and out, roll the mouse wheel up or down, respectively. The operations are illustrated in Figure 4.1.

Figure 4.1. Zooming and panning.

4.3 Trackball

A trackball is a virtual control that enables the rotation of three-dimensional objects using a standard two-dimensional mouse. To operate it, place the cursor over the object, hold down the left mouse button to "grab" a point, and move the cursor to relocate that point into a new, rotated position. This interaction style is widely used in 3D graphics because it mimics the natural action of holding and examining a physical object. It allows intuitive rotation around any axis and makes it easy to bring the object closer for detailed inspection.

This paradigm assumes a camera that rotates around the scene in response to mouse movement. The scene is enclosed within a virtual sphere, with the camera positioned at a point on the sphere's surface. The camera's location, denoted as point P, is defined using spherical coordinates (theta, phi, and rho), as shown in Figure 4.2.

However, for internal representation of the camera's orientation, quaternions are typically more convenient and effective. Quaternions represent rotations by storing both a rotation axis and an angle in a compact and efficient format. A quaternion consists of four components, [x y z a], where the first three values define the axis of rotation, and the fourth specifies the rotation angle. This structure simplifies the process of combining multiple rotations.

Manipulating the trackball works as follows:

- The trackball is active when the cursor appears as an open hand.
- Pressing and holding the left mouse button emulates the grasping of the sphere. The cursor changes to a closed hand, allowing the user to move the mouse and rotate the sphere.

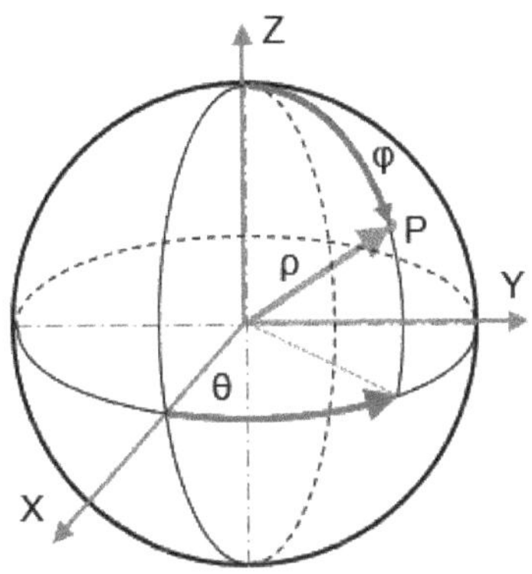

Figure 4.2. Representation of a trackball.

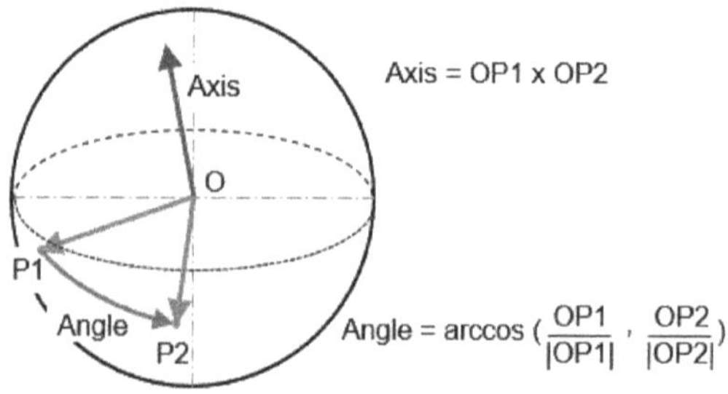

Figure 4.3. Computation of the quaternion.

- Rotation continues until the left mouse button is released, at which point the corresponding quaternion is computed. When the left mouse button is pressed, the current cursor position is recorded as point P1. When the button is released, the cursor's position is stored as point P2. The rotation axis for the quaternion is then computed as the cross product of vectors OP1 and OP2, where O is the center of the virtual sphere (coordinates 0, 0, 0). The angle of rotation is determined by the arc formed between P1 and P2, as shown in Figure 4.3.
- If the mouse is still in motion when the button is released, the action is interpreted as an impulse, causing the sphere to continue rotating. A subsequent click on the left mouse button, while the mouse is stationary, will stop the motion.

The trackball simulation is integrated with the zoom functionality controlled by the mouse wheel: scrolling the wheel adjusts the zoom level. This action can also be replicated by moving the mouse vertically while holding down the right mouse button, during which the cursor changes to a zoom icon.

4.4 Display 2D Sketch

4.4.1 *Display 2D sketch toolbar*

The *Display 2D Sketch* toolbar manages the visualization of two versions of the sketch: the original "input" sketch and the "working" version generated by the 2D refinement process. It also controls the display of three variants of the vectorized 2D drawing, along with most of the related

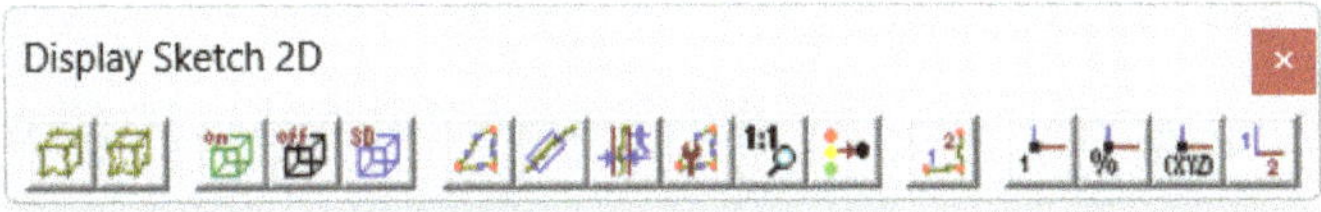

Figure 4.4. *Display 2D Sketch* toolbar.

information for both the sketch and the line drawing. The *Display 2D Sketch* toolbar, shown in Figure 4.4 is located in the *2D Sketch* window. The commands affect exclusively the information displayed in that window.

The toolbar contains the following commands:

- Display input sketch.
- Display working sketch.
- Display vectorized online Line-Drawing.
- Display 2D-refined offline Line-Drawing.
- Display BRep 3D model.
- Display stroke nodes and corners.
- Display sliding strips.
- Display tolerance bands.
- Display Stroke Configuration dialog.
- Zoom All, which resets the zoom and pan settings to restore the complete view of the drawing. It is also a convenient method to recover from unintended screen navigation.
- Display all lines in black. Included as a convenience for capturing black and white images.
- Display stroke numbers. A label displaying the stroke number is added and connected to the stroke's midpoint with a red reference line. A parameter to display stroke merits is included in the Display Strokes Configuration dialog.
- Display vertex numbers. A label with the vertex number is added to each vertex in the line-drawing.
- Display vertex merits. A label with the merit (in %) is added to each vertex in the line-drawing.
- Display vertex coordinates.
- Display edge numbers. A label with the edge number is added to each edge in the line-drawing and connected to the edge's midpoint with a red reference line.

4.4.2 *Display input sketch*

This command enables the user to display a sketch in the *2D Sketch* window. This sketch corresponds to the one described in the input modes section. The completed strokes are shown in a pale yellow-green color, while the stroke in progress is displayed as a thin blue line, as shown in Figure 4.5. Upon the user's request, the nodes and corners can be displayed as large dots.

4.4.3 *Display working sketch*

This command enables the user to display the current sketch in the *2D Sketch* window. The sketch is shown as being processed by the various tools active in the vectorization and refinement sections. Naturally, if no rebuilding is performed, the working sketch will remain identical to the input sketch. The main differences between the two sketches stem from the execution of the sketch refinement process, such as the grouping of dashed lines, as shown in Figure 4.6.

4.4.4 *Display vectorized online line drawing*

This command enables the user to display the line drawing (obtained interactively while sketching) in the 2D Sketch window. The lines of the drawing are shown in bright green, overlaid on the strokes of the sketch,

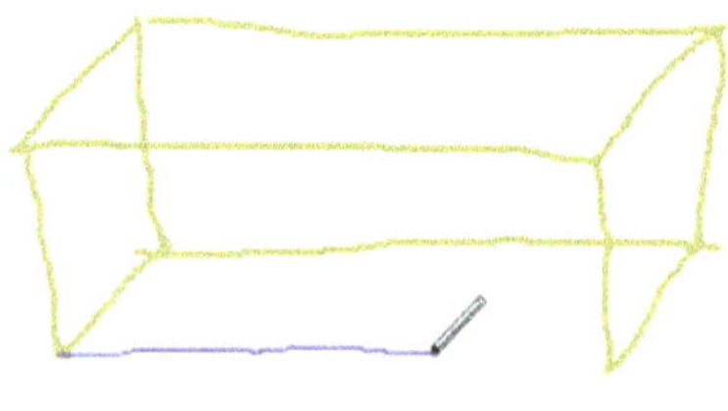

Figure 4.5. Display input sketch.

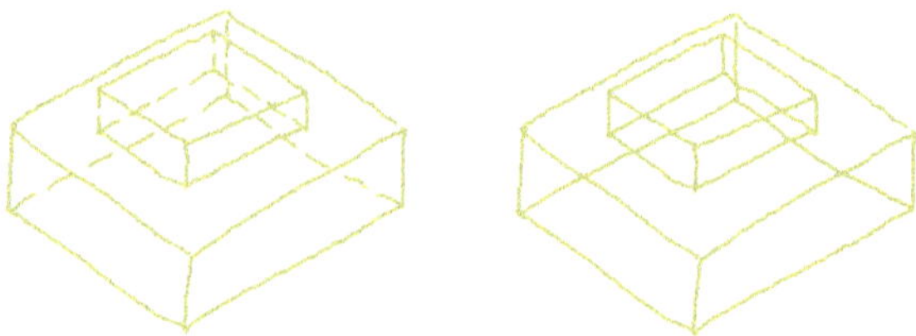

Figure 4.6. Display working sketch.

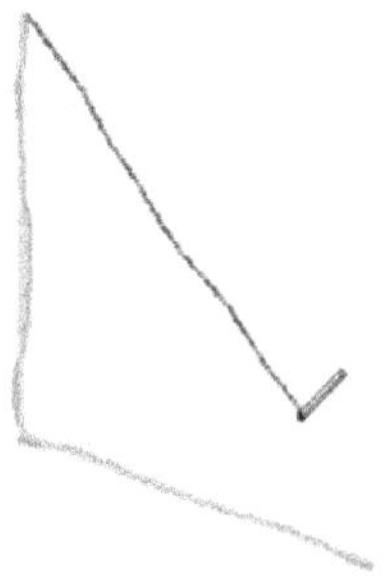

Figure 4.7. Display vectorized online line drawing.

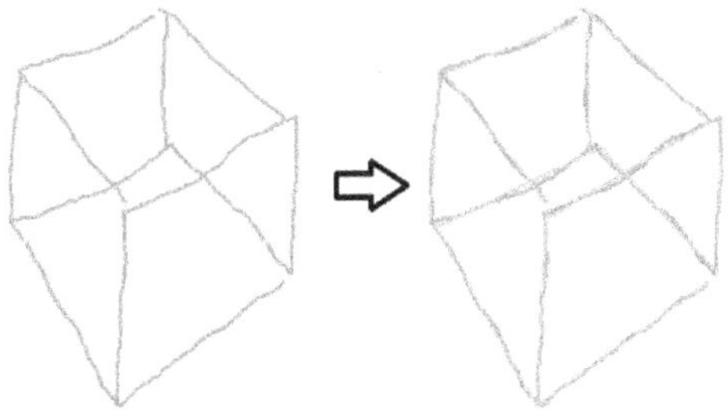

Figure 4.8. Online line drawing being populated with a copy of the offline line drawing.

as shown in Figure 4.7. The interactive line drawing is generated in real time as each stroke is drawn, but only when the interactive vectorization mode is enabled. This mode can be activated from the 2D Vectorization toolbar. When interactive vectorization is disabled, the "online" line drawing is populated with a copy of the offline line drawing, as illustrated in Figure 4.8. During online vectorization, no 2D refinements are applied, so the endpoints of lines are not yet grouped into vertices.

4.4.5 *Display vectorized offline line drawing*

This command enables the user to display the line drawing in the *2D Sketch* window. The line drawing corresponds to the version calculated offline during vectorization (only if the *batch vectorization* option is enabled in the *2D Vectorization* toolbar). The lines are displayed in black, overlaid on both the original sketch strokes and the interactive (online) line drawing, as shown in Figure 4.9.

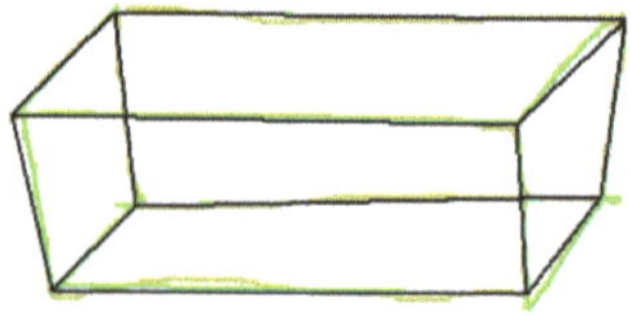

Figure 4.9. Display vectorized offline line drawing.

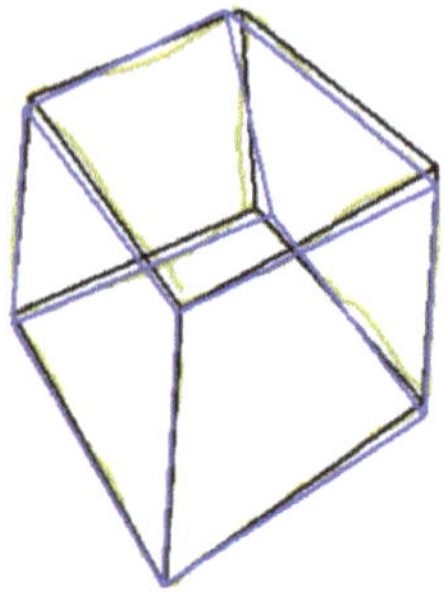

Figure 4.10. Display B-Rep 3D model.

4.4.6 *Display B-Rep 3D model*

This command allows the user to display the 2D projection of the current B-Rep 3D model in the *2D Sketch* window. The model must have been previously generated using one of the available inflation methods. Overlaying the projection of the 3D model (in blue) onto the 2D line drawing (in green or black) provides a clear visual comparison that highlights how the 2D coordinates are transformed during the inflation process, as shown in Figure 4.10.

In this window, the B-Rep model is always shown from the front view of the inflation coordinate system. To this end, the projection matrices described in the *B-Rep inflation coordinates* section are used. To view the 3D model from different angles, use the *Display B-Rep Model* window. Vertex coordinates can be displayed in either the inflation or sketch coordinate systems, as detailed in the *Display Vertex Coordinates* section.

4.4.7 *Display stroke nodes and corners*

Since sketches are composed of hand-drawn strokes, which consist of ordered nodes connected by segments that approximate the original scribbles, this command allows for the visualization of both the

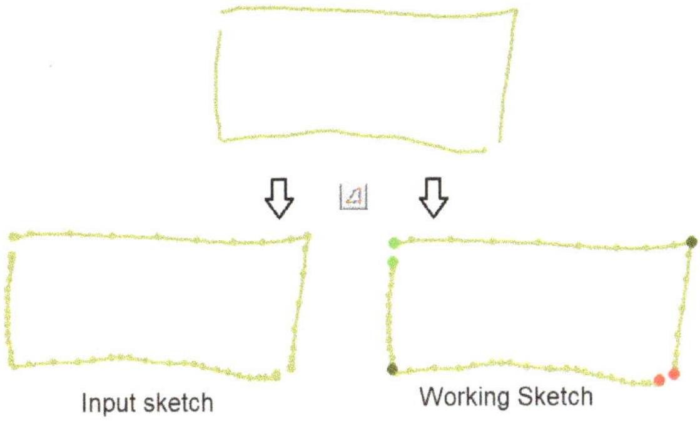

Figure 4.11. Display stroke nodes and corners.

segments and their corresponding nodes. In REFER, nodes are hidden by default when a sketch is displayed. When visible, nodes are shown as brown dots on the screen. If the *Display working sketch* option is active, corners (nodes where sharp changes in direction occur) are also shown as dark brown dots, unless they have been segmented during the sketch refinement stage. Stroke tips are highlighted as bright dots, with green indicating the first node and red indicating the last. An example is shown in Figure 4.11. The *Stroke Display Configuration* dialog provides a set of parameters that allow users to customize how strokes are visualized.

4.4.8 *Stroke display configuration dialog*

The *Stroke Display Configuration* dialog, shown in Figure 4.12, can be accessed via the *Display 2D Sketch* toolbar located in the *2D Sketch* win dow. The dialog includes various parameters that control how strokes are rendered.

When *stroke node display* is enabled, users can toggle between the original and the resampled nodes. The dialog also provides the option to display the convex hull of each stroke, offering helpful visual cues for analyzing and classifying curved strokes. An example is shown in Figure 4.13.

The *Display corners* option in this dialog allows stroke nodes to be shown independently of corner visibility. When the *Display Tolerance Bands* option is active, this dialog provides control over which types of tolerance bands are shown (for straight lines, elliptical arcs, both, or none). When the *Display Numbers in Strokes* option is enabled, the system

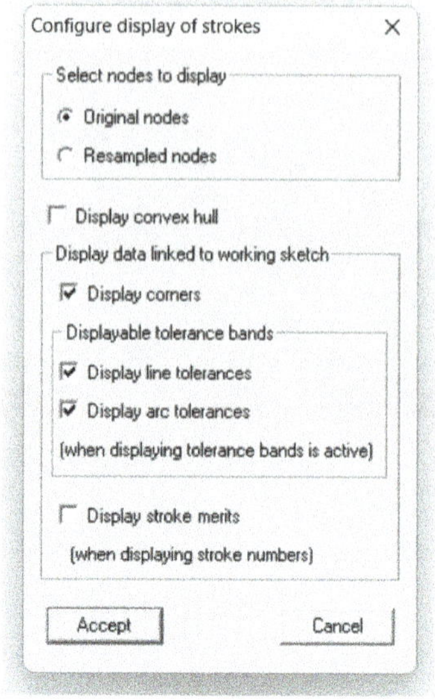

Figure 4.12. Stroke display configuration dialog.

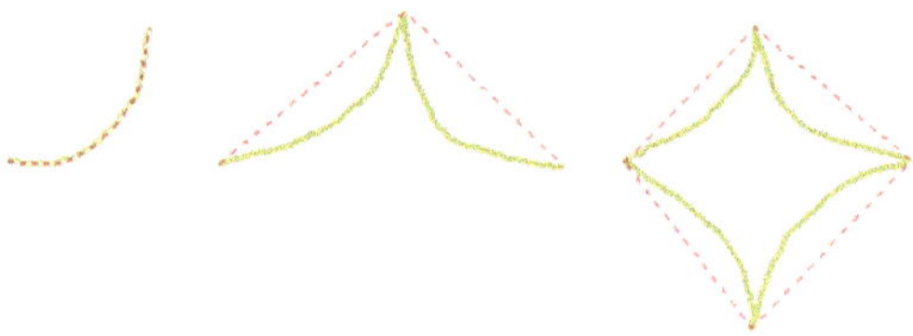

Figure 4.13. Convex hull display.

displays a label indicating the stroke number, connected to the stroke's midpoint by a red reference line. Additionally, if (1) the working sketch is visible and (2) the *Display Stroke Merits* option in this dialog is also enabled, each label is extended to include two merit values: the fit quality for a straight line and for an arc. These values range from 0 (poor fit) to 1 (perfect fit). An example is shown in Figure 4.14.

4.4.9 *Display sliding strips*

This command enables the display of sliding strips for all visible strokes in the *2D Sketch* window. Sliding strips are auxiliary geometric boundaries used in the method proposed by Masood and Sarfraz (2007), Sarfraz

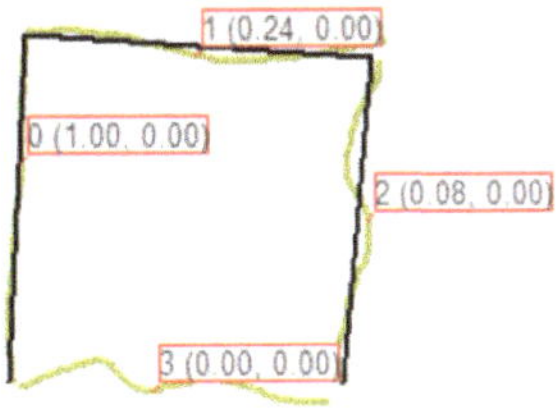

Figure 4.14. Display stroke merits.

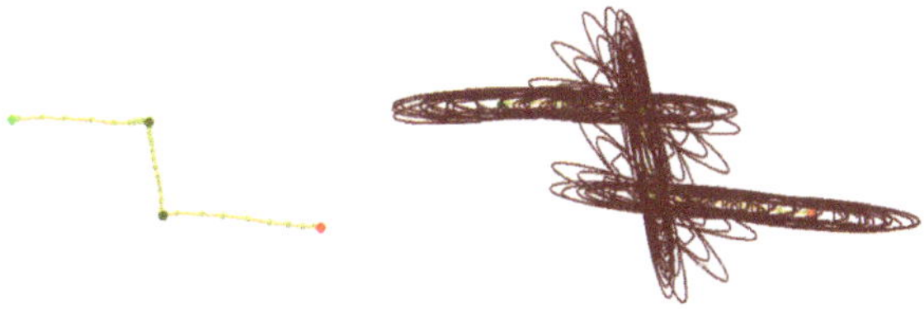

Figure 4.15. Display sliding strips.

et al. (2009), and Swati *et al.* (2009) for corner detection. The Sliding Strips approach identifies corners by analyzing variations in the number of stroke nodes contained within concentric rectangular and/or elliptical strips centered on potential corner locations.

This command allows the user to visualize the mathematical process used by the algorithm to identify corner locations. It is only available when the working sketch is displayed. It is important to note that this visualization is computationally intensive and may significantly clutter the screen. For this reason, it is recommended to enable it only when strictly necessary. An example is shown in Figure 4.15.

4.4.10 *Display tolerance bands*

This command enables the user to display the tolerance bands for all visible strokes in the 2D Sketch window. These tolerance bands are auxiliary geometric boundaries used in the merit calculation process during the fitting of straight segments (Plumed *et al.*, 2015) and elliptic arcs (Company *et al.*, 2015b). The *Stroke Display Configuration* dialog includes parameters that allow independent control over the visibility of straight-line tolerance bands and elliptical arc tolerance bands.

The command allows the user to visualize the mathematical process used by the algorithm to compute the merits of stroke fitting. It is only

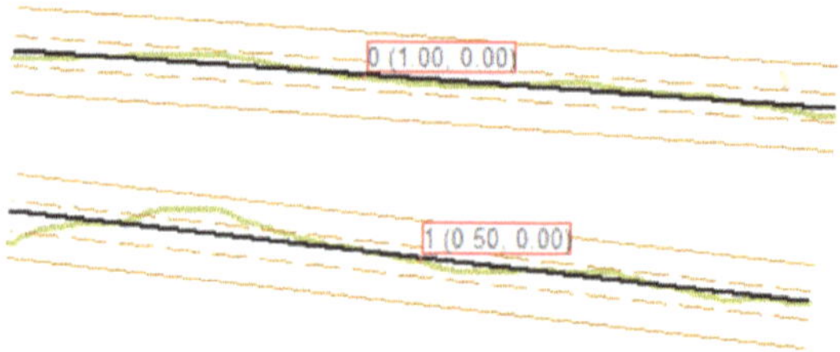

Figure 4.16. Display tolerance bands.

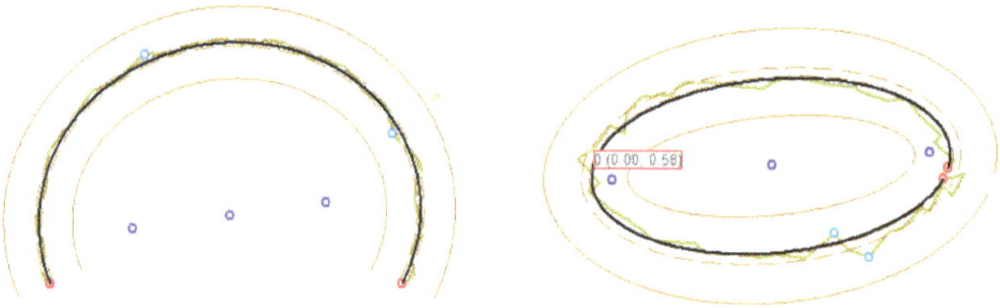

Figure 4.17. Display tolerance bands for arcs.

available when the working sketch is currently displayed. It is important to note that this visualization may significantly clutter the screen. For this reason, it is recommended to enable it only when strictly necessary. An example is shown in Figure 4.16.

Tolerance bands for arcs are supplemented with additional visual markers. The center and foci are shown as circles with bright blue borders and white fill, while the two points on the stroke with the poorest fit are marked with circles featuring pale blue borders and white fill, as shown in Figure 4.17.

4.4.11 *Display zoom all*

This command resets the zoom and pan settings to their default values, centering the scene in the window and scaling it to fill the available space, as shown in Figure 4.18. The command is especially helpful when accidentally navigating the scene outside the window's borders, allowing it to quickly reset the view and restore the scene to its default position.

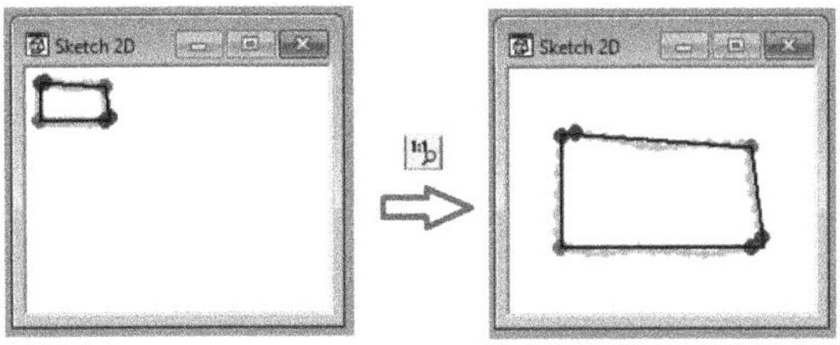

Figure 4.18. Display zoom all.

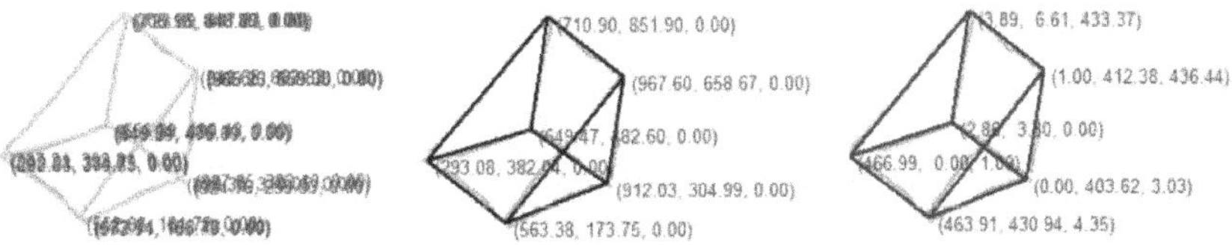

Figure 4.19. Display vertex coordinates.

4.4.12 *Display vertex coordinates*

This command displays the coordinates of the vertices in the line drawing by adding a label in brackets with the three Cartesian coordinates for each vertex. The third coordinate is set to zero for both the online and offline drawings, as well as for the B-Rep 3D model, until the CSG or inflation processes are calculated.

Coordinates are displayed only when one of the line-drawing views is active. Since there are three types of line drawings (online, offline, and B-Rep 3D), the coordinates shown correspond to the most recently active drawing. In other words, the B-Rep coordinates take priority over the offline coordinates, and the offline coordinates take priority over the online ones. An example is shown in Figure 4.19.

It is important to note that the left image in Figure 4.19 is not blurred; rather, it displays the coordinates of each tip of nearby (sometimes nearly overlapping) lines before the vertices are merged in the input line-drawing. Zooming in on these close tips will allow the coordinates to be read more clearly, as shown in Figure 4.20.

Note also that if the main axes have been calculated and are displayed, the B-Rep coordinates are relative to that coordinate system, rather than the inflation coordinate system, as shown in Figure 4.21.

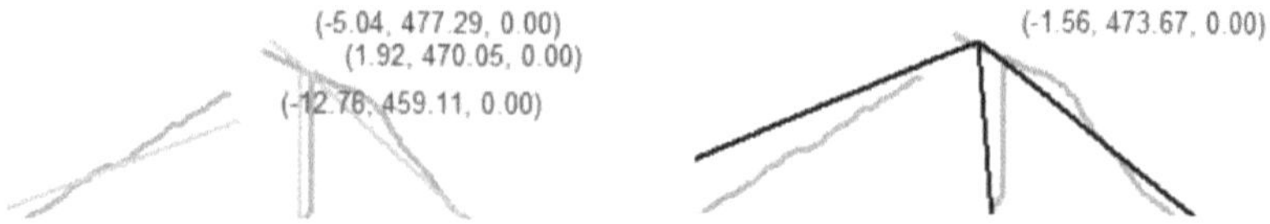

Figure 4.20. Zooming in on the line tips to facilitate readability.

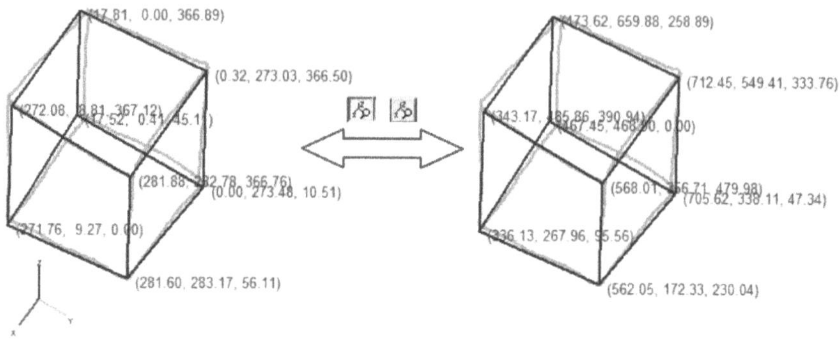

Figure 4.21. B-Rep coordinates are relative to main coordinate system (left) or inflation coordinate system (right).

4.5 Display Cues

4.5.1 *Introduction and toolbar*

The primary purpose of the *2D Sketch* window is to provide a framework for visual interaction during the sketching process. Additionally, it features the *Display Cues* toolbar, which grants access to a set of tools for interactively displaying various cues detected by the application during the reconstruction process. Naturally, the detection of cues must be activated, and the reconstruction process must have been completed before the user can access the visualization of the cues.

The *Display Cues* toolbar, shown in Figure 4.22, is located in the *2D Sketch* window.

The toolbar contains the following commands:

- Display dashed lines.
- Display vertex valences.
- Display parallel edges.
- Display parallel edges in sequence.

Figure 4.22. Display cues toolbar.

- Display vanishing points.
- Display vanishing points in sequence.
- Display subgraphs in sequence.
- Display perimeters in sequence.
- Display vertex labels.
- Display edge labels.
- Display main axes.
- Display faces in implosion.
- Display faces in sequence.
- Display hidden edges.
- Display vertex levels.
- Display symmetry planes.
- Display symmetry planes in sequence.
- Display symmetry subgraphs in sequence.
- Display symmetry axes.

4.5.2 *Display dashed lines*

This tool allows the user to display the dashed lines detected by the Grouping Non-Solid Strokes algorithm in the *2D Sketch* window. Dashed lines are highlighted with thick lines, with a unique shade of blue assigned to each dashed line. This display option is only active if dashed lines have been previously calculated. Once activated, the dashed lines remain highlighted until the option is turned off.

Dashed lines are visible in the input sketch (as shown in Figure 4.23), but they are merged into single lines in the working sketch (as shown in Figure 4.24). Dashes can be highlighted in both representations.

4.5.3 *Display vertex valences*

This tool allows the user to display the vertex valences detected by the *Cue Valences* command in the *2D Sketch* window. Once activated, a label is attached to each vertex or junction of the line-drawing. The label begins

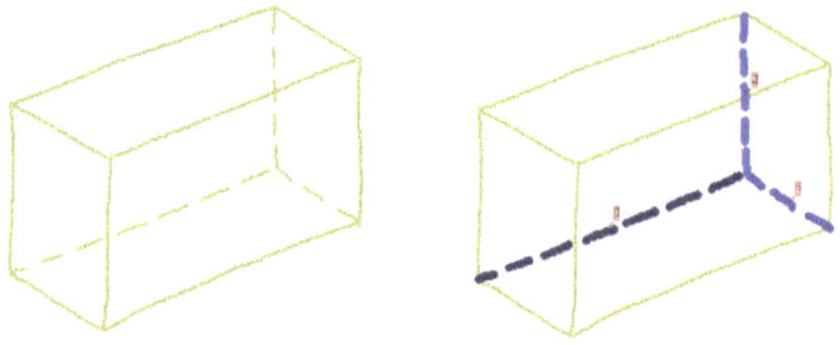

Figure 4.23. Display dashed lines (input sketch).

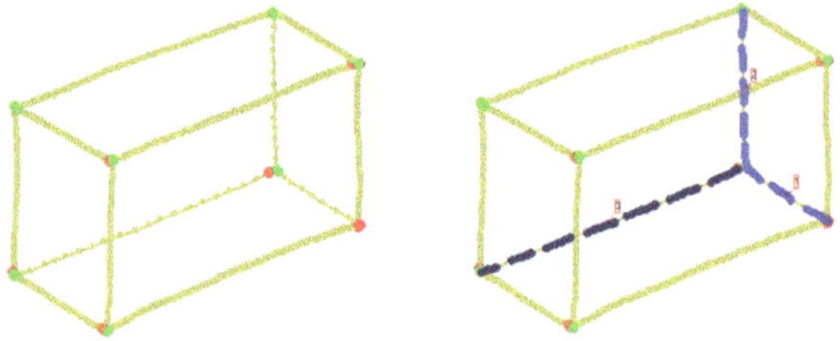

Figure 4.24. Display dashed lines (working sketch).

with the prefix "V" followed by the value of the valence. For example, "V2" indicates that exactly two edges meet at the junction. The labels of valences can be switched off at any time.

As shown in Figure 4.25, the two labels in the lower right corner indicate that the "expected" junction was not merged—likely because the tips were too far apart. This situation serves as a visual cue to the user that the corner may need to be redrawn, or that the parameters controlling the Refine 2D Merge Vertices process may require adjustment.

4.5.4 *Display parallel edges*

This tool allows the user to display the parallel edges detected by the *Cue Parallel Edges* command in the *2D Sketch* window. The lines grouped as parallel are not necessarily strictly parallel but perceived as intended to represent parallel edges. All bundles of parallel lines are simultaneously highlighted with thick green lines, with each bundle displayed in a different shade of green. An example is shown in Figure 4.26.

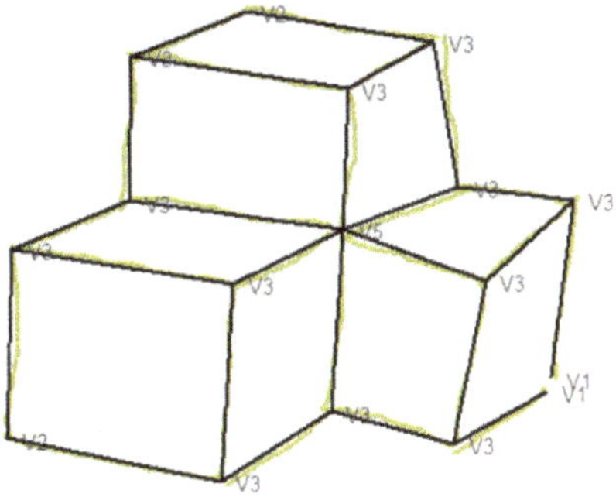

Figure 4.25. Display vertex valences.

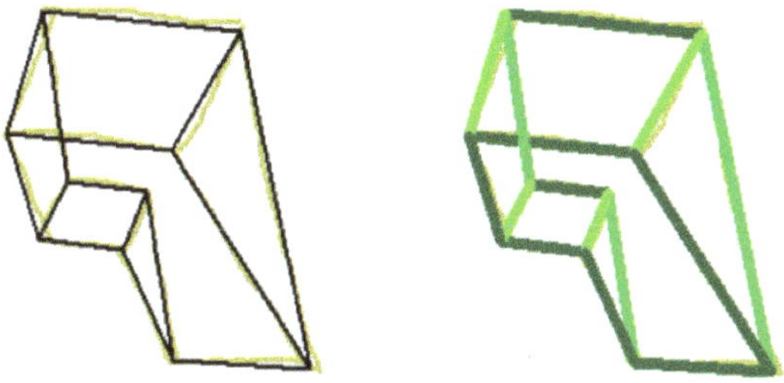

Figure 4.26. Display parallel edges.

4.5.5 *Display parallel edges in sequence*

This tool serves as an alternative *to Display Parallel Edges*. It allows the user to display the parallel edges detected by the *Cue Parallel Edges* command in the *2D Sketch* window. The bundles of parallel lines are highlighted sequentially with thick green lines. Once the command is activated, the first group of parallel lines is displayed, and an auxiliary dialog appears, allowing the user to select the next or previous group of parallel lines. The auxiliary dialog can be closed at any time by clicking the close button (located in the upper right corner) or by deactivating the command. An example is shown in Figure 4.27.

4.5.6 *Display vanishing points*

This tool enables the display of the vanishing points detected by the *Cue Vanishing Points* command in the 2D Sketch window. Alternatively, the groups of lines that converge to any single vanishing point can be

Figure 4.27. Display parallel edges in sequence.

displayed in sequence. It is important to note that the lines do not necessarily converge to a single vanishing point but are perceived as converging toward a common bundle of vanishing points.

All potential vanishing points are numbered and highlighted with orange dots. The centroid of the drawing is indicated by a hollow blue dot. The region of Valid Vanishing Points (RVP) is defined between an inner and an outer ring, both represented by thin blue lines. Clusters of vanishing points are marked with blue dots.

The entire group of vanishing points is framed within a rectangle, outlined with a thin blue line. When nearly parallel lines are present, the vanishing points can be quite distant from the center of gravity, causing the rectangle to appear very large. An example is shown in Figure 4.28.

4.5.7 *Display vanishing points in sequence*

This tool serves as an alternative to the *Display Vanishing Points* command, enabling the user to visualize the parallel edges associated with vanishing points, as detected by the *Cue Vanishing Points* algorithm, in the *2D Sketch* window.

The bundles of parallel lines are highlighted one at a time with thick red lines. All candidate vanishing points associated with the current bundle are shown as orange dots, while the corresponding cluster center is labeled with a number and marked as a thick blue dot. An example is shown in Figure 4.29.

Once the command is activated, the first group of parallel lines is displayed. An auxiliary dialog appears, allowing the user to navigate through the groups of vanishing points by selecting the next or previous group. The auxiliary dialog can be closed at any time by clicking the close button in the upper-right corner or by deactivating the command.

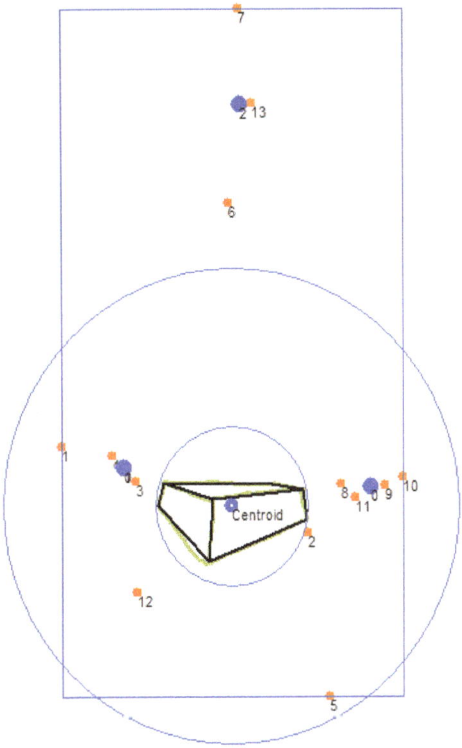

Figure 4.28. Display vanishing points.

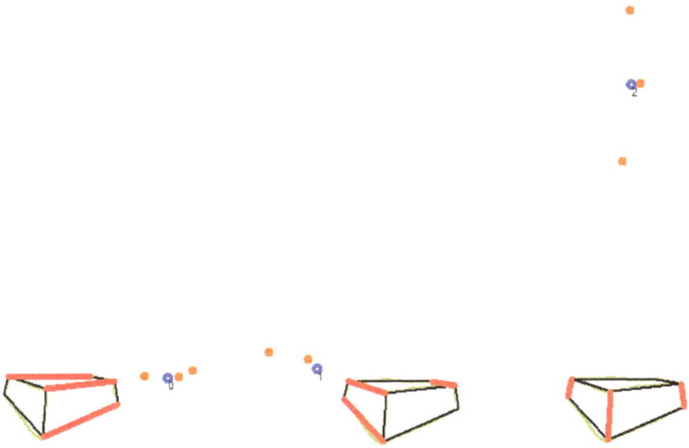

Figure 4.29. Display vanishing points in sequence.

4.5.8 Display subgraphs in sequence

This tool allows the user to display the subgraphs detected by the *Cue Subgraphs* algorithm in the *2D Sketch* window. Once the command is activated, the first subgraph is highlighted with thick orange lines (as shown in Figure 4.30), and an auxiliary dialog appears, allowing the user to navigate through the subgraphs using "Previous" and "Next" options. The auxiliary dialog can be closed at any time by clicking the close button in the upper right corner, or by deactivating the subgraph display command. As the subgraphs are visualized through highlighted lines in the line-drawing, the *Display Input Sketch* option must be active.

4.5.9 Display perimeters in sequence

This tool enables the user to display the contours identified by the *Cue Perimeters* algorithm in the *2D Sketch* window. Once the command is activated, the lines that form the first detected perimeter are highlighted with thick orange lines (as shown in Figure 4.31), and an auxiliary dialog appears, allowing the user to navigate through the sequence of detected perimeters. The auxiliary dialog can be closed at any time by clicking the close button in the upper-right corner, or by deactivating the "Display Perimeters in Sequence" command.

4.5.10 Display vertex and edge labels

The *Display Vertex Labels* tool allows the user to display the vertex labels detected by the *Cue Labels* algorithm in the *2D Sketch* window. All vertices in the line drawing are labeled by connecting each junction to its corresponding label with a red reference line. The label itself is

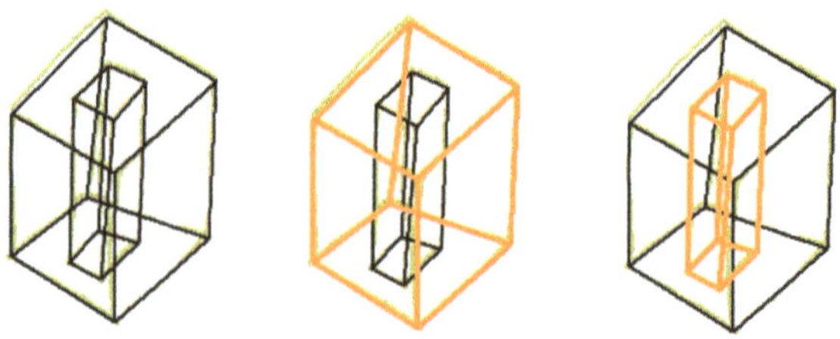

Figure 4.30. Display subgraphs in sequence.

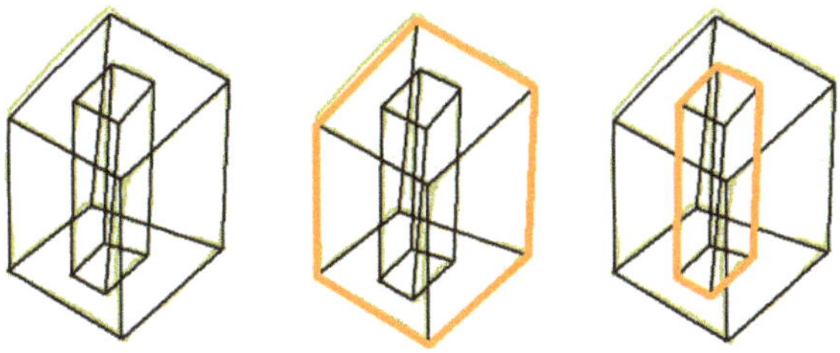

Figure 4.31. Display perimeters in sequence.

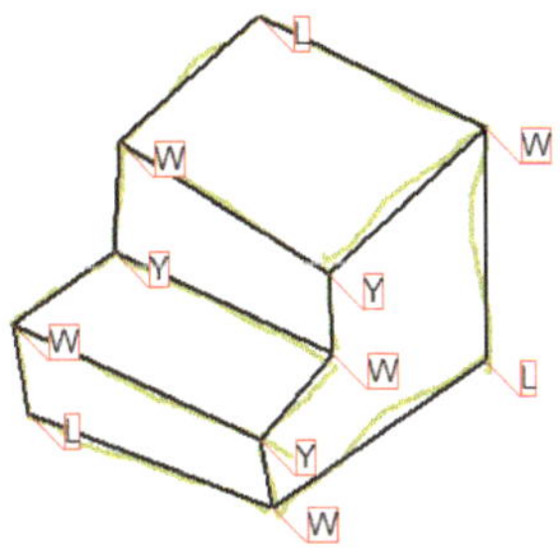

Figure 4.32. Display vertex labels.

enclosed within a red rectangle. The symbols used to label the vertices are explained in the *Cue Labels* section. An example is shown in Figure 4.32.

Similarly, the *Display Edge Labels* tool allows the user to display the edge labels detected by the *Cue Labels* algorithm in the *2D Sketch* window. Each edge in the line drawing is labeled with a tag near its midpoint. An example is shown in Figure 4.33. The symbols used to label the edges are as follows:

- > for boundary edges (The direction in which the contour is traversed is indicated by aligning a symbol along the path)
- + for convex edges
- − for concave edges
- ? for unlabeled edges

Vertex and edge labels may be displayed simultaneously, providing a comprehensive view of the labeling process.

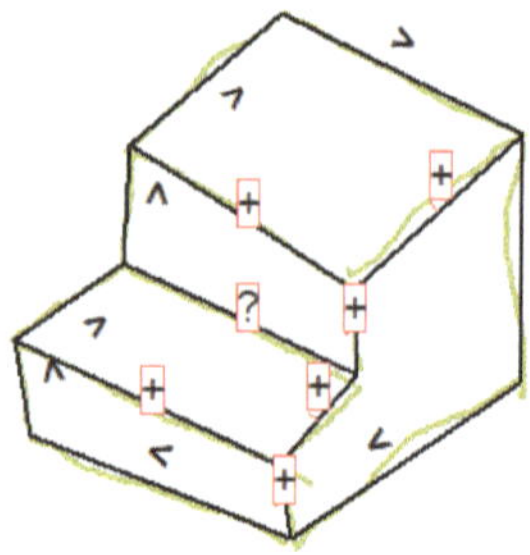

Figure 4.33. Display edge labels.

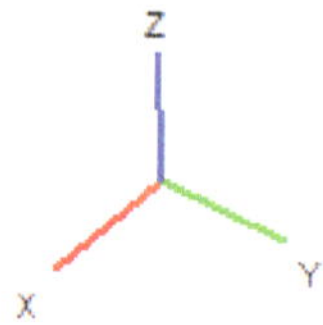

Figure 4.34. Display main axes.

4.5.11 *Display main axes*

This tool allows the user to display the main axes detected by the *Cue Main Axes* algorithm in the *2D Sketch* window. The display of the main axes is typically combined with the display of the normalon lines. An icon representing the three main axes is shown at a corner of the screen (see Figure 4.34). The X axis is shown in red, the Y axis in green, and the Z axis in blue.

It is important to note that this command indirectly controls the display of B-Rep model coordinates. As explained in the *Display Vertex Coordinates* section, when this command is active simultaneously with the coordinate display, the displayed coordinates correspond to the main axes. Otherwise, the coordinates are presented in the inflation coordinate system.

4.5.12 *Display faces in implosion*

This tool allows the user to display the faces detected by the *Cue Faces* algorithm in the *2D Sketch* window. All faces in the line drawing are displayed simultaneously in an “imploded” state, as if deflating by pulling their vertices inward toward their centroid. These faces are rendered with thin, pale blue lines overlaid on the original drawing, offering a quick visual summary of all detected faces. An example is shown in Figure 4.35.

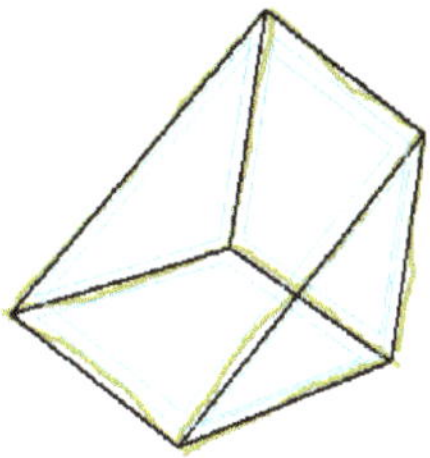

Figure 4.35. Display faces in implosion.

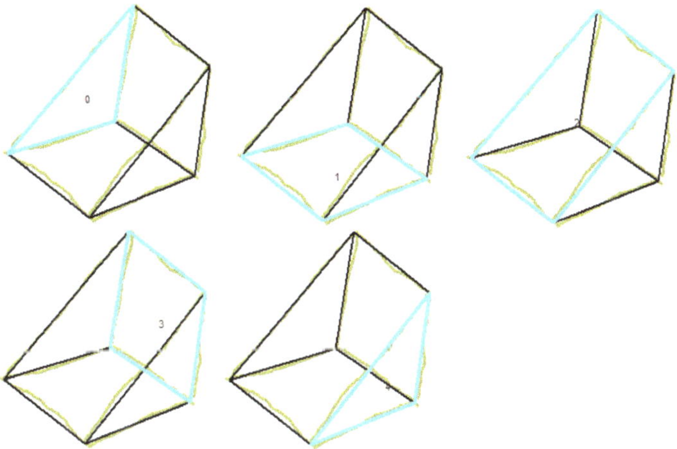

Figure 4.36. Display faces in sequence.

This approach can result in overloaded images when the object is complex. As an alternative, the option to *Display Faces In Sequence* allows for a more detailed analysis of each individual face.

4.5.13 *Display faces in sequence*

This tool allows the user to display the faces detected by the *Cue Faces* algorithm in the *2D Sketch* window. Once activated, the command displays the lines of the first face in thick pale blue lines. The face number is shown near the barycenter of the face. An example is shown in Figure 4.36.

An auxiliary dialog will appear when running this command, allowing the user to browse through the various faces. The dialog can be closed at any time by clicking the close button in the upper right corner or by deactivating the command to display faces in sequence.

4.5.14 *Display hidden edges*

This tool allows the user to display the hidden edges detected by the *Cue Hidden Edges* algorithm in the *2D Sketch* window. Edges labeled as hidden are displayed as thin dashed lines. An example is shown in Figure 4.37.

This feature only works if the hidden edges are labeled during the line-drawing input, or if hidden line detection is active and yields a positive result (which may not always be the case, as illustrated in Figure 4.38).

4.5.15 *Display vertex levels*

This tool allows the user to display the vertex levels detected by the *Cue Vertex Levels* command in the *2D Sketch* window. All vertices in the line drawing are labeled with their level rank, shown as an "L" followed by

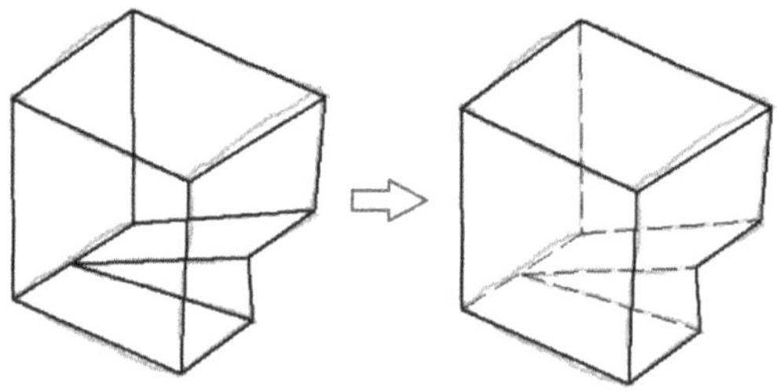

Figure 4.37. Display hidden edges.

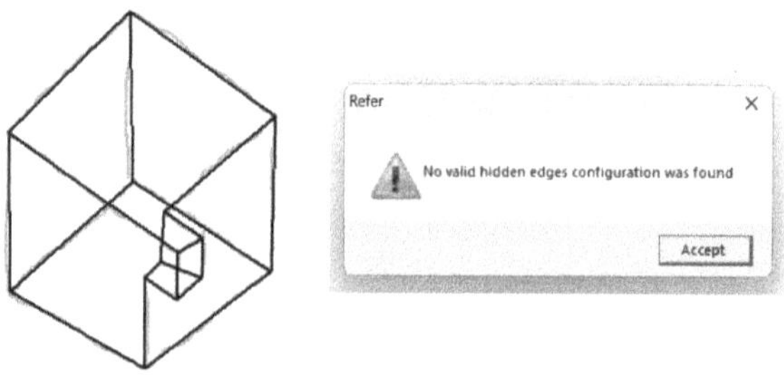

Figure 4.38. *Display hidden edges* when hidden line detection does not yield a positive result.

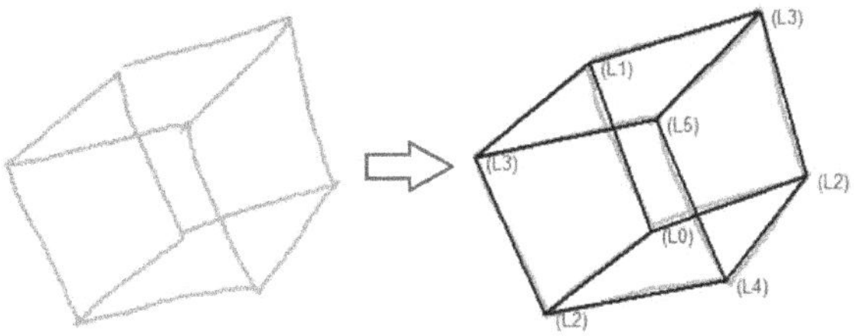

Figure 4.39. Display vertex labels.

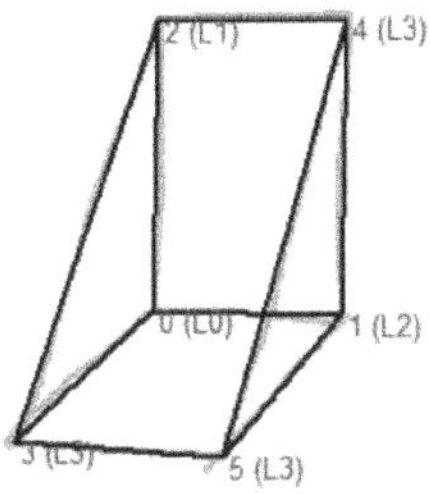

Figure 4.40. Vertex labels in a cavalier representation of a wedge.

the level number in parentheses. Higher levels indicate vertices presumed to be closer to the viewer, while Level 0 corresponds to the farthest vertex. An example is shown in Figure 4.39.

The visualization demonstrates that the heuristic approach does not accurately discriminate the depth of vertices, particularly for concave shapes or shapes depicted in oblique axonometry. In the cavalier representation of the wedge shown in Figure 4.40, the vertices closest to the viewer (3 and 5) are labeled with the same depth level as the upper-right vertex (4).

4.5.16 *Display symmetry planes*

This command enables the user to display all candidate symmetry planes detected by the *Cue Bilateral Symmetry* algorithm in the *2D Sketch* window. The application displays the perimeters of the intersections between the presumed 3D shape depicted in the 2D line-drawing and each of its candidate symmetry planes. An example is shown in Figure 4.41.

A great number of symmetry planes may be detected than one might intuitively expect, as the planes are identified based on topological rules

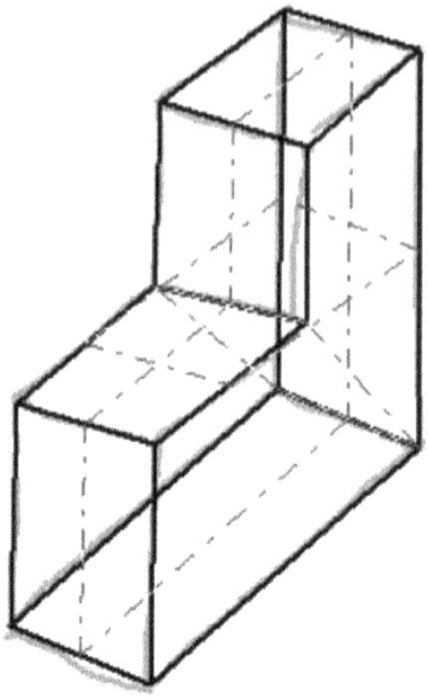

Figure 4.41. Display symmetry planes.

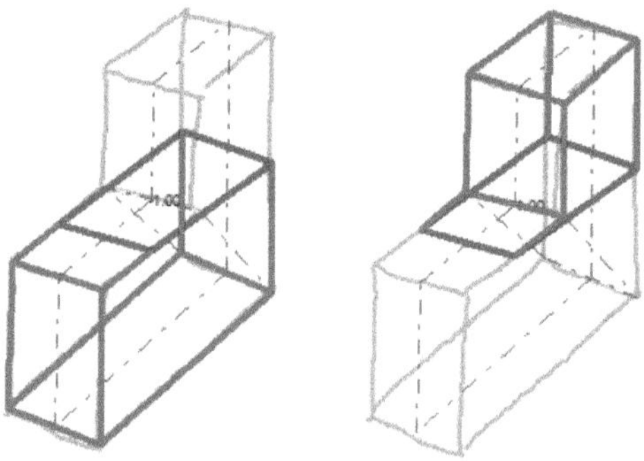

Figure 4.42. Symmetry plane that results in a negative merit score.

rather than geometric measurements or ergonomic assumptions. For example, the symmetry plane in Figure 4.42 is topologically valid, as it divides the shape into two "mirror" halves. However, it is geometrically absurd, resulting in a negative merit score.

The *Symmetry Planes in Sequence* command allows to cycle through candidate symmetry planes, enabling step-by-step inspection. In Figure 4.42, this feature was used in combination with the *Symmetry Subgraphs in Sequence* command to highlight a geometrically incorrect symmetry plane.

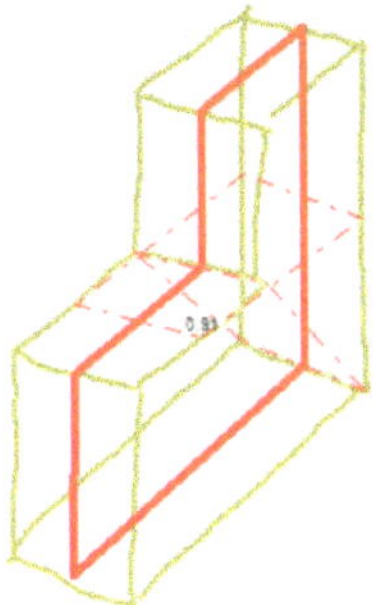

Figure 4.43. Display symmetry planes in sequence.

4.5.17 *Display symmetry planes in sequence*

This tool allows the user to display a single symmetry plane detected by the *Cue Bilateral Symmetry* algorithm in the *2D Sketch* window. The merit of the currently displayed symmetry plane is also shown, with values ranging from −1 to 1. A value of 1 indicates the highest confidence that the detected 2D symmetry plane corresponds to a true symmetry plane of the 3D model. An example is shown in Figure 4.43.

Once the command is activated, the first symmetry plane is highlighted with thick red lines, and an auxiliary dialog appears, allowing the user to select the next or previous symmetry plane. The auxiliary dialog can be closed at any time by clicking the close button (located in the upper right corner) or by deactivating the command to display symmetry planes.

4.5.18 *Display symmetry subgraphs in sequence*

This tool allows the user to display one of the two symmetric subgraphs, which result from dividing the 3D shape into two halves by the symmetry plane, as calculated by the *Cue Symmetry* algorithm, in the *2D Sketch* window. Once activated, the command highlights the first half in thick pale blue lines, and an auxiliary dialog appears to allow the user to select the other half. An example is shown in Figure 4.44.

To display any of the subgraphs, the *Symmetry Planes in Sequence* view must first be activated. Only the subgraphs associated with the currently active symmetry plane will be shown. Due to a current limitation in

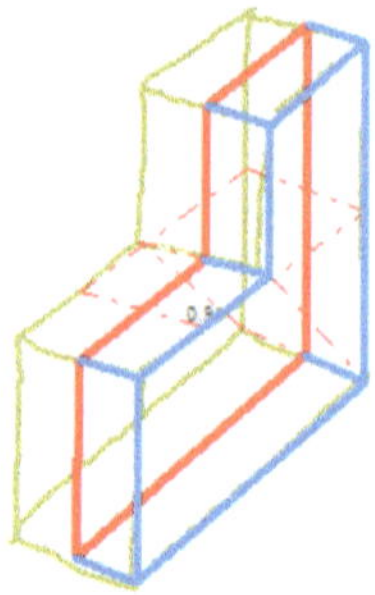

Figure 4.44. Display symmetry subgraphs in sequence.

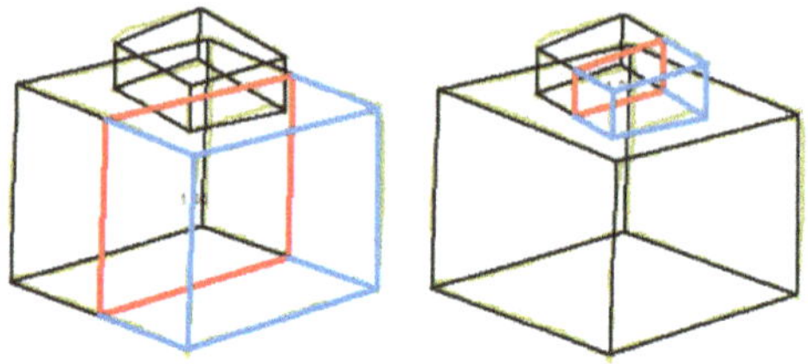

Figure 4.45. Visualization of subgraphs associated with the currently active symmetry plane.

handling symmetry planes shared across different subgraphs, multiple subgraphs cannot be displayed simultaneously. As a result, only the symmetry half subgraph corresponding to the active plane is shown at any given time. An example is shown in Figure 4.45.

4.5.19 *Display symmetry axes*

This command enables the display of the axes of symmetry of all the faces detected in the 2D line-drawing by the *Cue Faces* algorithm in the *2D Sketch* window. Candidate symmetry axes are computed for each face detected in the line drawing, representing the initial stage of the bilateral symmetry cue—specifically, the detection of skewed facial symmetry. When the corresponding visualization is activated, all identified axes are displayed simultaneously, as shown in Figure 4.46.

Candidate symmetry axes can be overlaid on the symmetry planes, enabling users to visualize that not all detected axes are ultimately incorporated into the final set of candidate symmetry planes, as shown in Figure 4.47.

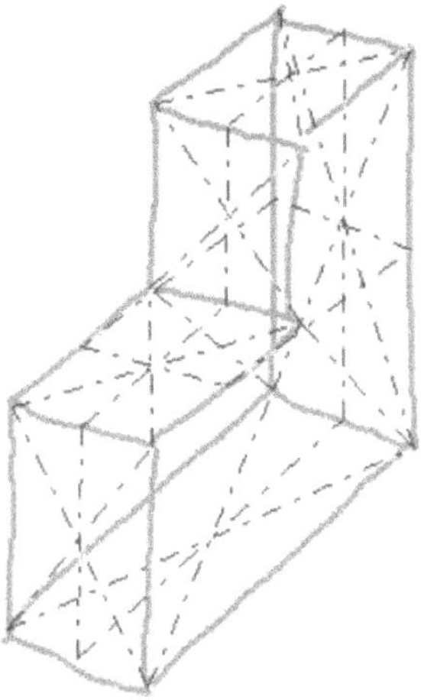

Figure 4.46. Display symmetry axes.

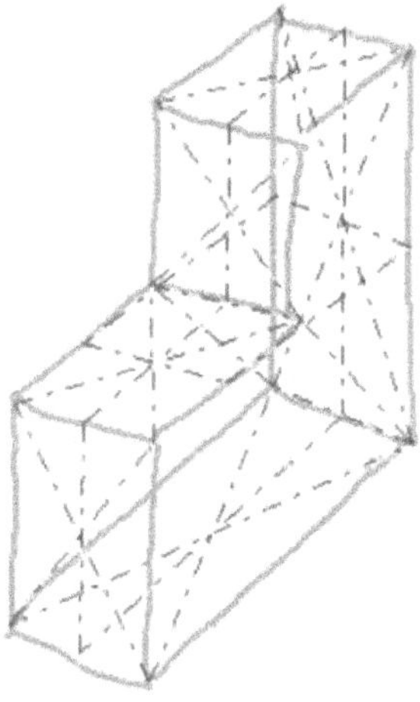

Figure 4.47. Candidate symmetry axes.

4.6 Display Features

4.6.1 *Introduction and toolbar*

The *Display Features* toolbar, shown in Figure 4.48, provides access to a range of display tools that enable the user to interactively visualize the various features detected by the application during the reconstruction process. The detection of features must be activated, and the reconstruction

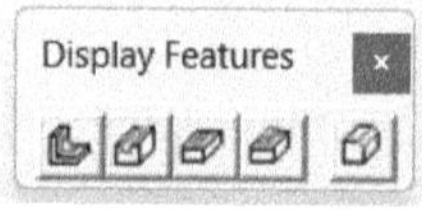

Figure 4.48. Display features toolbar.

must be completed before the user can access the visualization of the detected features.

The toolbar contains the following commands:

- Display ribs and dividers.
- Display rails and slots.
- Display bosses (Not yet implemented).
- Display steps and pockets.
- Display chamfers and rounds (Not yet implemented).

These display options are only available if the corresponding features have already been detected. Each highlighted subgraph is labeled to indicate whether it represents a particular feature, along with its corresponding figure of merit shown in parentheses. The figure of merit is a value between 0 and 1, where 1 indicates maximum confidence that the subgraph correctly represents the specified feature, and 0 indicates no confidence.

4.6.2 *Display ribs and dividers*

This tool allows the user to display the ribs and dividers detected by the *Feature Ribs And Dividers* algorithm in the *2D Sketch* window. The lines that form the subgraph representing either a rib or a divider are highlighted with thick strokes (green for ribs and red for dividers). The brightness of the color indicates the merit of the feature: brighter hues correspond to higher merits, while darker shades indicate lower ones. An example is shown in Figure 4.49.

4.6.3 *Display rails and slots*

This tool allows the user to display the rails and slots detected by the *Feature Rails And Slots* algorithm in the *2D Sketch* window. The lines that form the subgraph representing either a rail or a slot are highlighted with

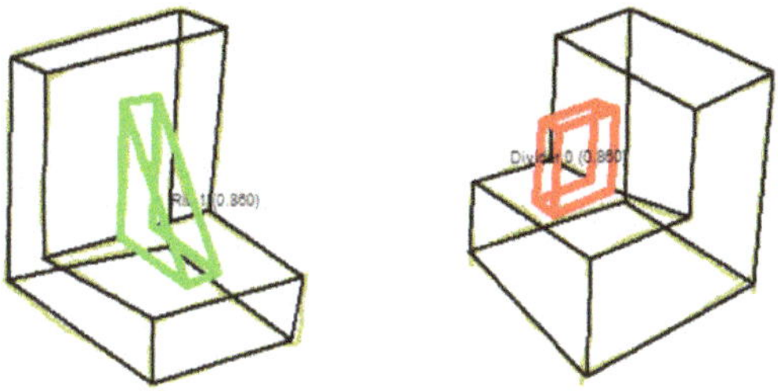

Figure 4.49. Display ribs and dividers.

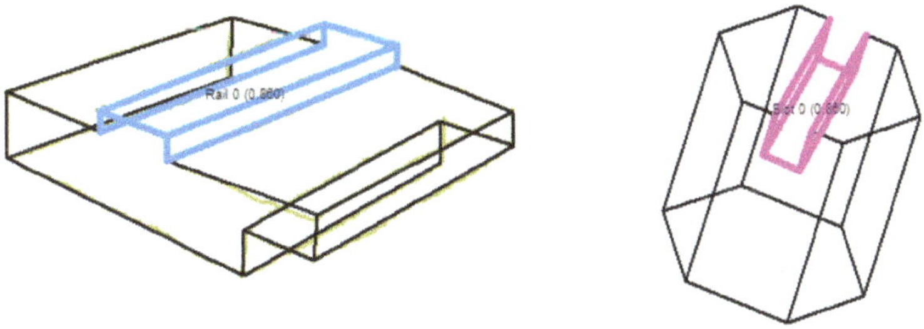

Figure 4.50. Display rails and slots.

thick strokes (pale blue for rails and pink for slots). The brightness of the color indicates the merit of the feature: brighter hues correspond to higher merits, while darker shades indicate lower ones. An example is shown in Figure 4.50.

4.6.4 *Display steps and pockets*

This tool allows the user to display the steps and pockets detected by the *Feature Steps And Pockets* algorithm in the *2D Sketch* window. The lines that form the subgraph representing either a step or a pocket are highlighted with thick strokes (blue for steps and purple for pockets). The brightness of the color indicates the merit of the feature: brighter hues correspond to higher merits, while darker shades indicate lower ones. An example is shown in Figure 4.51.

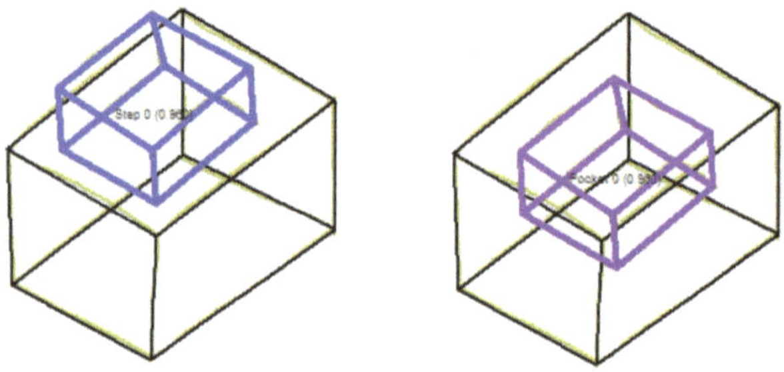

Figure 4.51. Display steps and pockets.

Figure 4.52. Display direct inflation toolbar.

4.7 Display Direct Inflation

4.7.1 *Introduction and toolbar*

The *Display Direct Inflation* toolbar, shown in Figure 4.52, provides access to a range of display tools that enable the user to interactively visualize the reconstruction process of direct inflation. The reconstruction must be completed before the user can access the visualization of the inflation process.

The toolbar contains the following commands:

- Display normalon.
- Display spanning tree.
- Configure spanning tree.

4.7.2 *Display normalon*

This tool allows the user to display the normalon shapes detected by the cue typology and parsed by the axonometric inflation in the *2D Sketch* window. Edges are color-coded based on their normalon directions, which can be visualized using the *Display Main Axes* command. An example is shown in Figure 4.53.

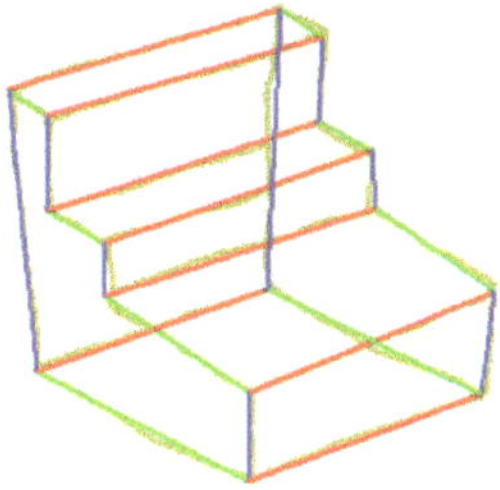

Figure 4.53. Display normalon.

Figure 4.54. Display spanning tree.

4.7.3 *Display spanning tree*

This tool allows the user to display the spanning tree used during axonometric inflation when the detected cue typology is either normalon or quali-normalon, and the direct inflation mode is set to automatic or normalon. The spanning tree is shown in the *2D Sketch* window. When the direct inflation mode is set to levels, the tool instead displays the spanning tree used for levels inflation.

Once the command is activated, the full spanning tree is displayed, with the most recently added branch highlighted in red and all previous branches in green. An auxiliary dialog appears, allowing the user to navigate through the branches of the sequence. An example is shown in Figure 4.54.

When navigating back to the starting point, the vertex from which the spanning tree originated is displayed first as a thick green dot, followed by the first branch shown as a thick red line, as shown in Figure 4.55.

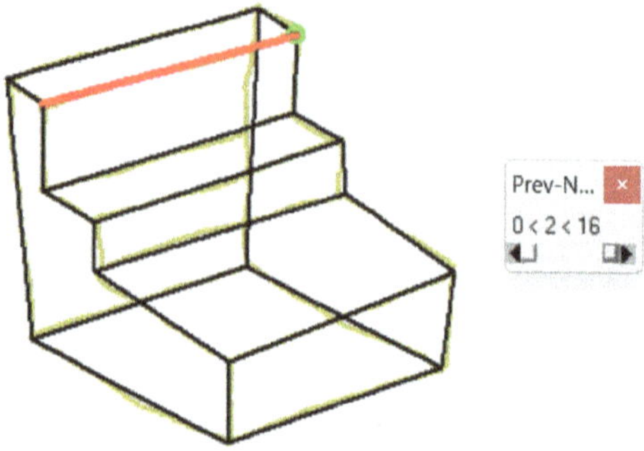

Figure 4.55. Starting point of spanning tree.

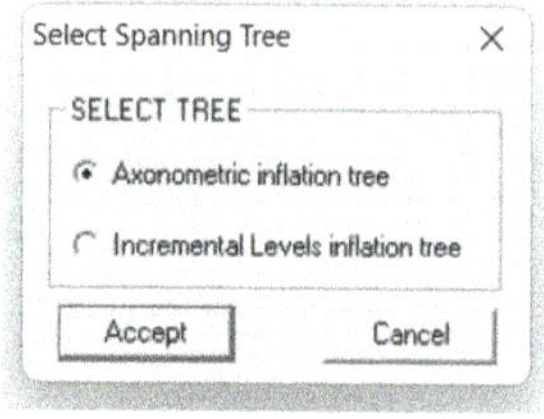

Figure 4.56. Spanning tree configuration.

4.7.4 *Spanning tree configuration*

REFER provides fine-tuning parameters to configure the details of the fitting procedure, which can be accessed through the *Details* button in the *2D Vectorization Configuration* Dialog box. The parameters are specific to each type of primitive.

REFER offers three direct inflation approaches, which are discussed in the *Direct Inflation Configuration* section. The random inflation approach does not require a spanning tree, whereas the other two approaches rely on different spanning trees. This dialog allows users to display either the axonometric or the level-based spanning tree, depending on the selected approach (see Figure 4.56).

There are three types of axonometric spanning trees. The original version, introduced in 2001, is retained for historical reference. The improved version from 2011 employs a recursive algorithm and allows the user to select either depth-first search (DFS) or breadth-first search (BFS) for vertex propagation. In this approach, incremental z-coordinates are assigned to "child" vertices connected to a "parent" vertex, provided they have not yet been visited. Child vertices are processed in order of decreasing distance from the parent.

The breadth-first search (BFS) approach is used exclusively for the level-based spanning tree. In this method, z-coordinates are assigned to all child vertices by either increasing or decreasing the z-coordinate of the parent vertex. As with the axonometric spanning tree, the child vertices are visited in order of their decreasing distance from the parent vertex.

4.8 Display Optimization Process

4.8.1 *Introduction and toolbar*

The purpose of the *Optimization* window is to monitor key visual information related to the optimization process. The *Optimization Process* toolbar, shown in Figure 4.57, provides access to a variety of diagrams and parameters for customizing them.

Before users can monitor the Optimization-Based Inflation process, it must first be executed. Note that the inflation optimization data overrides the drawing's original inflation data, i.e., this visualization is only available when the drawing has been refined, but the 3D model has not yet been inflated. The following commands are available to control the visualization of the optimization process:

- Display cost evolution and step length (for Hill-Climbing), or cost and temperature (for Simulated Annealing optimization)
- Display evolution of all Z variables
- Display evolution of some Z variables
- Display isolines of a design subspace

Figure 4.57. Optimization process toolbar.

The toolbar also includes commands to configure how the information is displayed:

- Display legends for regularities
- Display legends for isolines
- Open the Optimization Graphics Configuration dialog
- Open the configuration of the parameters to represent isolines of different design subspaces
- Zoom out
- Zoom in
- Zoom all
- Display pan controls.

4.8.2 *Display optimization cost evolution*

When configured accordingly, the *Optimization* window can display information related to the evolution of costs during the optimization-based inflation process. The cost diagrams are shown in Figure 4.58.

Two overlapping diagrams are displayed. The first diagram (corresponding to the left-side vertical axis) shows the cost evolution, typically

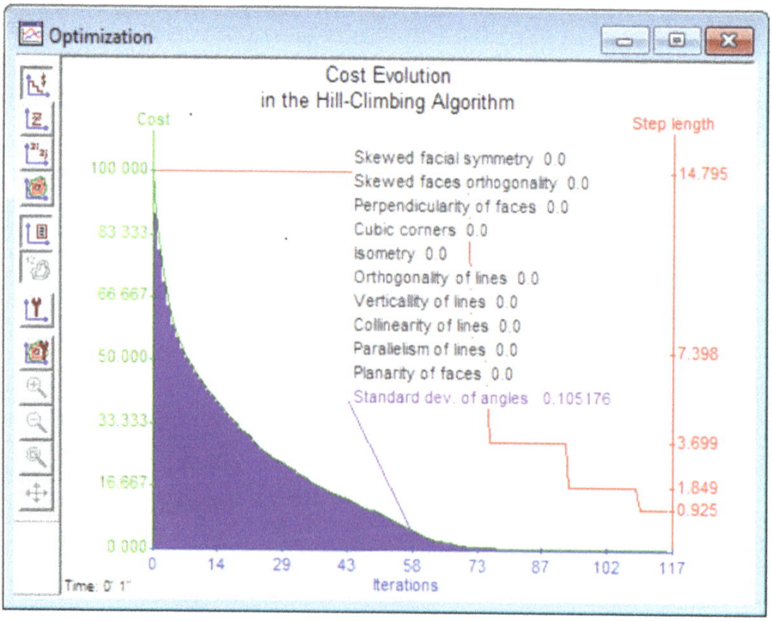

Figure 4.58. Display optimization cost evolution.

decreasing to zero or near zero. If multiple regularities are calculated, the cost for each is plotted, overlapping with the others.

In the second diagram (corresponding to the right-side vertical axis), red steps indicate the "step length," which represents the maximum allowed variation of the variables during each iteration. This step length gradually decreases over time. An example is shown in Figure 4.59.

4.8.3 *Display optimization evolution (all Z)*

When configured, the *Optimization* window can display the evolution of the Z coordinates for all the vertices in the line-drawing during the optimization-based inflation process, as shown in Figure 4.60. This diagram is typically used in conjunction with the View B-Rep Inflation feature.

4.8.4 *Display optimization evolution (some Z)*

When configured, the *Optimization* window can display the evolution of the Z coordinates for some of the vertices in the line-drawing during the optimization-based inflation process, as shown in Figure 4.61. The

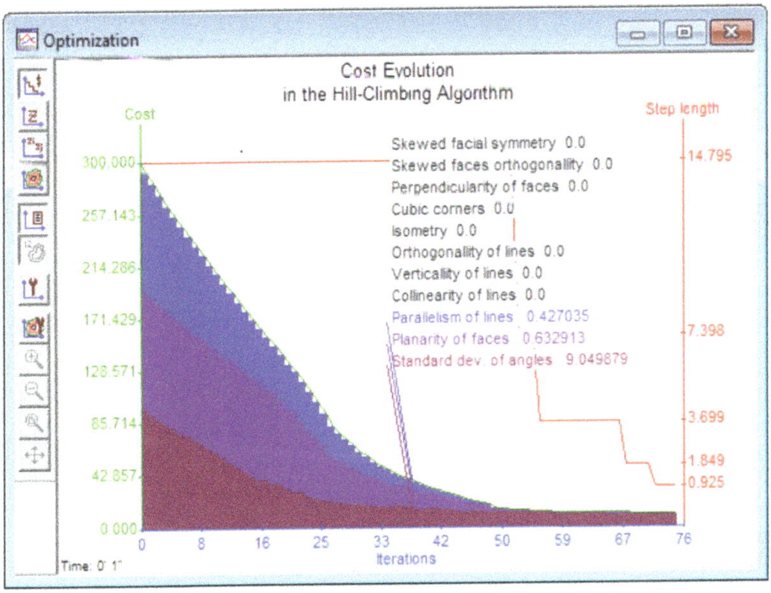

Figure 4.59. Display optimization cost evolution showing step length.

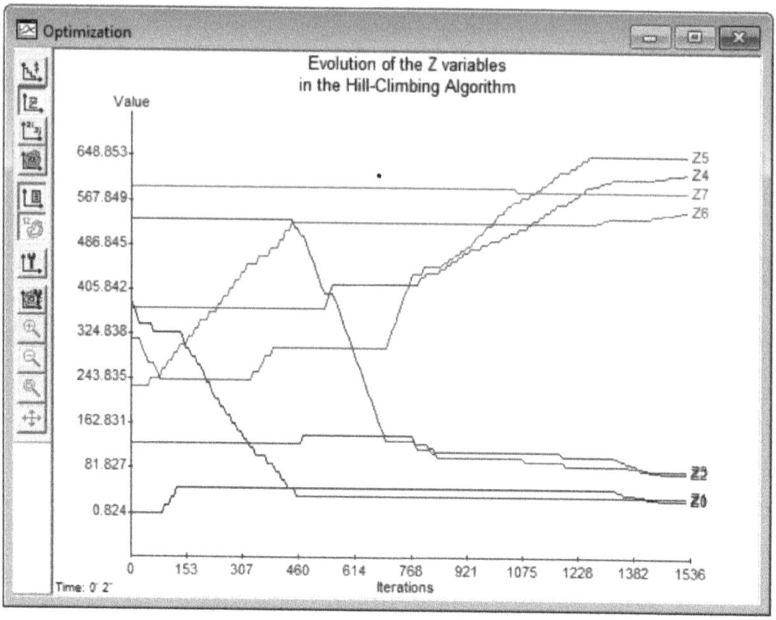

Figure 4.60. Display optimization evolution (all Z).

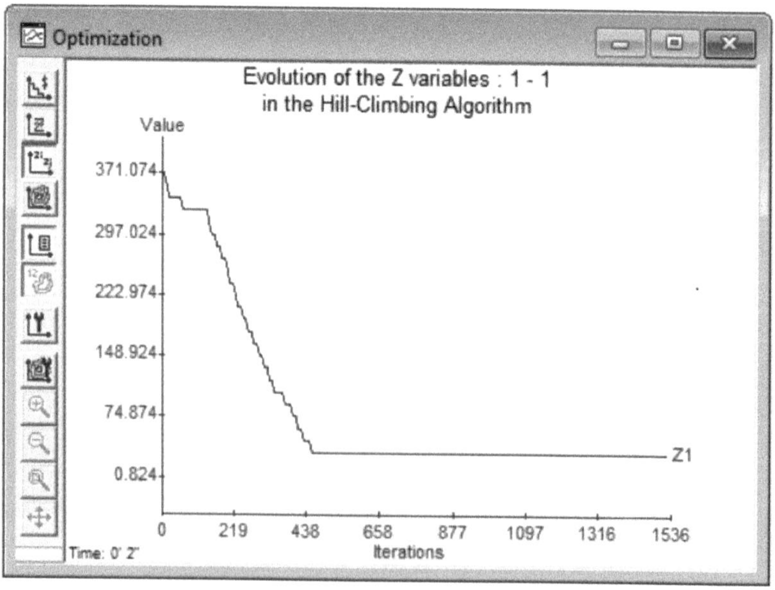

Figure 4.61. Display optimization evolution (some Z).

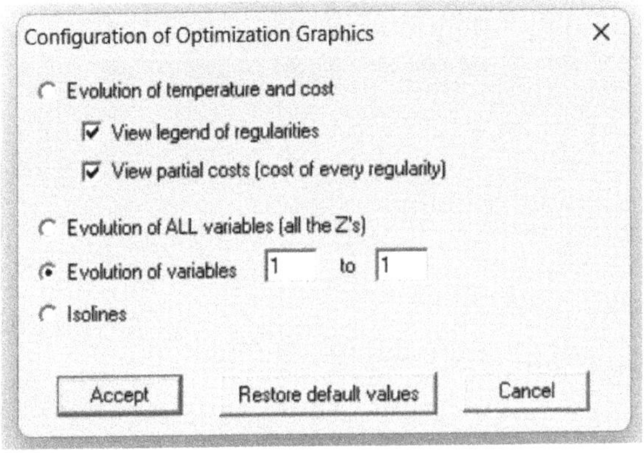

Figure 4.62. Optimization graphics configuration dialog.

Optimization Graphics Configuration command opens a dialog (see Figure 4.62) that enables the selection of the variables to visualize during the optimization process.

4.8.5 *Display optimization evolution isolines*

When configured, the *Optimization* window can display the isolines representation of a design sub-space during the optimization-based inflation process. The design space represents the set of feasible solutions to the optimization problem, specifically, the range of the objective function values as its dependent variables are varied.

Isolines are contour lines that connect points of equal value on a map. They provide a cartographic representation of numerical data, making it easier to identify geographic patterns and relationships. For this reason, isolines are well-suited for visualizing two-dimensional design subspaces. They help users intuitively recognize the maxima and minima of a function defined by two variables. However, when the function depends on more than two variables, only subspaces defined by a selected pair can be visualized at a time. An example of an isoline diagram of a cost function is shown in Figure 4.63.

Isolines illustrate how the value of the objective function changes as two controlling variables (specifically, the Z coordinates of vertices) vary within a defined range mapped on the horizontal and vertical axes. This

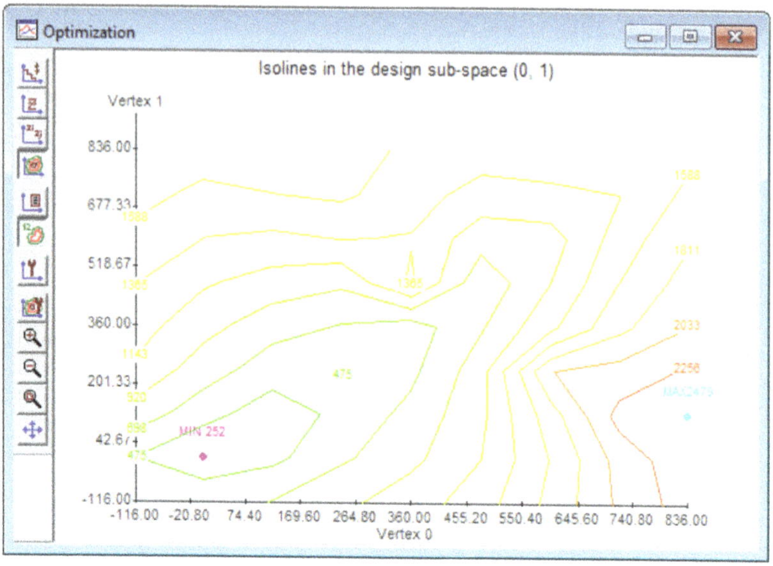

Figure 4.63. Display optimization evolution isolines.

visualization helps identify potential local minima that may exist between the current solution and the desired optimal solution.

The visualization can be adjusted by selecting a different variable pair to define the subspace, as well as modifying the range of values for each variable. Colors are used to indicate proximity to the optimum: isolines near the optimal value are shown in green, while those farther away appear in red. The display controls are located in the *Optimization Process* toolbar. It is important to note that computing design spaces is computationally intensive, so updating the visualization may result in significant delays.

4.8.6 *Optimization graphics configuration*

The *Optimization Graphics Configuration* dialog, shown in Figure 4.64 and available from the *Optimization Process* toolbar, allows the user to configure the type of information displayed in the *Optimization* window. It replicates the functionality of the commands that directly switch between displaying costs, variables, or isolines, while also offering additional configuration parameters:

- The cost display can be configured to include both the legend of regularities and the partial costs.

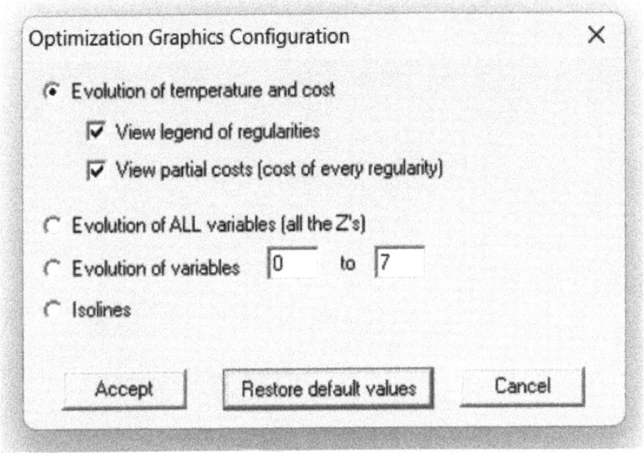

Figure 4.64. Optimization graphics configuration dialog.

- When displaying the evolution of variables, the specific variables to be shown can also be selected.

4.9 Display B-Rep Model

4.9.1 *Introduction and toolbar*

The primary purpose of the *B-Rep Model* window is to provide a visual interaction framework for the B-Rep model generated through inflation during the reconstruction process. This reconstruction must be completed before the B-Rep 3D model can be visualized. Additionally, the *B-Rep Model* toolbar, shown in Figure 4.65, enables the user to interactively display the various components of the B-Rep model and control the viewing perspective.

An alternative visualization of the B-Rep model is available via the *Display 2D Sketch* toolbar. Unlike the 3D view, this 2D representation does not support interactive viewpoint control through the Trackball. However, it provides access to additional model information, such as the 3D coordinates of the vertices.

The toolbar contains the following commands:

- View B-Rep Inflation.
- View B-Rep Inflation Dialog.
- View B-Rep Evolution.

Figure 4.65. B-Rep model toolbar.

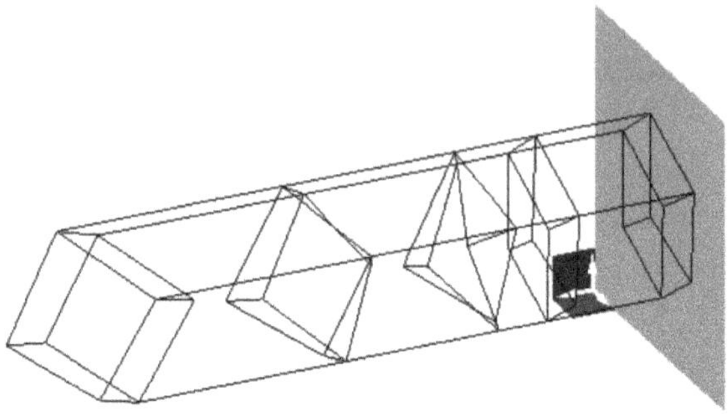

Figure 4.66. View exploded B-Rep inflation.

- Switch Necker Reversal.
- View B-Rep in Main Axes.
- B-Rep: FrontView.
- B-Rep: Top View.
- B-Rep: Side View.
- B-Rep: Isometric View.
- Display B-Rep Axes.
- Display B-Rep Edges.
- Switch To B-Rep with Thick Edges.
- Display B-Rep Faces.
- Display Triangulated Faces.
- Display Face Normals.

4.9.2 *View exploded B-Rep inflation*

This tool allows users to display some of the intermediate models produced during the iterative Inflation Optimization. These models are displayed in the B-Rep Model window, as shown in Figure 4.66.

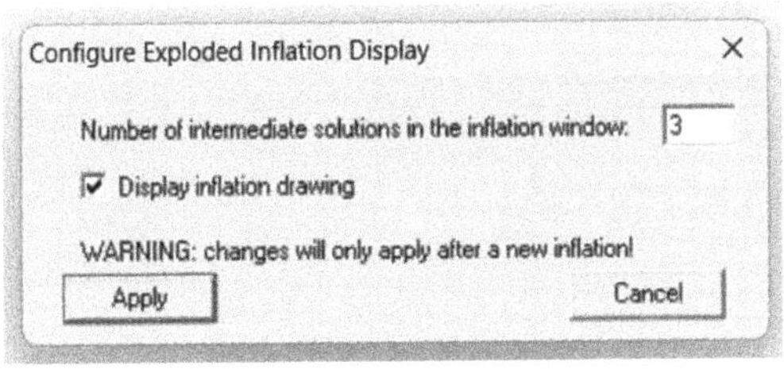

Figure 4.67. View exploded B-Rep inflation dialog.

Successive intermediate models are offset along the Z-axis to illustrate the progression of inflation—from the initial 2D drawing to the final 3D model—through several intermediate candidates. This visualization helps understand how the B-Rep model evolves during inflation and can reveal tangled shapes that may indicate local minima in the optimization process. The number of intermediate models displayed is controlled via the *View Exploded B-Rep Inflation* Dialog, shown in Figure 4.67.

4.9.3 *View B-Rep evolution*

This tool allows users to display the intermediate models produced while Optimization-Based Inflation is in progress. The intermediate models will be displayed in the *B-Rep Model* window

This visualization is optional, as users may consider it an unnecessary computational load during long or complex optimizations, given that it serves purely as a visual aid to illustrate the model's evolution. The command is only available when View B-Rep Inflation is inactive, as the two features are mutually exclusive.

4.9.4 *View necker reversal*

By inflating with positive Z coordinates, this approach assumes that the input sketch is not drawn from a worm's-eye perspective, effectively discarding one of the two geometrically valid Necker reversals. The *View Necker Reversal* option allows users to display this alternative geometrically valid interpretation. The reversal is achieved by inverting the sign of the Z coordinates for all vertices within the inflation reference system. An example is shown in Figure 4.68.

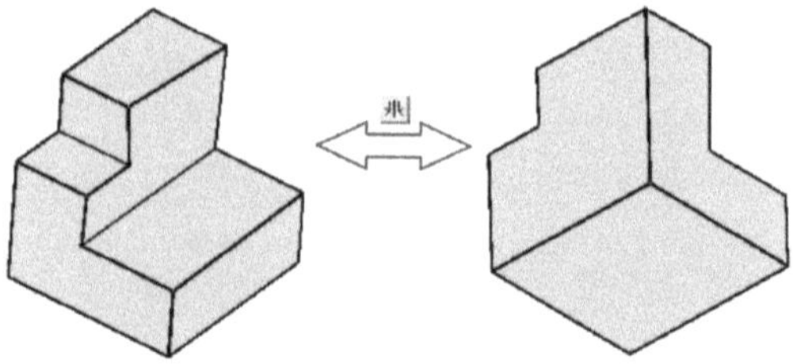

Figure 4.68. View Necker reversal.

Figure 4.69. View B-Rep in main axes.

4.9.5 *View B-Rep in main axes*

The orientation of the B-Rep model in the *B-Rep Model* window can be freely adjusted using the Trackball. Additionally, the user can select from four standard views: Front, Top, Side and Isometric. Since these views are defined relative to the inflation reference system (which can also be displayed), they may be of limited use for the viewer, except for the front view, which aligns the 3D shape with the original 2D line drawing. An example is shown in Figure 4.69.

If the main reference system has been calculated, the *View B-Rep In Main Axes* control allows switching the visualization of the B-Rep model

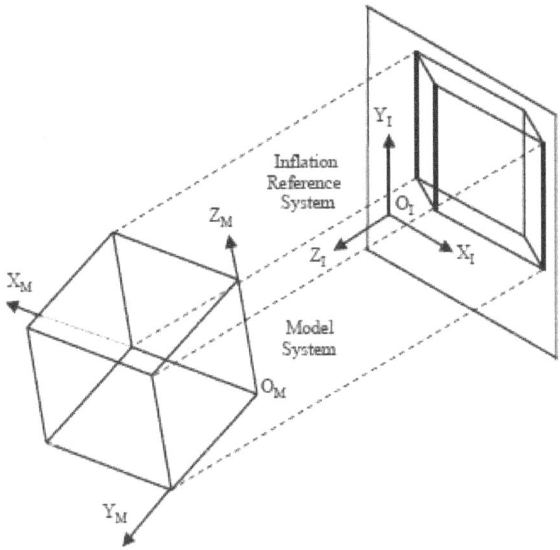

Figure 4.70. Switching between the inflation to the main reference system.

from the inflation reference system to the main reference system, as illustrated in Figure 4.70.

The standard orthographic views (see Figure 4.71) relative to the main reference system are generally more informative, as they often offer qualitative insights into the accuracy and quality of the reconstruction. The main coordinate system is displayed with the model, with the X-axis in red, the Y-axis in green, and the Z-axis in blue.

4.9.6 *B-Rep: Front, top, side, and isometric views*

These tools allow the user to display the front, top, side, and isometric views of the B-Rep Inflation Model in the *B-Rep Model* window. Note that the standard views are only meaningful after activating *View B-Rep In Main Axes*. Examples of these commands are shown in Figure 4.72.

4.9.7 *Display B-Rep axes*

Although humans can naturally perceive the volume depicted by the series of lines in a pictorial view (our brains are naturally wired to

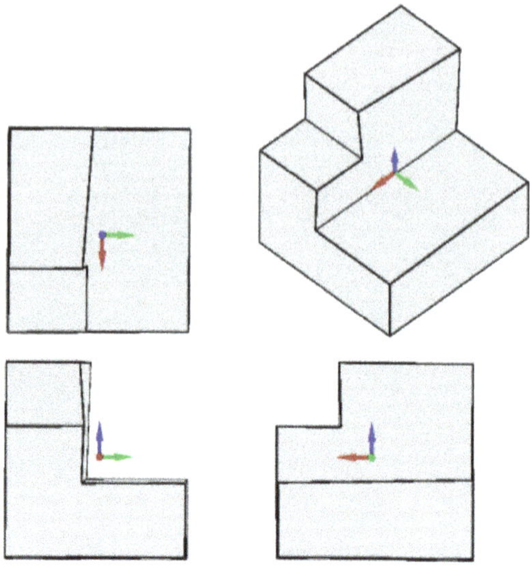

Figure 4.71. Standard orthographic views.

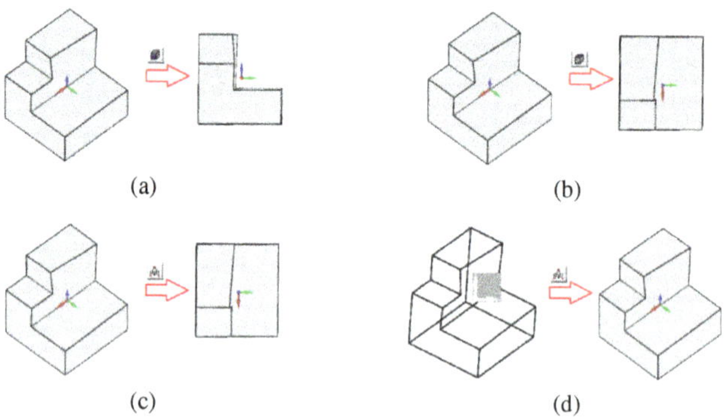

Figure 4.72. B-Rep: front (a), top (b), side (c), and isometric (d) views.

interpret drawings as depicting 3D shapes, even when we deliberately try not to), the image stored in the system is two-dimensional. Therefore, SBM systems typically work with up to four coordinate systems. The vectorization of the original sketch outputs a line-drawing that is still 2D and commonly referenced to the 2D drawing plane—the XY plane ($Z = 0$). In Figure 4.73, the plane is defined by the X_P and Y_P coordinates. In the subsequent reconstruction stage, geometric and perceptual

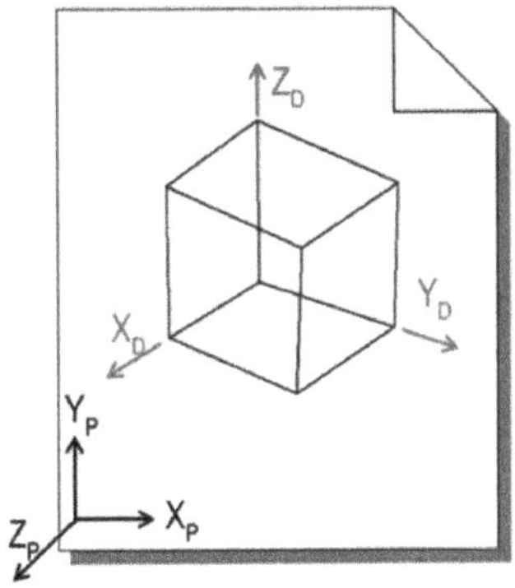

Figure 4.73. 2D drawing plane.

information is used to infer a consistent set of Z_P coordinates, thus inflating a 3D model (in reference to X_P, Y_P, Z_P).

Although we immediately perceive the three "main" directions indicated by the arrows (X_D, Y_D, Z_D), the computer can only infer this information after some successful reconstructions (see *Cue Main Axes*).

The *Display B-Rep Axes* command allows the user to display the coordinate system of the B-Rep Inflation Model in the *B-Rep Model window*. The coordinate axes display tool operates in two distinct modes:

- Inflation Axes are shown by default and represented as three orthogonal white arrows. They are accompanied by three reference planes: the XY plane (Z = 0) in grey, the XZ plane (Y = 0) in dark grey, and the YZ plane (X = 0) in blue, as shown in Figure 4.74.
- Main Axes are displayed when the *View B-Rep in Main Axes* option is enabled. In this mode, the orthogonal axes are color-coded: X-axis in red, Y-axis in green, and Z-axis in blue, as shown in Figure 4.75.

Because the origin of the coordinate system is located at the model's center of gravity, enabling face shading may result in part or all of the reference system symbol being obscured, as shown in Figure 4.76.

4.9.8 *Display B-Rep edges*

This tool allows the user to display the edges of the B-Rep Inflation Model in the *B-Rep Model* window, as shown in Figure 4.77. Disabling edge display has no effect when faces are not visible, as it would render

Figure 4.74. Inflation axes.

Figure 4.75. Main axes.

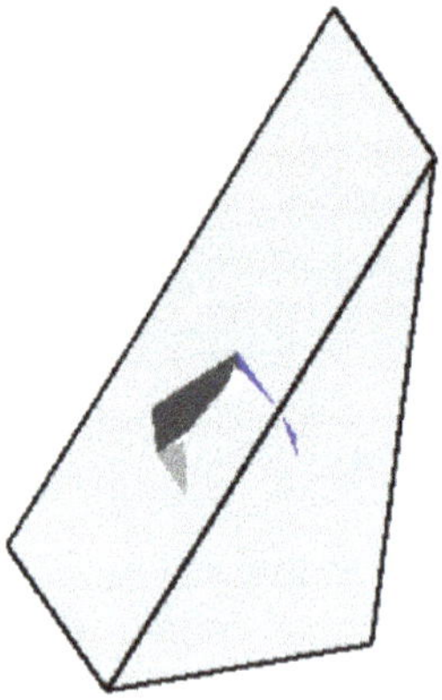

Figure 4.76. Obscured symbol due to face shading.

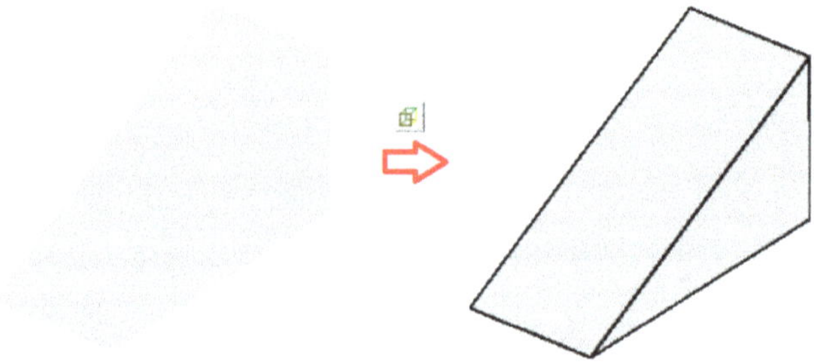

Figure 4.77. Display B-Rep edges.

the entire model invisible. This tool is primarily useful for hiding edges when shaded faces of the B-Rep model are shown, or more commonly, when viewing the triangulation without emphasizing individual face boundaries (see Figure 4.78).

4.9.9 *Display B-Rep with thick edges*

This tool allows the user to display the edges of the B-Rep model as thick lines in the *B-Rep Model* window, as illustrated in Figure 4.79. Edges are rendered as thin cylinders to prevent them from being fully occluded by the shaded faces when viewed from certain angles.

4.9.10 *Display B-Rep faces*

This tool allows the user to activate face shading mode in the B-Rep Inflation Model in the *B-Rep Model* window, as shown in Figure 4.80. The shading procedure relies on OpenGL and ensures that faces are flat and suitable for shading by automatically triangulating them. This process is typically unnoticeable to the user. The triangulation used for shading is

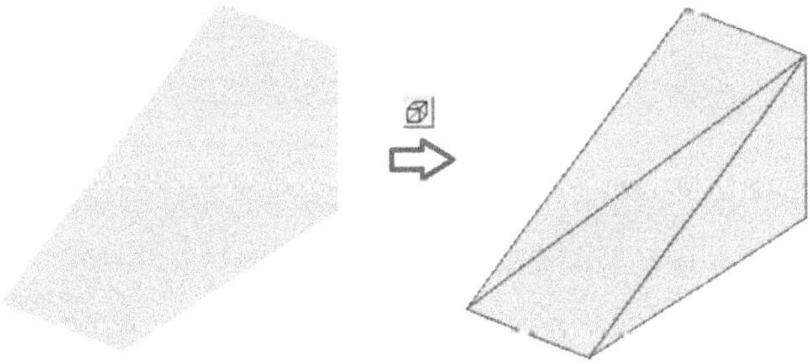

Figure 4.78. Display B-Rep edges to view triangulation.

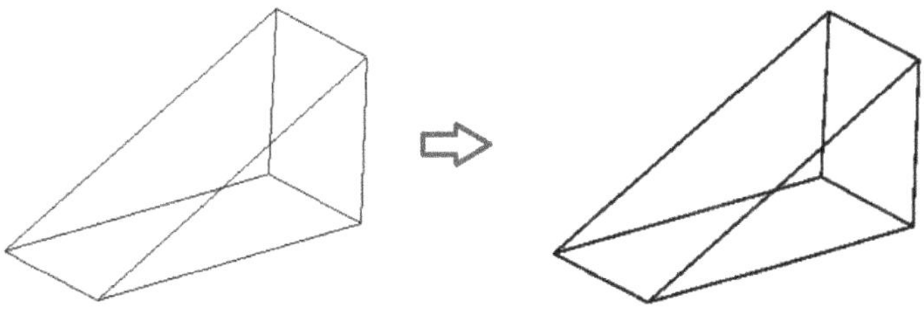

Figure 4.79. Display B-Rep with thick edges.

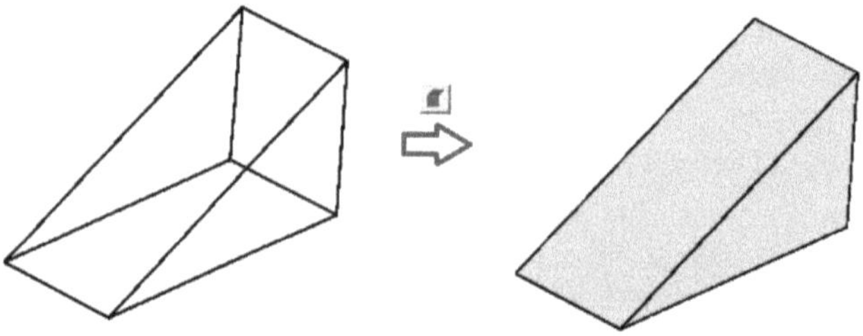

Figure 4.80. Display B-Rep faces.

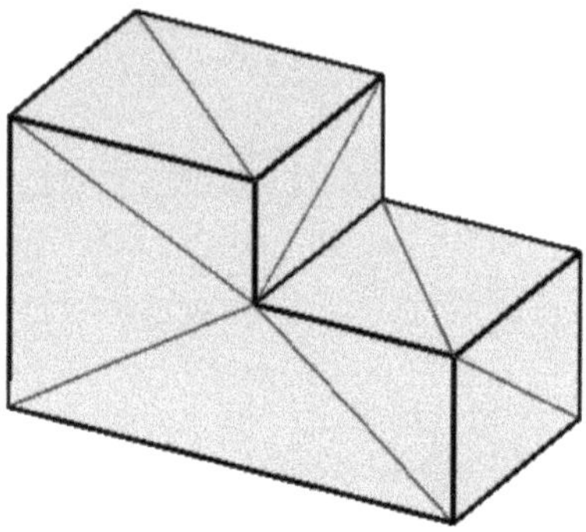

Figure 4.81. Display triangulated faces.

handled internally by OpenGL and may differ from the triangulation of the faces in the B-Rep model, which can be displayed separately at the user's discretion.

4.9.11 *Display triangulated faces*

The algorithm triggered by this command triangulates the faces of the B-Rep model, as shown in Figure 4.81. The method tessellates polygonal faces, converting them into sets of triangular facets, and returns triplets of vertex indices for each triangular facet. Faces are assumed to be planar, though they may have a slight tilt.

The approach triangulates simple polygons (both convex and concave) located in a two-dimensional plane, with vertices defined by two-dimensional points. The triangulation is performed on the XY projection of the face, which simplifies the process. The result remains invariant to projections, except in cases of degenerate projections, which should not occur since the faces were originally detected from a flat image represented in the XY plane. The approach requires the vertices of the polygon to be arranged in counterclockwise (CCW) order. REFER's approach is adapted from Moreno's C++ implementation (Moreno, 1998c).

The algorithm identifies convex vertices within the input polygon. For each detected convex vertex, it forms a triangle using the following points:

- V_{prev}: the vertex immediately preceding the convex vertex;
- V_{curr}: the convex vertex itself;
- V_{next}: the vertex immediately following the convex vertex.

If the triangle formed does not enclose any other vertex of the input polygon, it is guaranteed that the segment (V_{prev}, V_{next}) does not intersect any polygon edge. In that case, the triangle is added to the output, and the vertex V_{curr} is removed from the polygon. If the triangle does contain other polygon vertices, the algorithm searches for one such vertex V_s that can be connected to V_{curr} without intersecting any edges. The segment (V_{curr}, V_s) then splits the polygon into two simpler sub-polygons. The triangulation function is recursively applied to each of these sub-polygons. The recursion stops when a polygon has only three vertices, i.e., it is already a triangle.

Identifying a suitable vertex that can be connected to V_{curr} without intersecting the input polygon is challenging. A common but flawed strategy is to simply select the vertex closest to V_{curr}. This intuitive approach often fails, particularly when the resulting triangle intersects the polygon. A more robust method is to identify the vertex with the greatest perpendicular distance from the line defined by the segment (V_{prev}, V_{next}). As noted by Moreno (1998c), "to obtain this point the algorithm does not directly compute a single distance (it would be too hard, since we are talking about perpendicular distance). Instead, the algorithm finds the point P for which the area of the triangle (V_{prev}, V_{next}, P) is maximized."

4.9.12 *Display face normals*

REFER provides the capability to display the 3D normals of the B-Rep model, once it has been successfully inflated into a 3D representation. An example is shown in Figure 4.82.

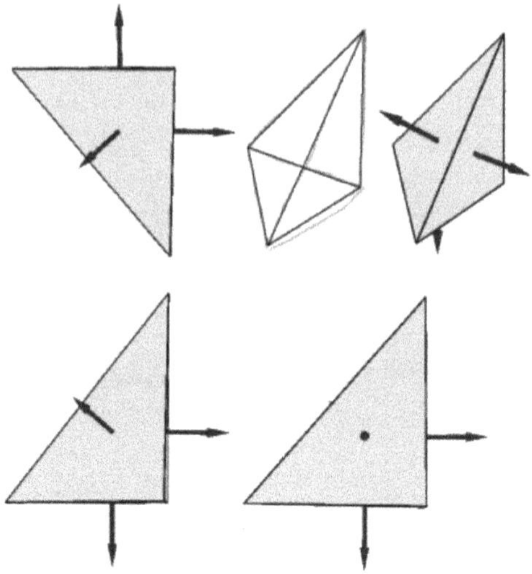

Figure 4.82. Display face normals.

The display of normals can be enabled whenever the model is visualized in the inflation coordinate system. This option is automatically disabled when switching to the main coordinate system. If triangulation has been computed and its display is enabled, a normal vector is shown for each triangular facet, as illustrated in Figure 4.83.

4.10 Display CSG Model

4.10.1 *Introduction and toolbar*

The primary function of the *CSG Model* window is to provide a framework for visual interaction with the CSG model reconstructed during the reconstruction phase. Access to the 3D CSG model visualization is only possible once this reconstruction has been completed. To support this interaction, the window includes the *Display CSG Model* toolbar, which enables the user to selectively visualize different components of the CSG model, along with complementary information such as the datum graph.

The *Display CSG Model* toolbar, shown in Figure 4.84, provides access to a variety of diagrams and configuration parameters.

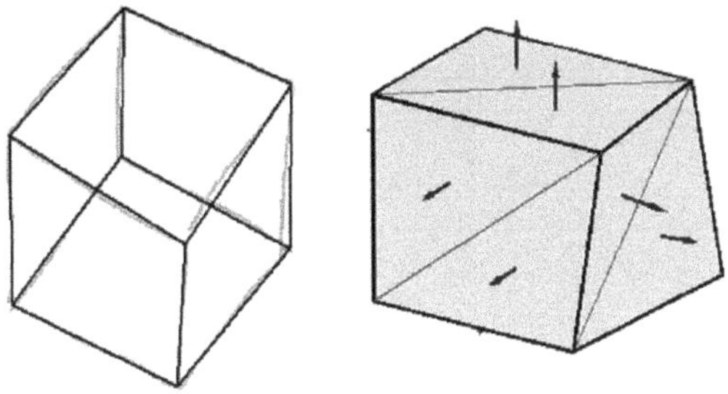

Figure 4.83. Normal vectors in triangulated facets.

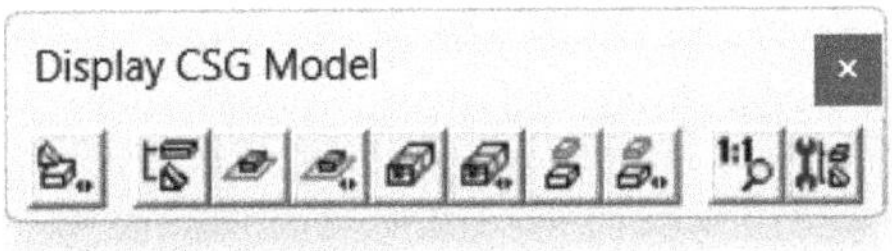

Figure 4.84. Display CSG model toolbar.

The toolbar contains the following commands:

- Display CSG features in sequence (When this button is activated, all other visualization options in the toolbar are turned off. To use the other options again, the button must first be deactivated.)
- Display CSG Tree.
- Display CSG Datums.
- Display CSG Datums in Sequence.
- Display CSG Profiles.
- Display CSG Profiles in Sequence.
- Display CSG Deconstruction.
- Display CSG Deconstruction in Sequence.
- Display CSG Zoom All: This option resets zoom and pan, returning the drawing or model to its full view.
- Display CSG Configuration Dialog.

4.10.2 *Display CSG features*

The *Display CSG Model Features* command toggles the visualization of the features automatically reconstructed as 3D B-Rep models using axonometric inflation. When the command is activated, the first feature is displayed, and a dialog appears to allow the user to navigate to the next or previous feature. The dialog can be closed at any time by clicking the close button in the upper-right corner or by deactivating the command. An example is shown in Figure 4.85. Each inflated B-Rep model can be viewed in 3D, allowing the user to rotate the model with the trackball, as well as zoom and pan.

If CSG automatic inflation is inactive (or fails to execute successfully) only the 2D drawing of the current feature is displayed. The purpose of this visualization is to verify that the CSG deconstruction produces features that can be reconstructed and match the shape perceived by humans in the 2D drawing. Additionally, CSG feature information can be saved as DXF files, which can then be exported to other applications or reimported individually into REFER for detailed checks or further reconstruction operations.

4.10.3 *Display CSG tree*

The *Display CSG Tree* command toggles the visualization of the model tree, which reflects the datums and features identified during deconstruction of the input drawing. The purpose of this visualization is to show the parent–child relationships established during reconstruction. It sequentially displays all the features (including the core) extracted during the deconstruction of the 2D drawing and automatically reconstructed as 3D B-Rep models using axonometric inflation.

Features, which are recognized during the feature reconstruction stage, can also be displayed individually using the *Feature Display*

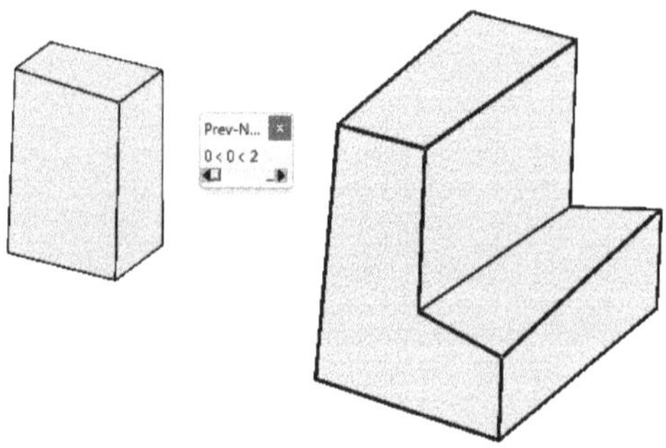

Figure 4.85. Navigating the features of a CSG model.

commands in the *2D Sketch* window. Once identified during the CSG modeling stage, features are organized according to their parent–child relationships and the branches to which they belong. This tree structure can be visualized in this window. An example is shown in Figures 4.86 and 4.87.

The default configuration arranges the information in two columns, the core and nodes in the left column and the features in the right column. Users can change this arrangement in the configuration dialog, which also controls the labels, which provide information such as the number of branchcs, feature names, and the merit assigned to each feature during detection.

4.10.4 *Display CSG datums*

The *Display CSG Datums* command toggles the visualization of the datums extracted during the deconstruction of the input drawing to generate the embedded CSG model. All datums identified during reconstruction are displayed in the window, overlaid on the datum graph. An example is shown in Figure 4.88.

The configuration dialog allows users to control the display of labels, which provide information such as datum number or whether it is a symmetry datum. This visualization is intended to show the relationships between datums. To analyze individual datums in detail, it is recommended to use *the Display CSG Datums in Sequence* command.

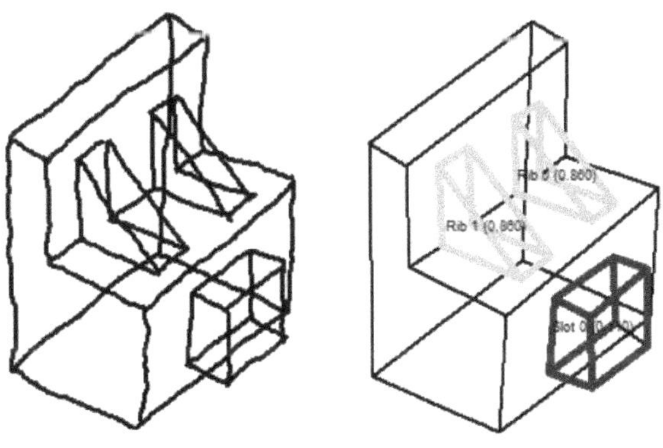

Figure 4.86. Initial sketch (left) and sketch with recognized features highlighted (right).

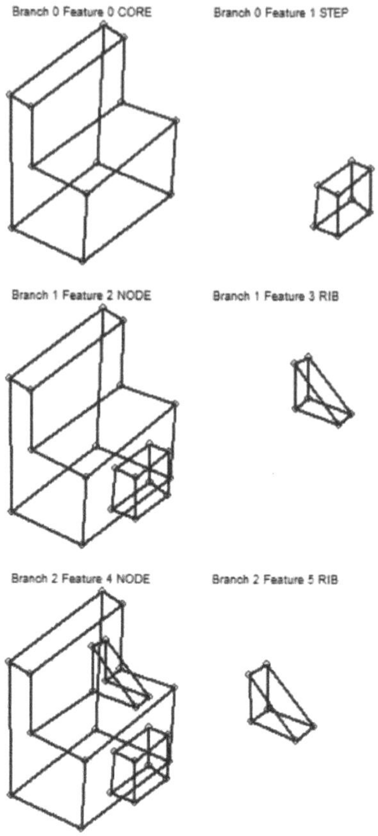

Figure 4.87. CSG model tree visualization.

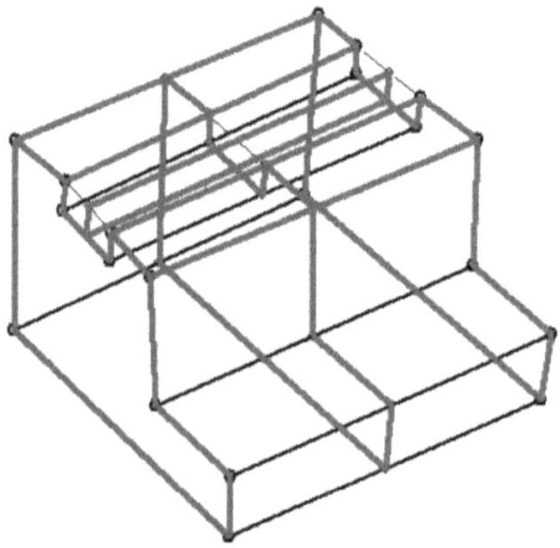

Figure 4.88. Display CSG datums.

4.10.5 *Display CSG datums in sequence*

The *Display CSG Datums in Sequence* command toggles the visualization of the datums extracted during the deconstruction of the input drawing to generate the embedded CSG model.

While the *Display CSG Datums* command shows all datums at once, which can cause overlapping and make the drawing hard to read (see Figure 4.89), the sequential display solves this issue by presenting each datum individually. When this command is activated, only the first datum is displayed, and an auxiliary dialog appears to allow navigation to the next or previous datum, as shown in Figure 4.90. This visualization is intended to display the datums produced during the reconstruction process, as illustrated in Figure 4.91.

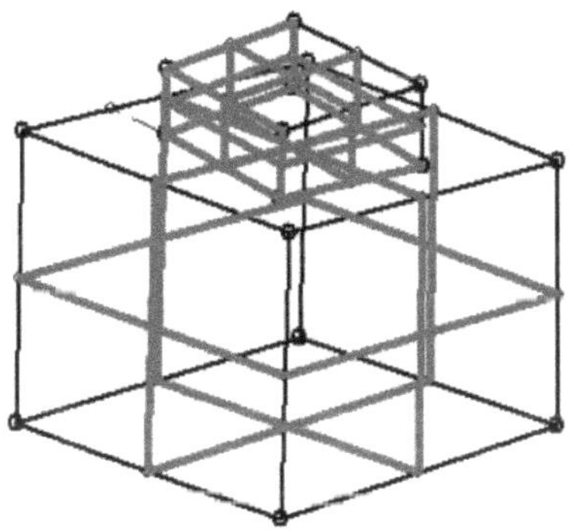

Figure 4.89. Overlapping of datums when shown all at once.

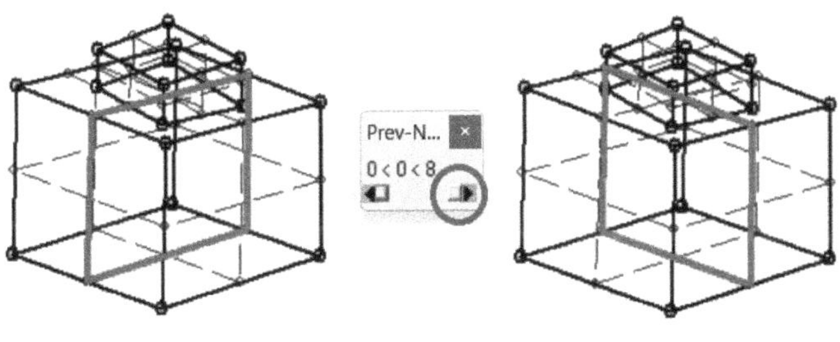

Figure 4.90. Display CSG datums in sequence.

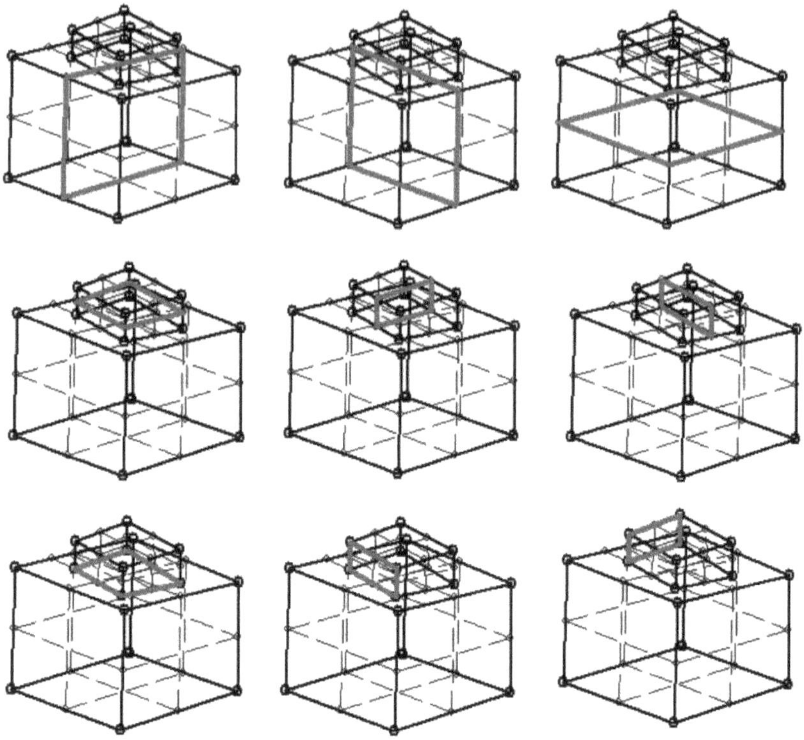

Figure 4.91. Visualization of datums.

This type of visualization is also useful to examine the datum graph used internally to determine the relative positions of the datums. For example, Figure 4.92 shows a "bridge" in the datum graph connecting two disjoint subgraphs in a representation of a block with a step.

4.10.6 *Display CSG profiles*

The *Display CSG Profiles* command toggles the visualization of the profiles extracted during the deconstruction of the input drawing to generate the embedded CSG model. All profiles identified during reconstruction are displayed in the window, overlaid on the datum graph. An example is shown in Figure 4.93. This visualization shows the relationships between profiles. To analyze individual profiles in detail, it is recommended to use the *Display CSG Profiles in Sequence* command.

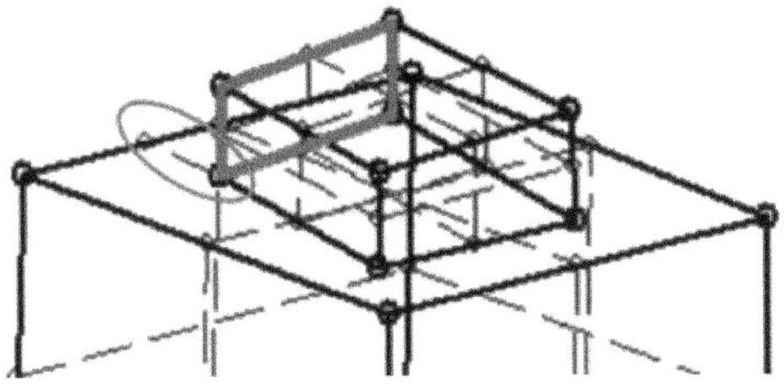

Figure 4.92. "Bridge" in the datum graph connecting two disjoint subgraphs of a block with a step.

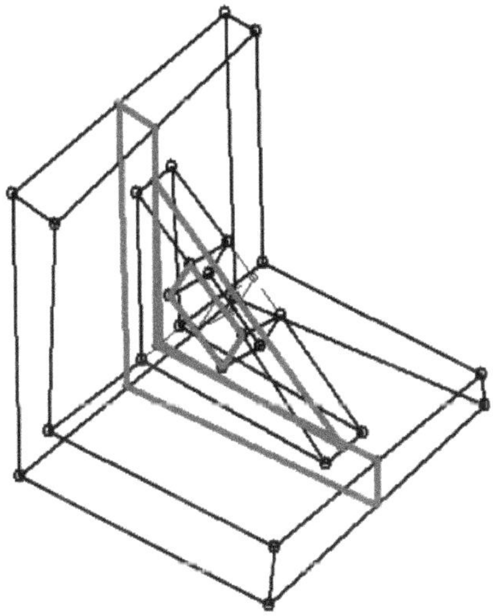

Figure 4.93. Display CSG profiles.

4.10.7 *Display CSG profiles in sequence*

The *Display CSG Profiles in Sequence* command toggles the visualization of the profiles extracted during the deconstruction of the input drawing to generate the embedded CSG model. When activated, only the first profile is displayed, and a dialog provides navigation to the next or previous profile. This visualization shows the profiles generated during reconstruction. It is also useful for reviewing the datum graph, which is used internally to determine the relative positions of profiles and datums. An example is shown in Figure 4.94.

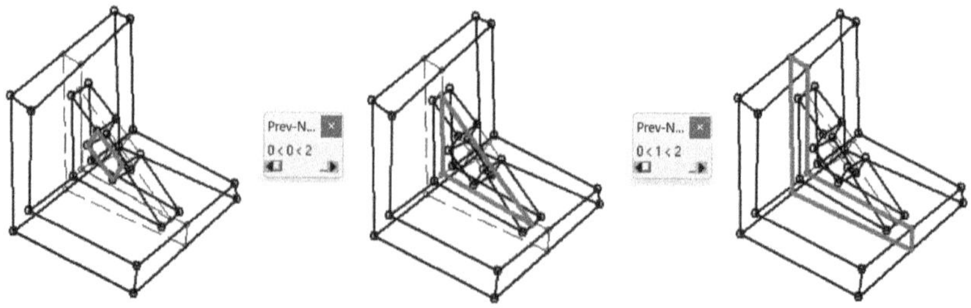

Figure 4.94. Display CSG profiles in sequence.

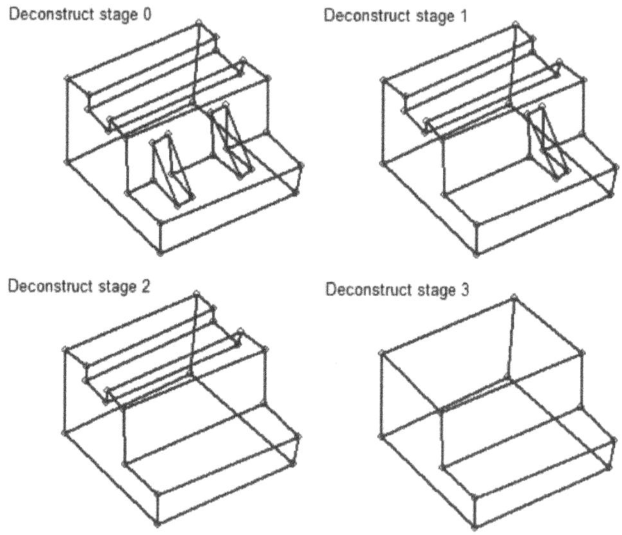

Figure 4.95. Display CSG deconstruction.

4.10.8 *Display CSG deconstruction*

This visualization shows the sequence of 2D drawings generated during the CSG deconstruction process. The *Display CSG Deconstruction* command toggles the visualization of the successive deconstructed 2D drawings generated by removing features from the input drawing to create the CSG model. All deconstructed drawings are displayed in the window, arranged as successive "vignettes," as shown in Figure 4.95.

4.10.9 *Display CSG deconstruction in sequence*

The *Display CSG Deconstruction in Sequence* command toggles the visualization of the deconstructed drawings generated by successively removing features from the input drawing to create the CSG model. When activated, only the first drawing is displayed, and a dialog allows the user to navigate to the next or previous drawing, as illustrated in Figure 4.96.

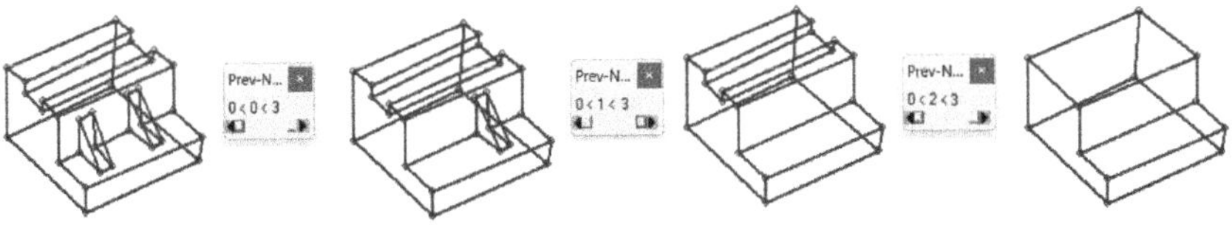

Figure 4.96. Display CSG deconstruction in sequence.

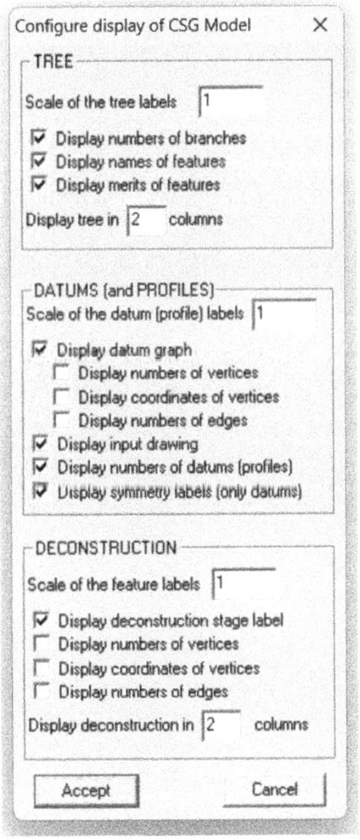

Figure 4.97. Display CSG configuration dialog.

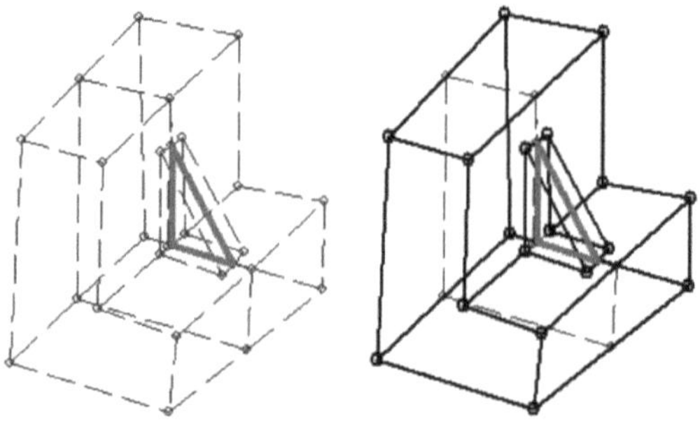

Figure 4.98. Datums and profiles superimposed on the datum graph.

4.10.10 *Display CSG configuration dialog*

The *Display CSG Configuration* dialog, shown in Figure 4.97, includes various parameters that control the display of CSG information. The dialog is divided into three sections, each dedicated to controlling the display of the model tree, the datums and profiles, and the deconstruction process.

The parameters allow users to control the size and type of labels displayed for each visualization type. For the display of the model tree and deconstruction sequence, additional parameters are provided to select the number of rows and columns for arranging the displayed images.

Datums and profiles are displayed over a representation of the datum graph. By default, the input drawing is also superimposed in the foreground of the graph. An example is shown in Figure 4.98.

Chapter 5
Output

5.1 Introduction

REFER is designed as a stand-alone application for generating 3D models reconstructed from 2D sketches. However, its primary purpose is to serve as a testbed for exploring ideas in Sketch-Based Modeling (SBM). As such, the reconstruction flow is highly configurable and requires close monitoring.

The resulting 3D models are visualized using simplified renderings that facilitate quality assessment. A wide range of display options is available to support experimentation with different reconstruction configurations.

In addition to saving final 3D models, most intermediate outputs, such as sketches and vectorized line drawings, can also be saved in different formats. The *Save* command is used for exporting 2D information, whereas *Save As* allows for saving both 2D and 3D data. The requested file format implicitly determines the type of information to be exported.

5.2 Save

The *Save* command is available in the REFER's main toolbar, as shown in Figure 5.1.

The command is used to automatically output information from REFER, based on the following criteria:

(1) If the current pathname (the internal variable that stores the path and filename of the most recently opened file) includes the name of a previously opened file, REFER updates that file with the current

Figure 5.1. *Save* command.

information. Depending on the file extension, the command operates in one of three modes:

(a) If the extension is *.stk*, the current sketch is saved.
(b) If the extension is *.Sketch*, the vectorized line drawing is saved in RIBALD format.
(c) For any other extension, the current line drawing is saved as a *DXF* file. DXF files saved by REFER include the string *(Rf)* immediately before the file extension to help identify them as output generated by the application.

(2) If the current pathname is empty, typically because a new sketch was created directly within REFER, a default name is assigned (e.g., "Refer" followed by a natural number), and the data is saved according to the current mode. If the mode is sketching, the current content is saved as a sketch file (*stk*). Otherwise, the content is assumed to be a line drawing in the reconstruction stage and saved as a *DXF* file.

5.3 Save As

The *Save As* command is available in the REFER's main toolbar, as shown in Figure 5.2.

The command opens an interactive dialog that enables the user to specify the file name and location, and the type of information to be saved, as shown in Figure 5.3.

Figure 5.2. *Save As* command.

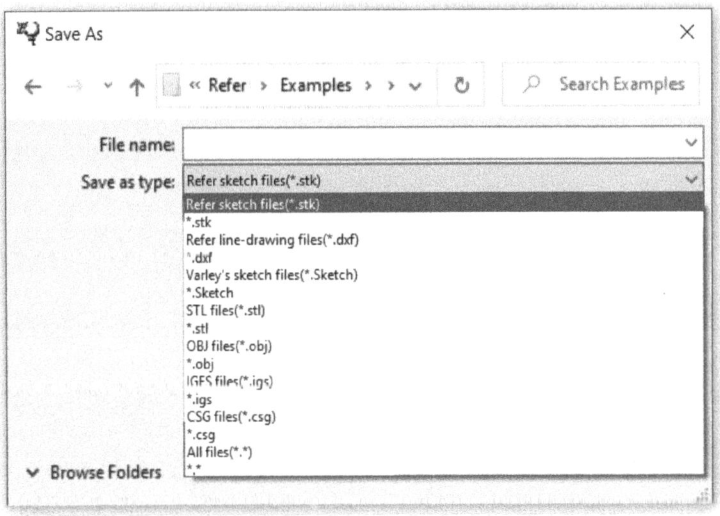

Figure 5.3. *Save As* dialog.

By selecting a specific file extension, users can save the current information available in REFER, corresponding to the various input file formats associated with the application:

- If the file extension is *.stk*, the sketch is saved.
- If the file extension is *.Sketch*, the vectorized line-drawing is saved as a RIBALD file.
- If the file extension is *.dxf*, the line-drawing is saved in *DXF* format.

The *Save As* command is also used to save reconstructed 3D models as polygonal meshes (STL or OBJ formats), B-Rep models (IGES format), and CSG (REFER's custom format).

5.4 3D Model

A CAD model is a mathematical representation of an object, used to describe and predict its behavior before it is physically built. According to ISO 10303-43, such representations are structured collections of interrelated data elements, organized for specific purposes. CAD models are expressed using formal languages that must conform to an underlying representation scheme. In the context of Mechanical CAD, two representation schemes dominate contemporary systems, as defined in ISO 10303-108: explicit and procedural.

Hybrid models result from combining both explicit and procedural representation methods. Hoffmann (1989) refers to these models as *dual-representation* models. He argues that the tendency to combine both representations stems from the complementary strengths and limitations of each approach. CSG models are procedural and naturally define solid volumes. In contrast, B-Rep models explicitly define surfaces and edges, allowing for efficient rendering and geometric computations.

A CAD model is stored within a document, defined in ISO 11442 as a fixed and structured collection of information that can be managed and exchanged as a unit between users and systems. Representation schemes give rise to *file formats*, which are standardized structures for encoding information suitable for computer storage. However, the relationship between schema and format is not one-to-one: depending on specific goals and implementation strategies, multiple grammars may be used to represent the same schema, leading to the development of alternative file formats.

5.5 Explicit 3D Models

Models are considered explicit (also known as *declarative* or *evaluated*) when all geometric and topological details are directly available without requiring further computation. These models are sometimes referred to as "dumb" models, underscoring the absence of embedded information about design intent or the procedural history used to create or modify the shapes.

Explicit models can be exported from REFER in a variety of formats, including STL, OBJ, and IGS. These formats are widely supported by commercial CAD applications and can serve as input for downstream design, analysis, or manufacturing processes.

For formats that support only triangular faces, such as the STL representation, face triangulation is required. This process reduces high-level, semantically rich polygonal faces to low-level tessellated approximations, resulting in a loss of geometric and topological intent. To retain face information, the representation must meet at least the minimum topological requirements of a polygonal mesh.

Finally, to enhance both compatibility and generality, REFER can output reconstructed 3D models using the B-Rep representation. This is a specific type of explicit model in which connected sets of geometric elements define the vertices, edges, and faces that form the boundary of a solid. While the geometric data specifies the precise shape and spatial location of each element, the topological data captures the relationships between them.

Generally speaking, B-Rep models are composed of patches, which are segments of curved surfaces bounded by curved contours. These contours consist of curve segments that are delimited by vertices, which in turn correspond to specific points in space. For example, a simple cylinder can be represented using three patches: a cylindrical surface patch for the side, and two planar circular patches for the top and bottom faces, as shown in Figure 5.4.

There seems to be a notable terminological inconsistency in the literature regarding the use of the terms patch and face. In some sources, faces refer broadly to any bounded portion of a surface—whether planar, cylindrical, or arbitrarily curved. In contrast, other authors reserve the term *face* strictly for planar regions, using patch to describe curved surface segments. This discrepancy can lead to confusion, particularly in cases where the boundary of a surface region appears polygonal, but its vertices do not lie exactly in the same plane. Such configurations result in "tangled" faces, which are technically curved patches, even if they resemble planar polygons. An example is shown in Figure 5.5.

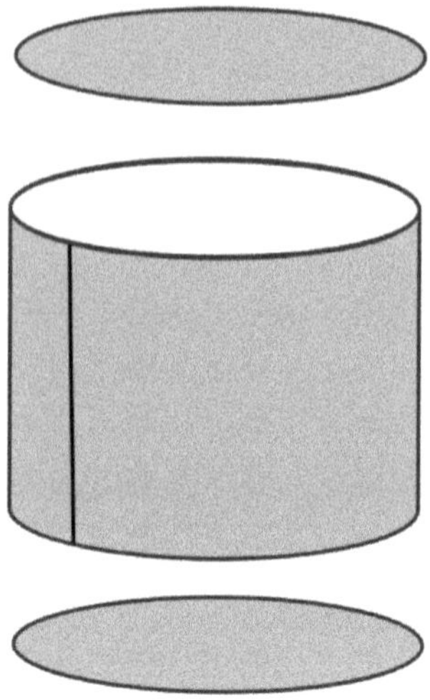

Figure 5.4. Patches that make a B-Rep cylinder.

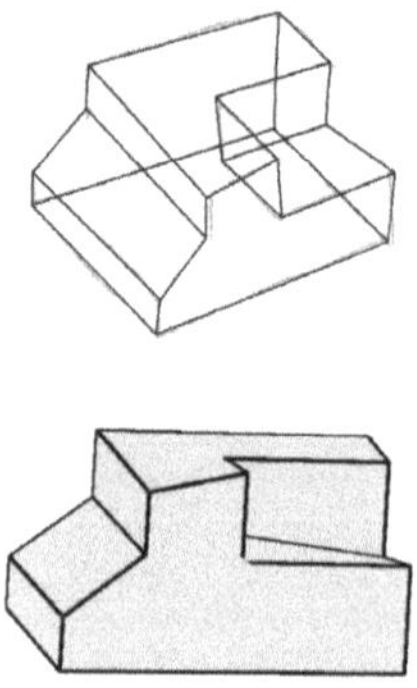

Figure 5.5. Polygon with non-planar patches.

5.6 Polygonal Mesh Output

A polygonal mesh is composed of two or more surface patches bounded by polygonal contours and joined together to form a coherent shape. REFER supports two alternative file formats for exporting polygonal-mesh models:

```
solid
   facet normal  0.000000  0.000000  -1.000000
      outer loop
         vertex  0.000000  0.000000  0.000000
         vertex  0.000000  10.000000  0.000000
         vertex  10.000000  0.000000  0.000000
      endloop
   endfacet
   facet normal  -1.000000  0.000000  0.000000
      outer loop
         vertex  0.000000  0.000000  0.000000
         vertex  0.000000  0.000000  10.000000
         vertex  0.000000  10.000000  0.000000
      endloop
   endfacet
   facet normal  0.000000  -1.000000  0.000000
      outer loop
         vertex  0.000000  0.000000  0.000000
         vertex  10.000000  0.000000  0.000000
         vertex  0.000000  0.000000  10.000000
      endloop
   endfacet
   facet normal  1.000000  1.000000  1.000000
      outer loop
         vertex  10.000000  0.000000  0.000000
         vertex  0.000000  10.000000  0.000000
         vertex  0.000000  0.000000  10.000000
      endloop
   endfacet
endsolid
```

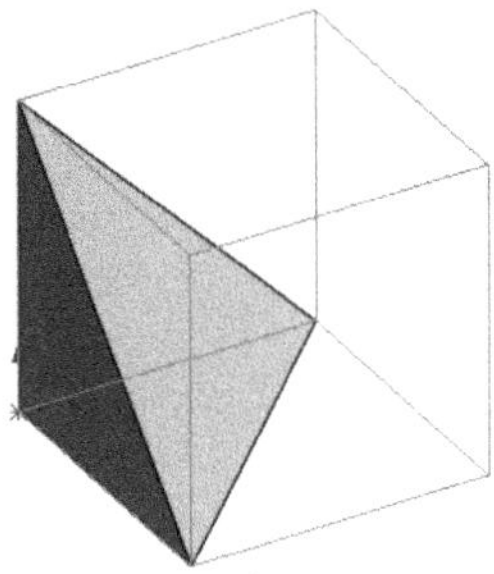

Figure 5.6. STL representation of a tetrahedron.

- STL files are limited to representing surfaces through triangles only (and not polygonal meshes directly). Therefore, REFER's B-Rep models must first be triangulated before they can be exported in STL format. The structure of the STL file is described as follows:

facet normal i j k
    outer loop
        vertex $x_i\ y_i\ z_i$
        vertex $x_j\ y_j\ z_j$
        vertex $x_k\ y_k\ z_k$
    endloop
endfacet

An example of the STL representation of a tetrahedron is shown in Figure 5.6.

- OBJ files support the representation of polygonal faces. By default, vertices are stored in counterclockwise order, which makes the explicit declaration of face normal optional.

Models are exported exactly as produced by REFER, including any errors or imperfections resulting from the reconstruction process. The example shown in Figure 5.7 illustrates a wedge-shaped input sketch with numbered vertices and edges, which was automatically reconstructed into a refined polyhedral model. The resulting polygonal model contains minor geometric imperfections, which are evident in the corresponding ASCII representation of the OBJ file, as shown in Figure 5.8.

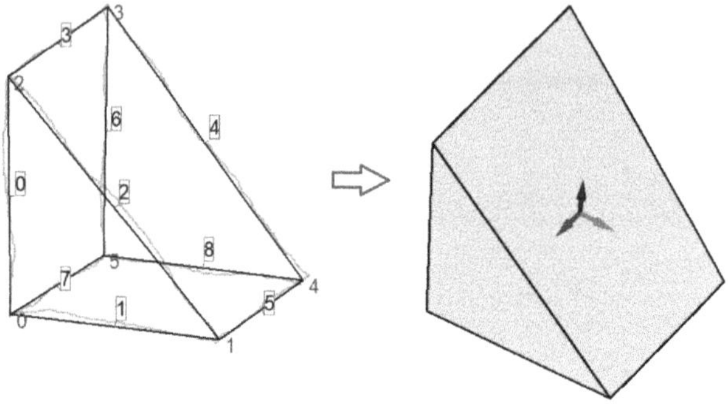

Figure 5.7. A wedge-shaped input reconstructed in 3D.

```
#solid (obj format)
#vertices
v  226.157838     0.000000     0.000000
v  226.157838   195.144451     0.000000
v  226.157838     0.000000   217.600568
v    0.000000     0.000000   216.842443
v    0.000000   195.281019     0.000000
v    0.000000     0.000000     0.000000
#normals (one per face)
vn  1.000000      0.000000    -0.000000
vn  0.000000     -1.000000     0.000000
vn -0.000000      0.000000    -1.000000
vn -0.002238      0.744475     0.667646
vn -1.000000      0.000000     0.000000
#faces
f     3//1     1//1     2//1
f     6//2     1//2     3//2     4//2
f     6//3     5//3     2//3     1//3
f     4//4     3//4     2//4     5//4
f     6//5     4//5     5//5
#endsolid
```

Figure 5.8. ASCII representation of the reconstructed 3D model shown in the previous figure.

5.7 B-Rep Output

B-Rep is a generic term that refers to various mathematical representations of surfaces, all sharing the common goal of using high-semantic geometric information to describe shape. This means avoiding breakdown, approximation, or oversimplification of geometry. B-Rep models are typically constructed from polysurfaces—sets of adjacent surface patches sewn together along their perimeters. This strategy reduces mathematical complexity while enhancing design intent by allowing independent control over the shape of each patch and the relationships between neighboring patches. The acronym *B-Rep* stands for "boundary representation," since solid objects are defined as the space enclosed by a closed surface or polysurface. This approach offers a compact and efficient way to represent volumes by storing only boundary data and omitting interior details.

Various mathematical representations exist for surfaces, each with its own strengths, variants, and configuration options. Non-Rational B-Spline Surfaces (NURBS) are widely regarded as a powerful and broadly accepted standard. For readers interested in learning more, we recommend the manual by Issa (2019).

B-Rep representations can be encoded in multiple file formats. Some legacy formats remain prevalent, as newer formats have not always been able to fully replace them (Stroud and Nagy, 2011). When saving to a B-Rep model, REFER first converts the polygonal meshes into NURBS surfaces. The resulting B-Rep models are then exported to the IGES file format.

IGES was originally developed to facilitate the exchange of technical drawings containing multiple views and annotations. It was not initially designed to represent full three-dimensional models. Instead, its primary focus was on conveying 2D geometry along with a broad range of complementary information, such as dimensions, tolerances, and manufacturing specifications. IGES was created during a time when data needed to be kept minimal to accommodate limited computer memory and peripheral constraints such as punched cards. These two design constraints have resulted in a format that is both complex and excessively verbose by modern standards.

A typical IGES B-Rep file is organized into five rigidly structured sections:

- *Start Section*: Contains a free-text identifier for the file.
- *Global Section*: Stores general information required to interpret the file (e.g., units, author, timestamps).

- *Directory Entry Section*: Encodes topological information about the model, such as entity types and references. Although this section can also include annotations and attributes, these are outside the scope of B-Rep modeling and are not discussed here.
- *Parameter Data Section*: Contains the detailed geometric data such as curves and surfaces associated with the model.
- *Terminate Section*: Marks the end of the file and finalizes the structure.

The syntax within each section is highly rigid and varies from one section to another. Rather than using human-readable labels, IGES uses numeric entity codes (e.g., 110 for a line) to indicate the type of data being represented. Furthermore, a dense network of references and pointers between the topological and geometric data is required to maintain the coherence of the model. Despite these drawbacks, IGES remains a widely supported format for exchanging B-Rep models, primarily because of its long-standing adoption and ability to support advanced geometry, including NURBS surfaces.

According to the detailed explanations provided by Neto *et al.* (2013) regarding the "direct link between the mathematical parameters and the information included in the IGES file format," curves are defined as entities 126 in the parametric space, while they are also defined as entities 110 in the Euclidean space. Both types of curves are grouped in entities 102 that, in turn, are linked to the 128 patch through entities 142. Finally, an entity 144 is used to trim the patch. This is more like IGES files produced by other CAD applications such as SolidWorks. Alternatively, B-Rep models may arise by mapping each polygonal face to a corresponding Rational B-Spline Surface (IGES entity 128). The edges that define the boundary of each patch are represented as Rational B-Spline Curves (entity 126). These curves are then linked together to form a Boundary (entity 141), which is used to trim the original surface, yielding a Bounded Surface (entity 143). Since the "143 approach" is common in some CAD applications like Rhinoceros, and the "144 approach" is used by other systems like SolidWorks, REFER implements both "flavors" for translating their 3D models into IGES B-Rep models. Users can select the approach via the corresponding parameter under *General Settings*.

It is worth noting that triangular polygons are trivially planar and can be easily represented as plane polygons. Alternatively, they may also be represented as NURBS patches, which is a convenient way to homogenize their representation with other polygonal faces. However, when mapped to untrimmed NURBS surfaces, they may introduce singularities due to the repeated fourth control point (e.g., P1, P2, P3, P3). Quadrilateral polygons, in contrast, can often be represented using ruled surfaces, which interpolate between two boundary curves and are more geometrically stable in NURBS form. Consequently, a more effective approach to representing triangular faces is to define a quadrilateral surface and then trim it to the desired triangular shape. REFER converts the triangular polygon (P1, P2, P3) into a quadrilateral patch by adding PS, a "rotational symmetric" of P1 relative to the midpoint M of P2-P3 inserted in the contour between them, as illustrated in Figure 5.9.

In the example shown in Figure 5.10, the four surfaces are intended to be flat faces. However, due to imperfections in the input sketch that are not corrected during the inflation process, the resulting surfaces are actually ruled surfaces. Also known as bilinear surfaces, these are the simplest type of non-flat surfaces (Salomon, 2011). As such, they are readily represented in IGES using the specific form 8 of the type 128 entity, which defines a Rational B-spline Surface, as shown in Figure 5.11.

In REFER, polygonal faces are defined by an ordered sequence of vertices and edges. When computing face normals, the vertex/edge order is adjusted as needed to ensure that counterclockwise (CCW) winding produces an outward-pointing normal. The example below shows the list of four contour points (P1, P2, P3, P4) provided by REFER to define a quadrilateral face, which is then represented as a ruled surface. The corresponding IGES parameters for this entity type are as follows:

- Upper index K1 of first sum (Number of control points less one in U) x9 = 1
- Upper index K2 of second sum (Number of control points less one in V)= 1
- Degree M1 of first basis functions (Surface degree in U)= 1
- Degree M2 of second basis functions (Surface degree in V)= 1
- Flag1 0= open in first direction
- Flag2 0= open in second direction
- Flag3 1= polynomial
- Flag4 0= nonperiodic in first direction
- Flag5 0= nonperiodic in second direction
- First knot sequence (0., 0., 1., 1.), alternatively, reparametrize to (0., 0., |P2-P1|, |P2-P1|)
- Second knot sequence (0., 0., 1., 1.), alternatively, reparametrize to (0., 0., |P3-P4|, |P3-P4|)
- Weights associated to each control point (1., 1., 1., 1.)

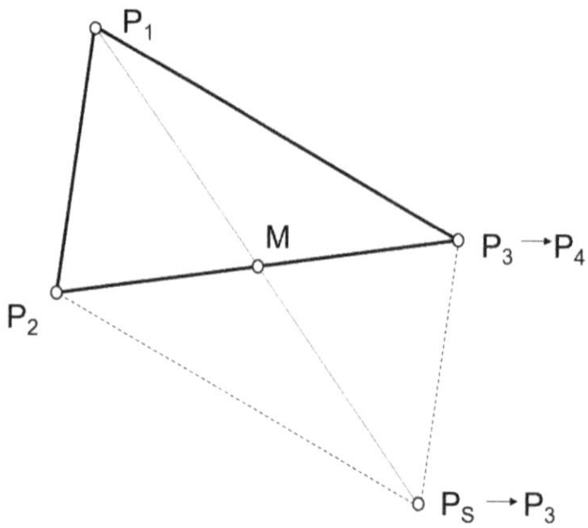

Figure 5.9. Conversion of a triangular polygon (P1, P2, P3) into a quadrilateral patch.

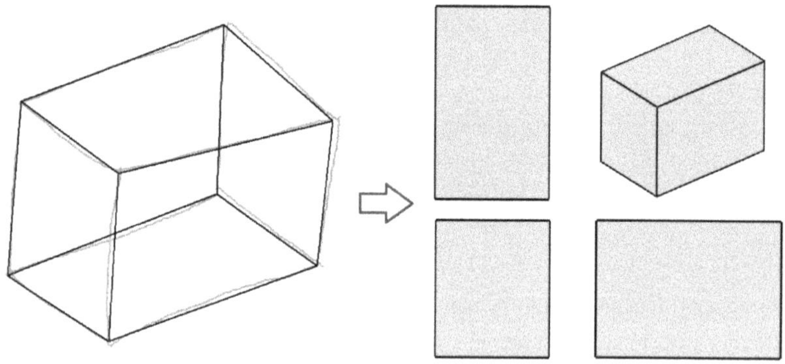

Figure 5.10. Ruled (non-flat) surfaces in the final model caused by imperfections in the input sketch.

- Control point (P1.x, P1.y, P1.z)
- Control point (P2.x, P2.y, P2.z)
- Control point (P4.x, P4.y, P4.z)
- Control point (P3.x, P3.y, P3.z)
- Start first parameter value U0 = 0.
- End first parameter value, U1 = |P2-P1|
- Start second parameter value V0 = 0.
- End second parameter value, V1 = |P3-P4|

General n-gon faces can also be represented using IGES type 128 entities. However, the required parameters differ from those used for

```
#REFER 7.0 BRep solid (IGES NURBS faces type 143)                       S      1
1H,,1H;,15HPrism REFER.igs,56HC:\D\Refer 7.0\Examples\5.0 Output\Prism\PG      1
rism REFER.igs,9HREFER 7.0,14HREFER IGES 5.3,32,38,6,308,15,15HPrism REFG      2
ER.igs,1.0,2,2HMM,1,0.250,15H20250611.100553,0.001D0,575.727815,,,10,0,1G      3
5H20250611.100553;                                                      G      4
     314       1       0       0       0       0       0       000000200D      1
     314       0       1       1       0                  Color       0D      2
     406       2       0       0       1       0       0       000000300D      3
     406       0      -1       1       3                  Layer       0D      4
     128       3       0       0       1       0       0       000000000D      5
     128       0      -1       4       8                  Q_gon       1D      6
     128       7       0       0       1       0       0       000000000D      7
     128       0      -1       4       8                  Q_gon       2D      8
     128      11       0       0       1       0       0       000000000D      9
     128       0      -1       4       8                  Q_gon       3D     10
     128      15       0       0       1       0       0       000000000D     11
     128       0      -1       4       8                  Q_gon       4D     12
     128      19       0       0       1       0       0       000000000D     13
     128       0      -1       4       8                  Q_gon       5D     14
     128      23       0       0       1       0       0       000000000D     15
     128       0      -1       4       8                  Q_gon       6D     16
314,80.0,80.0,90.0,19HRGB( 080, 080, 090);                            1P      1
406,2,1,8HLayer  1;                                                   3P      2
128,1,1,1,1,0,0,1,0,0,0.,0.,378.905520,378.905520,0.,0.,577.3925      5P      3
78,577.392578,1.,1.,1.,1.,0.000000,378.587076,387.754005,8.30595      5P      4
5,0.000000,400.877643,575.727815,344.090867,414.766660,567.24798      5P      5
9,6.105769,412.240663,0.,378.905520,0.,577.392578;                    5P      6
128,1,1,1,1,0,0,1,0,0,0.,0.,335.889070,335.889070,0.,0.,411.7552      7P      7
05,411.755205,1.,1.,1.,1.,570.115185,11.808491,0.534934,575.1431      7P      8
12,347.659501,0.000000,567.247989,6.105769,412.240663,575.727815      7P      9
,344.090867,414.766660,0.,335.889070,0.,411.755205;                   7P     10
128,1,1,1,1,0,0,1,0,0,0.,0.,559.729267,559.729267,0.,0.,414.7824      9P     11
24,414.782424,1.,1.,1.,1.,575.143112,347.659501,0.000000,15.4354      9P     12
94,342.858468,1.088515,575.727815,344.090867,414.766660,0.000000      9P     13
,378.587076,387.754005,0.,559.729267,0.,414.782424;                   9P     14
128,1,1,1,1,0,0,1,0,0,0.,0.,411.755205,411.755205,0.,0.,557.3955     11P     15
52,557.395552,1.,1.,1.,1.,570.115185,11.808491,0.534934,567.2479     11P     16
89,6.105769,412.240663,12.743708,7.711837,3.706099,8.305955,0.00     11P     17
0000,400.877643,0.,411.755205,0.,557.395552;                         11P     18
128,1,1,1,1,0,0,1,0,0,0.,0.,397.271194,397.271194,0.,0.,335.1676     13P     19
62,335.167662,1.,1.,1.,1.,12.743708,7.711837,3.706099,8.305955,0     13P     20
.000000,400.877643,15.435494,342.858468,1.088515,0.000000,378.58     13P     21
7076,387.754005,0.,397.271194,0.,335.167662;                         13P     22
128,1,1,1,1,0,0,1,0,0,0.,0.,335.167662,335.167662,0.,0.,557.3955     15P     23
52,557.395552,1.,1.,1.,1.,12.743708,7.711837,3.706099,15.435494,     15P     24
342.858468,1.088515,570.115185,11.808491,0.534934,575.143112,347     15P     25
.659501,0.000000,0.,335.167662,0.,557.395552;                        15P     26
S      1G      4D     16P     26                                      T      1
```

Figure 5.11. IGES representation of the model shown in the previous figure.

triangular or quadrilateral faces. In addition, the surface must be trimmed by a Boundary (entity 141) in order to produce a Bounded Surface (entity 143), which restricts the surface domain to the region defined by the face contour.

If face geometry has not yet been calculated, the function responsible for outputting the B-Rep model will default to exporting a wireframe representation. This wireframe is generated using IGES type 116 entities to represent vertices and type 110 entities to represent edges. The resulting file can be imported into most commercial CAD systems; however, some applications may require specific IGES translation settings. For example, in SolidWorks (see Figure 5.12), the "Free point/curve entities" option must be enabled to correctly import the wireframe geometry.

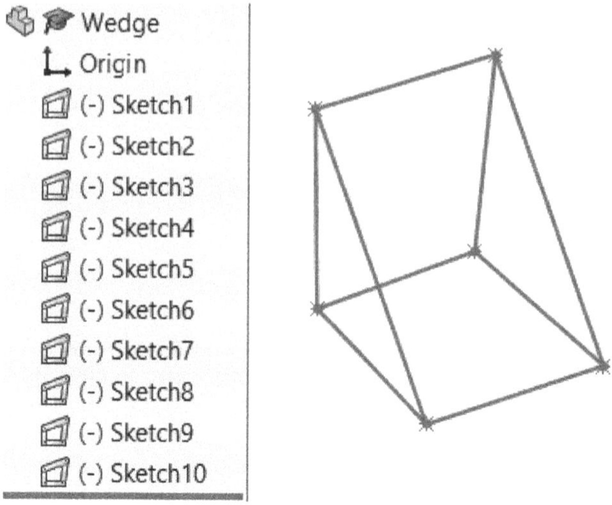

Figure 5.12. Imported geometry in SolidWorks.

As a potential future development, adopting STEP AP214 (or similar B-Rep formats) would enable the explicit representation of all faces, edges, and vertices. However, STEP 214 is not straightforward. A critical challenge lies in maintaining consistency between topological and geometrical information while minimizing redundancy. If this balance is not carefully managed, geometric errors can quickly propagate. Figure 5.13 illustrates such redundancy: each *advanced_face* is defined by both its *face_outer_bound* and an associated *plane*, introducing topological duplication. Similar duplication occurs with *edge_curve* entities, which are defined by two *vertex_point* entities and a *line*. This topological redundancy often results in geometric incompatibility, as different *cartesian_point* entities may be used to represent what is, in the SBM model, a single vertex.

5.8 Procedural 3D Models

Models are procedural (also known as *generative* or *history-based*) when they are defined by a sequence of operations or procedures, which may include the resolution of geometric and dimensional constraints. Procedural modeling techniques such as history-based, parametric, and feature-based modeling, generate 3D models from a defined set of rules or operations. Unlike explicit models, procedural models capture all or part of the design

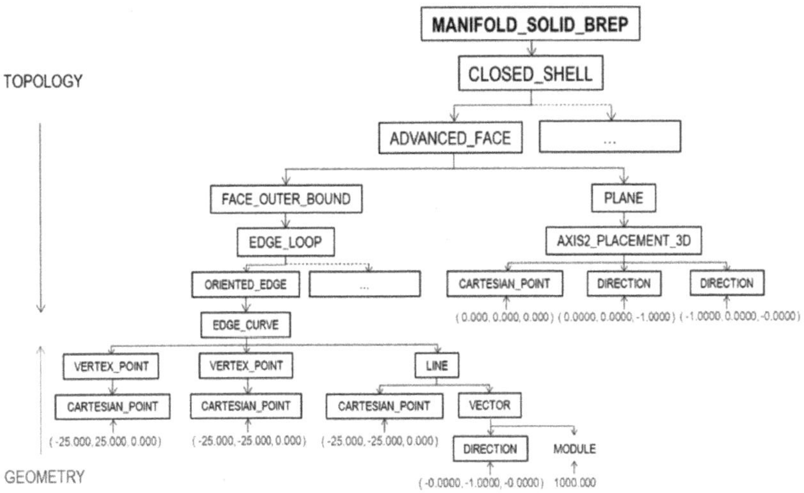

Figure 5.13. Topological representation in STEP 214.

intent. They also facilitate easier editing, as modifications can be made by adjusting parameters or reordering the modeling steps.

The procedural approach implemented in REFER aims to extract high-level information from a 2D sketch and represent it using a Constructive Solid Geometry (CSG) model. This CSG representation serves as a model tree for generating a procedural 3D CAD model (as shown in Figure 5.14). The model tree of a procedural model typically consists of the following elements:

(1) Modeling features, which are operations that follow the paradigm of sketching a 2D profile and sweeping it to generate a 3D shape.
(2) Datums, which are reference elements (planes, axes, coordinate systems, etc.) that serve as scaffolding to define the relative positions of features and to support future editing operations.
(3) Parent-child relationships among features, along with the implicit Boolean operations that define how those features interact and combine within the model hierarchy.

5.9 CSG Output

Once it was recognized that a mere collection of faces, edges, and vertices is insufficient to capture the conceptual and structural information implied by the method of construction, the need for a higher-level representation

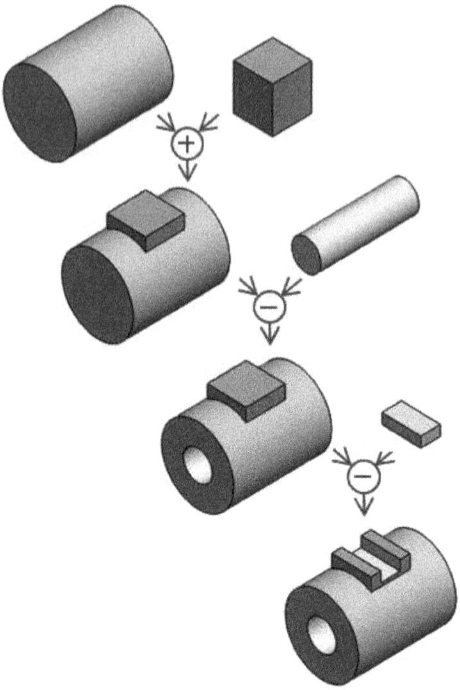

Figure 5.14. Example of a procedural model as a sequence of modeling steps.

became evident. Beyond the various proprietary formats introduced by early 3D procedural CAD applications, a notable early effort to define an open 3D model representation that satisfies the minimal requirements for archival, fidelity, and constraint support was the *Editable Representation* (Erep) proposed by Hoffmann and Arinyo (1992).

In a recent paper, Company *et al.* (2023) reviewed the state of the art in procedural CAD model exchange and discussed an approach for conveying procedural information during model exchange. The authors concluded that ISO 10303 AP 242 standard is theoretically feasible, although not available in practice, and advocated for defining "Conformance Classes to facilitate the practical development of the ISO 10303 AP 242 standard, thus enabling official implementations that can provide full exchange capabilities for procedural CAD models."

Given the current state of the art, REFER adopts an in-house CSG representation. The output is structured as a text-based list of datums and features. Features are organized into branches, which in turn constitute the

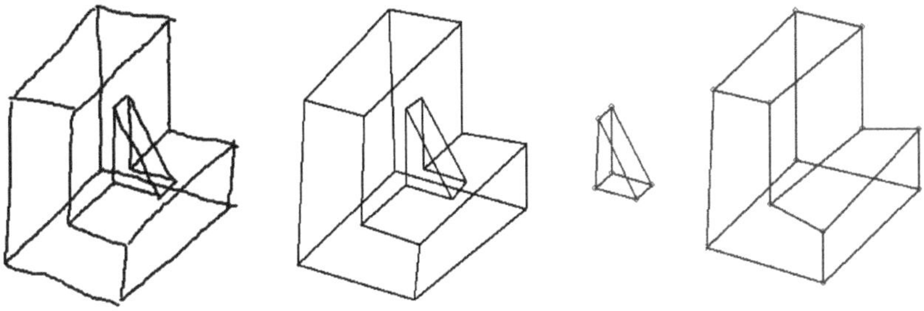

Figure 5.15. L-shaped block with a rib.

model tree, thereby encoding the parent–child relationships among features. Each feature is defined by its extrusion parameters and a triplet of datum planes that locate it within the main reference system. This representation has been shown to provide sufficient information for reconstructing the 3D model in any commercial CAD environment.

The example shown in Figure 5.15 illustrates an L-shaped block with a rib. Starting from the sketch, the 2D drawing is first vectorized and then deconstructed by detecting and removing the rib, thereby yielding the 2D representation of the core shape.

The corresponding CSG output file is organized into two main sections. The first specifies the datum planes, while the second describes the features that constitute the model, grouped into branches. The information is self-descriptive and arranged into structured blocks. Indentation is employed to clearly convey the hierarchical relationships among the data. An example is shown in Figure 5.16.

The primary objective is to ensure that any user can verify that the file contains all the information necessary to construct the procedural model. With minimal effort, the file could also be converted to a mark-up language such as XML. The main drawback of the current method is the lack of geometric precision. Even a brief inspection reveals discrepancies and imperfections that would inevitably produce models with inaccurate dimensions and shapes, and, more critically, with gaps or other errors in the Boolean combinations between features. Addressing this limitation should therefore be a key objective for future improvements, specifically by ensuring that both the features and their defining datums are geometrically exact and mutually consistent. It is important to emphasize that these output errors are not introduced by the reconstruction process itself but are inherited from inaccuracies in the input sketch. The reconstruction algorithm does not introduce new geometric errors or correct errors already

```
CSG model by REFER
  DATUMS
    Datum Planes =4
      Datum Plane Idd            = 0
        Reference subgraph       = 0
        Plane orientation        = YZ
        Distance in X            = 150.330847
      Datum Plane Idd            = 1
        Reference subgraph       = 0
        Plane orientation        = YZ
        Distance in X            = 150.330847
      Datum Plane Idd            = 2
        Reference subgraph       = 0
        Plane orientation        = XY
        Distance in Z            =  84.815215
      Datum Plane Idd            = 3
        Reference subgraph       = 0
        Plane orientation        = XZ
        Distance in Y            =  92.013807
    End Datum Planes
  END DATUMS
  TREE
    Branches =1
      Branch 0
        Feature 0
          CORE:
          Extrusion both sides        = YES
          Subgraph                    = 0
          Branch                      = 0
          Merit                       = 0
          Profile orientation         = YZ
          Profile vertices            = 6
          {
            V0 [110.427801,240.060312]
            V1 [  0.000000,240.060312]
            V2 [  0.000000,  0.000000]
            V3 [204.147488,  0.000000]
            V4 [204.147488, 97.142617]
            V5 [109.576160, 97.142617]
          }
          Extrusion direction         = X
          Linear_extrusion length     =304.427713
          Datums referencing feature =1
            Datum 0= 0
        Feature 1
          RIB:
          Extrusion both sides        = YES
          Subgraph                    = 0
          Branch                      = 0
          Merit                       = 0.86
          Profile orientation         = YZ
          Profile vertices            = 3
          {
            V0 [ 18.413994,104.427638]
            V1 [ 91.598230, 12.327402]
            V2 [ 18.413994, -9.265456]
          }
          Extrusion direction         = X
          Linear_extrusion length     = 48.937981
          Datums referencing feature =3
            Datum 0= 1
            Datum 1= 2
            Datum 2= 3
      End Branch 0
    End Branches
  END TREE
End CSG model by REFER
```

Figure 5.16. Output model.

present. In contrast, humans naturally "interpret" imperfect drawings and mentally construct error-free representations.

After generating the file, the user will be asked to specify whether corresponding 2D drawing files should also be produced for each detected feature (including the core shape). If the answer is "Yes," the system will output DXF files containing these 2D representations. The user can also generate a DXF file containing the datum graph, which corresponds to the internal representation used by the deconstructor to identify and store the geometry of datums, features, and their cross-sections. These dialog boxes are shown in Figure 5.17. These supplementary files provide detailed

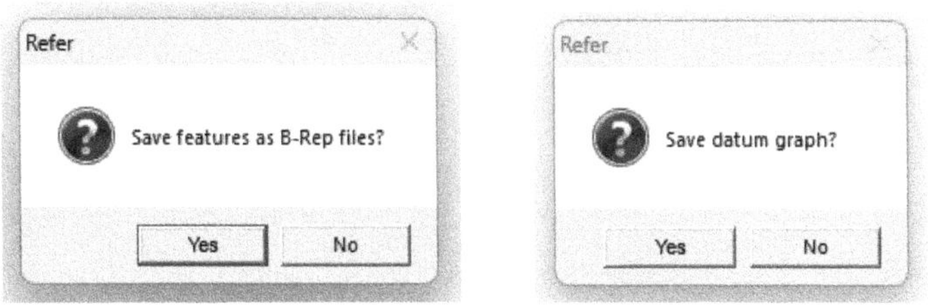

Figure 5.17. System dialogs for CSG outputs.

documentation of the CSG reconstruction process. They can be imported into external CAD applications (e.g., AutoCAD®) or reintroduced into REFER for a detailed examination of the deconstruction steps performed by the application to detect datums and features.

Chapter 6

Implementation

6.1 Introduction

The source code of REFER is available under terms similar (but not identical) to copyleft. In particular, the source code is not bundled with the executable. You are permitted to study, use, distribute, and modify the code, provided it remains free and open. However, you may not claim authorship of the original work. If you use REFER in academic or research contexts, please cite REGEO or the appropriate reference corresponding to the module or algorithm utilized.

REFER is written in C++ and is distributed as a Microsoft Visual Studio Solution. It uses Microsoft Foundation Classes (MFC) for the interactive environment, and can be compiled using Microsoft Visual Studio 2008 (©Microsoft Corporation 2007). Alternatively, the solution can be migrated to Microsoft Visual Studio Community 2015 or 2019. Warnings that appear during the migration process can generally be safely ignored. If you choose to compile REFER with an alternative to Microsoft Visual Studio 2008, the CALI library (Fonseca *et al.*, 2002) must be rebuilt accordingly. To this end, open the *Recognizer.sln* solution file located in the *RECOGNIZER* folder. Once the compilation is complete and the *Recognizer.lib* library is generated, it must be copied into the *lib* folder (where the *freeglut.lib* file is also located).

The first version of REFER was implemented in C for the Windows environment. Version 3 introduced a transition to C++ and the Microsoft Foundation Classes (MFC), along with a redesigned database structure. As a result, traces of the original C-based architecture remain visible in the current codebase. For this reason, it is often more effective to understand the code by examining the structure of its files, rather than focusing solely on its class hierarchy.

This chapter outlines the current code organization of the application, the design principles behind parameter management, and the key features of its databases.

6.2 Code Organization

When viewed through the *Solution Explorer* in *Microsoft Visual Studio*, the code is organized into three main folders, which respectively contain the *.cpp* source files, the *.h* header files, and the application resources (see Figure 6.1).

REFER's source code is organized in ten numbered folders (or "filters"), which contain the functions that implement the different stages of the SBM process:

- 0 Main
- 1 Sketching
- 2 Vectorization
- 3 Refinement 2D
- 4 Detection
- 5 Direct Inflation

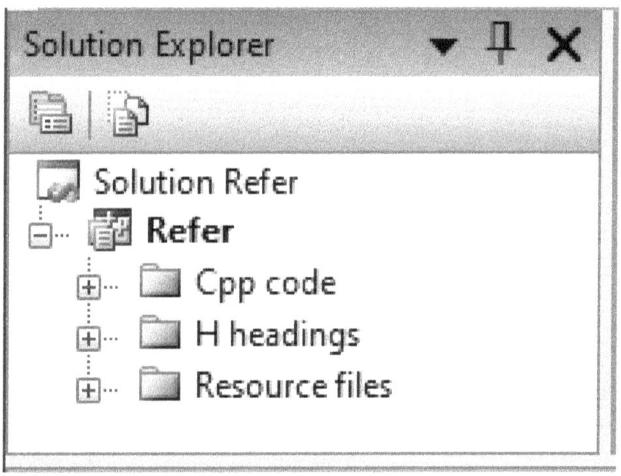

Figure 6.1. Main folders in REFER's code.

- 6 Optimization
- 7 CSG
- 8 Refinement 3D
- 9 Tessellation

The folder labeled "0 Main" contains the files that manage the MFC application (REFER), the documents opened by the application (*Refer Doc*), and the main flow of the SBM process (Reconstruction Manager), as shown in Figure 6.2.

The contents of the folders "1 Sketching" (which implements gesture recognition), "2 Vectorization," and "3 Refinement 2D" are shown in Figure 6.3.

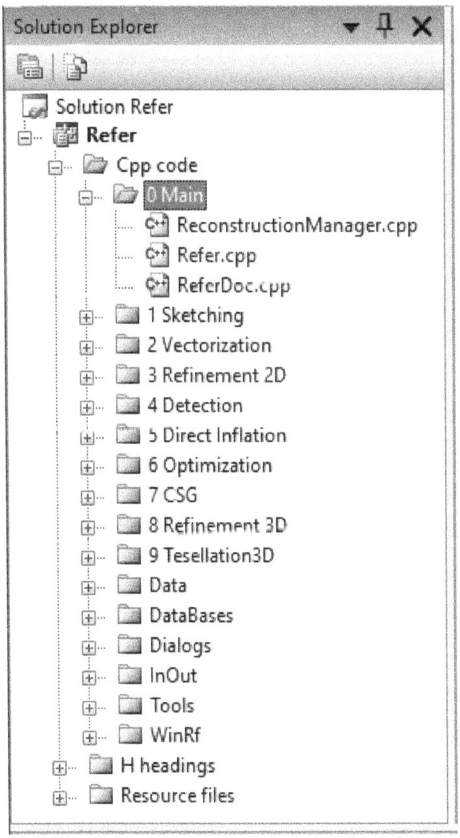

Figure 6.2. Contents of "0 Main" folder.

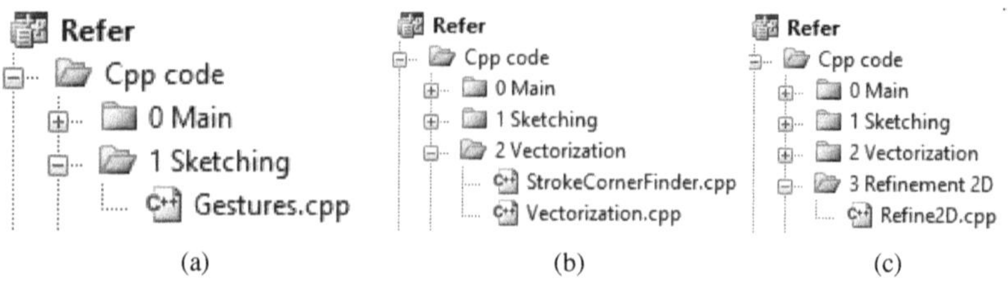

Figure 6.3. Contents of the folders "1 Sketching" (a), "2 Vectorization" (b), and "3 Refinement 2D" (c).

Figure 6.4. Contents of folder "4 Detection. " Cues (a) and Features (b).

The folder labeled "4 Detection" is further subdivided into two subfolders, which contain the functions aimed at finding cues and features in the vectorized sketch, as shown in Figure 6.4.

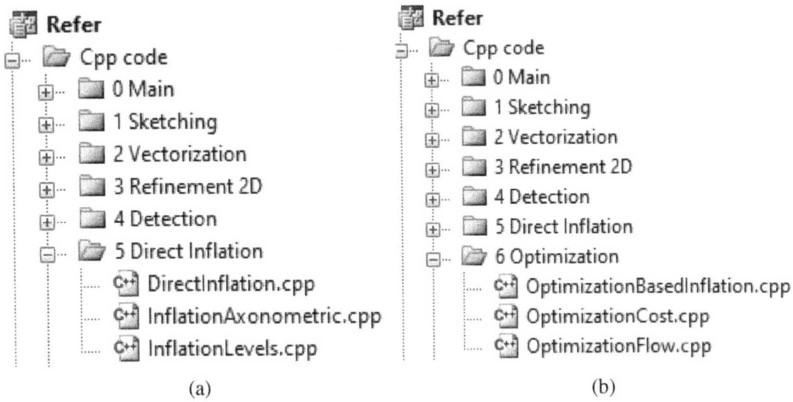

Figure 6.5. Contents of the folders "5 Direct Inflation" (a), and "6 Optimization" (b).

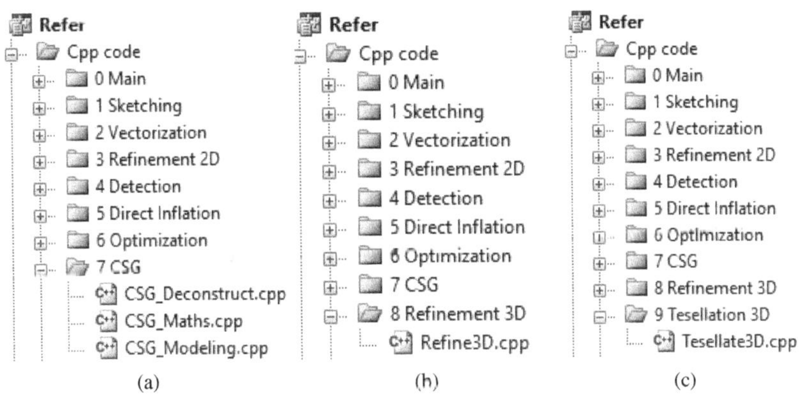

Figure 6.6. Contents of the folders "7 CSG" (a), "8 Refinement 3D" (b), and "9 Tessellation 3D" (c).

The contents of the folders "5 Direct Inflation" and "6 Optimization" are shown in Figure 6.5.

The contents of the folders "7 CSG," "8 Refinement 3D," and "9 Tessellation 3D" are shown in Figure 6.6.

The folder labeled "Data" includes the subfolders with the functions that are called to initialize and reset the parameters that govern the behavior of each of the stages of the SBM process. The contents are shown in Figure 6.7. These functions are where low-level adjustments or performance tuning of the application can be performed. Note that no "data" subfolder is included for the sketching stage.

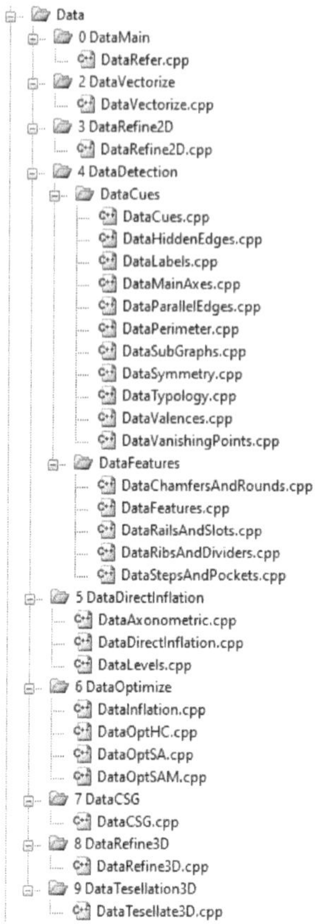

Figure 6.7. Contents of the “Data” folder.

The next folder contains the implementation of the databases, as shown in Figure 6.8.

The source files that implement the dialogs used to interact with the menus are grouped in the “Dialogs” folder (see Figure 6.9). Each file is labeled following the pattern *Dialog_task.cpp*, where task is replaced by a descriptive name corresponding to the specific functionality.

The functions responsible for managing the input and output of information (whether retrieved from or saved to files) are grouped in the “InOut” folder (see Figure 6.10). This is where information is bulk loaded into or extracted from the database.

The “Tools” folder contains a collection of utility functions that have proven useful across various tasks throughout the codebase (see Figure 6.11). These utilities are categorized into three specific groups:

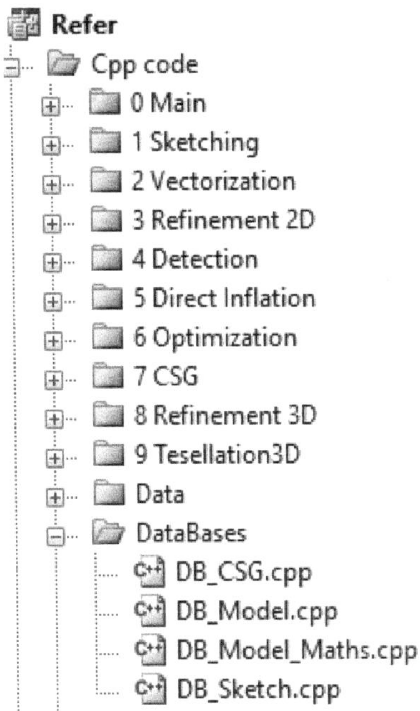

Figure 6.8. Contents of the "Databases" folder.

geometry, matrix, and vector. However, not all commonly used functions are located here. In many cases, tasks that might appear generic are instead implemented individually, with slight variations tailored to specific contexts. This approach was sometimes intentional, aiming to optimize performance. More often than not, it reflects the incremental development of the codebase, where different programmers addressed similar problems using their own solutions.

The information processed by the application is displayed in specific types of child windows, which are managed through dedicated MFC functions. These functions are implemented in files named *WinRf_task.cpp*, where task is replaced by a descriptive label indicating the purpose of each file. All such files are organized within the "WinRf" folder (see Figure 6.12). This is the core area where the application's interactive behavior is handled and coordinated.

As a result of the application's long evolution, and for reasons of convenience, most of the functions related to sketching assistance have been concentrated in the file *WinRf_ViewSketch2D.cpp*. Only the functions specifically related to gesture management have been placed

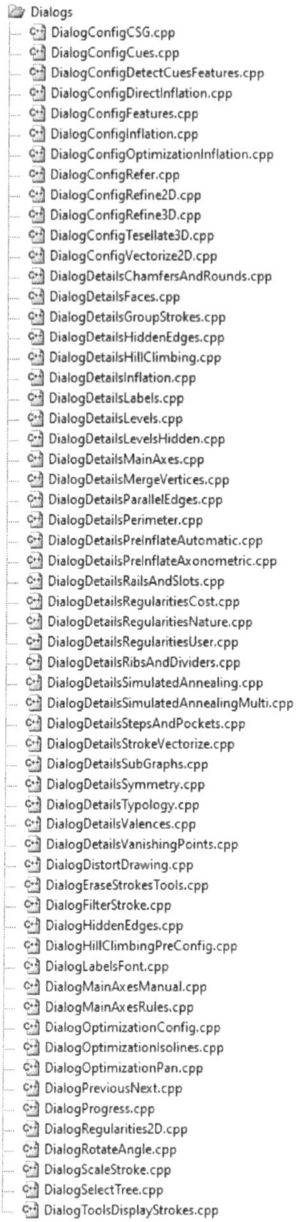

Figure 6.9. Contents of the "Dialogs" folder.

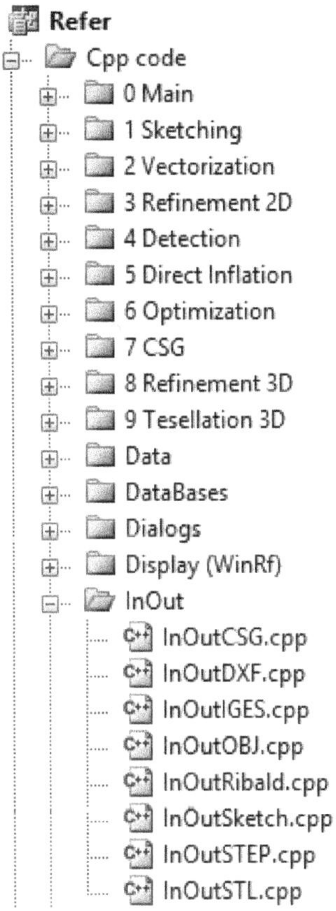

Figure 6.10. Contents of the "InOut" folder.

separately in the "Sketching" folder. Consequently, the lengthy *WinRf_ViewSketch2D.cpp* file not only manages sketch visualization but also handles sketch creation and editing, vectorized line-drawing, and the interaction required to display the cues and features detected in the vectorized drawing.

Some final remarks about the code:

- Most of the MFC code was written manually, rather than generated using the built-in assistants or "wizards." This was a conscious decision to avoid the excessive boilerplate code typically introduced by these tools in their effort to create generic frameworks, and to keep the codebase as concise and readable as possible.

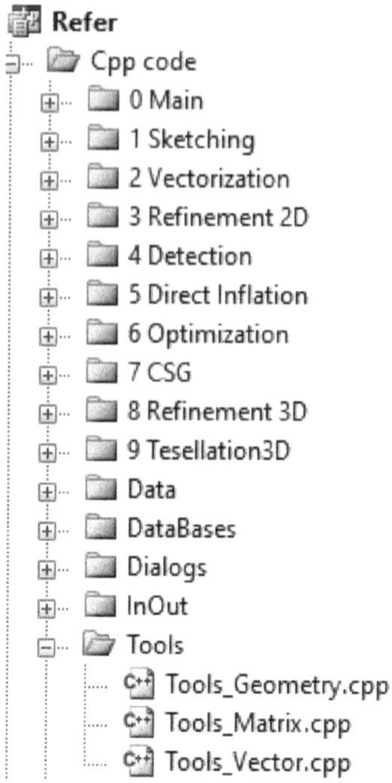

Figure 6.11. Contents of the "Tools" folder.

- The code has been thoroughly annotated with comments to enhance clarity and readability. Most comments and variable names are in English, though some remnants of the original implementation remain in Spanish, both in comments and naming, due to the early development stages.
- The codebase was not cleaned prior to release. As a result, it includes bypassed sections for methods that were never fully implemented, as well as approaches that were eventually discarded in favor of other alternatives.

6.3 Parameter Management

As explained in the previous section, the *Data* folder includes subfolders that contain the functions responsible for initializing and resetting the parameters governing each stage of the SBM process. These functions are where fine-tuning of the application can be performed.

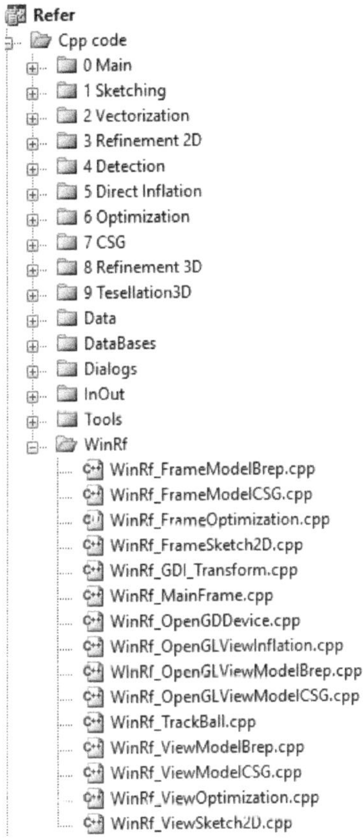

Figure 6.12. Contents of the "WinRf" folder.

Most, if not all, of the approaches implemented in REFER rely on numerous adjustment parameters, as many of the algorithms are based on heuristic principles or on complex and still poorly understood psychological foundations. To account for a wide variety of specific cases while closely approximating human perception, various algorithms requiring a large number of parameters have been developed.

To support the use of REFER as a testbed of ideas, all parameters are accessible to users through dialog windows, most of which can be opened via the corresponding toolbars. Recommended values are readily available via the *Default* buttons provided in each dialog window. Future developers should take note of the deliberate parameter management strategy: all parameters are accessible via dedicated *Data* functions.

Additionally, the places where these parameters are used have also been made explicit. Parameters are typically passed as arguments to functions, rather than being accessed globally. Furthermore, in an effort to encapsulate different approaches, most algorithms operate on local copies of general parameters. These copies are updated at runtime when the corresponding approach is executed.

Clearly, these design choices prioritize code clarity, often at the cost of computational efficiency. For commercial applications, performance-focused refactoring may be required. Nonetheless, the current implementation can be a valuable resource for developers interested in improving the algorithms and reducing the number of required parameters. Undoubtedly, some parameters may prove to be nonessential or redundant.

6.4 Databases

REFER does not rely on a single unified database. Instead, separate databases are used to store the information generated at each stage of the SBM process. However, these databases are not entirely independent. They share a similar structure and are sequentially linked: the final state of the database produced by one stage serves as the input for the database associated with the next stage.

Following the standard execution flow, the following databases are sequentially created during the operation of REFER:

- DB_Sketch. Contains the sketch information drawn by the user (either manually or imported) and possibly edited afterward.
- DB_Drawing. Stores the vectorized drawing (or imported line drawing). REFER uses two versions: *DB_DrawingOn* for online drawing and *DB_DrawingOff* for offline vectorized drawing.
- DB_BRep_Model. Holds the B-Rep 3D model data generated by inflating the vectorized drawing.
- DB_CSG. Contains the procedural CAD model, reconstructed by identifying features and datums within the drawing and/or the B-Rep model.

The code that implements the process can be found in the *ReconstructionManager.cpp* file, located in the folder *0 Main*, as outlined in the *Code Organization* section.

The use of separate databases is not aimed at optimizing information management efficiency, but rather at preserving all intermediate data generated during the various stages of the process. This design choice enables users to interactively visualize results at any stage. A secondary advantage of this architecture is the flexibility it offers in redefining the flow of the SBM process. Users can interactively modify the pipeline by activating or deactivating specific stages, enabling tailored experimentation and development.

At the core of the application lies the principle: "query the 2D geometry to perceive the 3D shape." Consequently, the databases are extensively queried throughout the application. Basic query implementations are embedded directly within the database source code, whereas more complex queries that require additional computation are handled in a separate file named *DB_Model_Maths.cpp*. High-level semantic queries—those at the heart of inferring meaningful 3D structures from drawings—are implemented independently, as detailed in the *Detection* chapter.

The database design predates the first implementation of the reconstruction application. As a large number of queries were added over time, various aspects of the databases were redesigned, while ensuring backward compatibility. The current database architecture, however, is not ideal for all modern query requirements. A detailed analysis of existing queries is strongly recommended, as it could inform a more effective database design for future reconstruction applications.

6.5 Sketch Database

The sketch database ("DB_Sketch") contains the information of the sketch drawn by the user (manually or imported) and possibly edited afterward. To capture all relevant information processed by the application, two sketch databases are created and populated sequentially during the execution of the SBM workflow: "DB_InputSketch" and "DB_WorkingSketch." The former stores the sketch as initially input by the user; the latter contains the refined sketch produced after grouping and segmenting operations.

Three core entities are defined in the sketch database, reflecting the hierarchical nature of sketches, which are composed of strokes, and strokes of points:

- POINTS IN STROKE. This entity stores the individual points that make up a stroke. Each point is characterized by its position (given as a pair of real coordinates relative to an implicit global 2D coordinate system) and by a time parameter that records the elapsed time since the previous point was entered.
- STROKE. This entity represents a single stroke. The *StrokePoints* attribute is a vector containing pointers to the sequence of points comprising the stroke. Additional parameters related to its vectorization, such as fitting merits, are also stored in this structure.
- SKETCH. This entity encapsulates a set of strokes belonging to the same sketch. The *SketchStrokes* attribute is a vector of vectors, allowing the grouping of related strokes. This includes strokes generated by segmenting a polyline-like stroke, as well as those resulting from grouping overtraced or non-solid strokes.

The following queries have been implemented to compute various metrics on individual strokes:

- Number of points in the stroke.
- Bounding box of the stroke, relative to the implicit global 2D coordinate system in which point coordinates are defined.
- Total length of the stroke.
- Approximate stroke length, excluding small-scale oscillations.
- Slope of the stroke, relative to the global 2D coordinate system.
- Length of the longest segment within the stroke.
- Centroid (geometric center) of the stroke.
- Stroke's orientation (clockwise or counterclockwise).
- Convex hull enclosing the stroke.
- Estimated positions of the stroke's corners.

The following queries are available to extract information about full sketches:

- The overall bounds of the sketch.
- The centroid of the entire sketch.
- The minimum valence of any stroke in the sketch. Valence is defined as the number of strokes with endpoints located within a given threshold distance of an endpoint of the current stroke.

6.6 Drawing Database

The drawing databases store information from drawings created externally by the user or vectorized directly by the application. Notably, REFER does not employ dedicated databases specifically for 2D drawings. Instead, it repurposes the databases originally designed to store information about the B-Rep model, simply disregarding the third coordinate of the vertices when used in the 2D context. To capture the full set of drawing-related information during the execution of the SBM process, two drawing databases are created and populated sequentially: "DB_DrawingOn" and "DB_DrawingOff."

Four entities are defined in the drawing database, as valid drawings may include:

- VERTICES. This entity stores points assumed to represent the vertices of the drawing. These vertices serve as the endpoints of edges and define their connectivity. Each vertex includes a pair of real coordinates, relative to the "screen" or "paper" bidimensional coordinate system. Aside from arcs, this is the only geometric information explicitly stored about the drawing. Although a third coordinate z is included in the database, it is not used when representing 2D drawings.
- EDGES. This entity stores straight segments that are assumed to represent straight edges of polyhedral 3D models. Each edge is topologically defined by the indices of its two vertices (Tail → Head). No geometric representation is included directly within the edge entity, as the coordinates of the endpoints are retrieved from the vertex data.
- ARCS. This entity stores elliptic or circular arcs, which may correspond to circular or elliptic edges in a 3D model. Each arc is defined by:
    - Center point (x, y).
    - Radius (for circular arcs).
    - Lengths of the semi-axes (a, b).
    - Counterclockwise angle of rotation from the x-axis to the major axis.
- FACES. This entity stores candidate faces of B-Rep models. Topologically, a face is represented as a cycle (a closed path of alternating vertices and edges) where no vertex or edge appears more than once. Since edges are defined by referencing their connected vertices, each face is represented by the list of edge indices forming the cycle.

Additional information about the faces may be computed during later stages of the SBM process, including symmetry axes, triangulation data, and surface normals.

It is important to clarify that the 2D drawing does not contain actual faces, edges, or vertices in the strict geometric sense used for polyhedral 3D models. Rather, REFER's drawing database stores 2D points that serve as candidate vertices, 2D segments that act as candidate edges, and 2D line cycles that are interpreted as candidate contours of polygonal faces.

Nonetheless, for simplicity, we refer to an entity called DRAWING (internally stored as a 2D MODEL) as being composed of a collection of such "vertices," "edges," "arcs," and "faces." Since the representation is topologically centric, the MODEL entity can encapsulate a 2D drawing by simply ignoring the third coordinate of each vertex and of the points defining arcs.

Additionally, the inclusion of elliptic arcs, alongside or in place of straight edges, broadens the range of representable shapes. This enhancement allows the database to support drawings that depict not only polyhedral models, but also models incorporating curved surfaces.

A vast number of functions have been implemented to assist users in maintaining the geometric integrity of the 2D drawings. However, future developers should exercise care in preserving data consistency when modifying the database code.

The drawing data is extensively queried to extract the information necessary for inferring the 3D model it represents. These high-level queries are detailed in the *Detection* chapter. They rely on a variety of low-level queries that are directly supported by the database (e.g., determining which faces are connected to each vertex).

6.7 B-Rep Model Database

The model databases store the information of the B-Rep models inflated by REFER. To capture all relevant data processed during the SBM workflow, two B-Rep model databases are sequentially created and populated: "DB_Inflation" and "DB_Brep_Model." Additionally, a third database, "DB_Brep_MainAxes," stores the model referenced to its principal coordinate system. This third database is generated automatically when certain SBM modules or visualization commands (those requiring a main reference system) are invoked by the user.

Four entities are defined in the model database, as valid B-Rep models may include the following:

- The VERTICES entity stores points that represent the vertices of the model. These are the endpoints of edges and serve to connect them. Each vertex is stored with three real-valued coordinates, defined relative to the inflation reference system. Apart from arcs, this is the only geometric information explicitly stored in the model. For the database "DB_Brep_MainAxes" (which is automatically generated when required by certain modules) the coordinates are re-referenced to the main reference system.
- The EDGES entity contains information about straight segments that represent straight edges of polyhedral 3D models. Each edge is topologically defined by the indices of its two vertices (Tail → Head). No separate geometric data is stored, as the edge geometry is implicitly defined by its endpoints.
- The ARCS entity stores elliptic or circular arcs, which represent curved edges in the model. Each arc is defined by the following parameters:
    - Center point (x, y).
    - Radius (for circles).
    - Lengths of the semi-axes (a, b).
    - Counterclockwise rotation angle from the x-axis to the major axis.
- The FACES entity contains information about polygonal faces in the B-Rep model. Topologically, each face is defined as a cycle—an alternating sequence of vertices and edges with no repetitions. Since REFER encodes edges via their connected vertex indices, a polygonal face can be stored simply as an ordered list of its constituent edges. The edges and vertices are assumed to be listed in counterclockwise (CCW) order as viewed from outside the face. However, this convention is only valid after face normal vectors are computed. When tentative faces are initially detected from the 2D drawing, normal information is not yet available, so an arbitrary orientation is assigned. Additional topological data may be appended in later stages of the SBM process, such as information about symmetry axes or triangulation.

It is worth noting that the inclusion of elliptic arcs—either in addition to or in place of straight edges—enables the database to represent B-Rep models that feature curved surfaces. However, the current inflation procedure does not yet propagate or convert the parameters defining these arcs from the 2D drawing into the 3D model during reconstruction.

A wide range of functions has been implemented to assist users in preserving the integrity of the entities that define a 3D B-Rep model. Nevertheless, future developers must exercise care to maintain the consistency of the stored information when modifying the database-related code.

6.8 CSG Model Database

Two key aspects must be clarified before introducing the CSG database, as they significant impact the design:

- Custom database structure. REFER implements a proprietary structure because commercial implementations of standard CSG model formats are not yet available (Company *et al.*, 2023).
- Practical yet non-optimal implementation. The current version of the CSG database is functional but not optimized. Its limitations in efficiency stem from two primary considerations. First, clarity has been prioritized over execution speed. Since the database is still under development, maintaining transparency in the implementation is essential to support future extensions. Rather than providing direct access to data (e.g., through pointers), a series of "call" functions have been introduced to mediate data access. This approach requires downloading data for local manipulation before uploading it back into the database. Second, clarity also contributes to robustness, which is equally critical. Because the database is used in an iterative deconstruction/reconstruction process, information is accessed and modified by various code fragments across different stages of execution. REFER's design helps ensure that no code segment attempts to access CSG model data that has already been deconstructed, and deleted, by another part of the code.

REFER's CSG database (DB_CSG) stores the Constructive Solid Geometry reconstructed from an input sketch or 2D line drawing. Within this database, two principal classes of entities are defined, reflecting the procedural nature of models, which are represented as features organized in a tree structure and located relative to datums:

- Class DATUMS: This class contains the set of datum planes that form the structural framework of the model. These planes provide the geometric references necessary to uniquely position the various features that constitute the complete CSG model. While future versions of the system are intended to incorporate additional types of datums such as points, axes, and full reference systems, the current implementation relies exclusively on datum planes.

- Class TREE: This class stores the hierarchical information of feature branches. Branches are sequences of successive features linked via parent—child relationships. Within a branch, each feature is defined as the child of its predecessor and the parent of its successor. By definition, the first feature in a branch has no parent within that branch. In the current implementation, the parent of each branch is defined as the model's core. At present, REFER supports only linear, 2D features that interact via simple one-to-one parent—child relationships. More complex configurations, such as multiple-child-to-parent or child-to-multiple-parent interactions, are not yet supported. Future versions are expected to address these cases by splitting branches. Specifically, when a feature in a branch has more than one child, it will be designated a *local core*, from which new branches will originate.

Features are complex, heterogeneous entities, and are therefore managed by a dedicated class. The specific information required for each feature depends on its type. Accordingly, specialized managers have been implemented for CORE, RIB, DIVIDER, RAIL, SLOT, STEP, and POCKET. Each feature type is managed independently, as no common database structure has been identified that could accommodate all feature categories. Given the intrinsic diversity of features, it is unlikely that such a unified structure can ever be defined. It should be noted, however, that certain feature types are logical complements or reversals of one another, and in these cases some information is managed jointly. In general, feature-related information is organized into three main blocks.

The first block of data stores information about the subgraph to which the feature belongs, its container faces, the merit assigned to it, and the labels (pointers) of the datums linked to the feature. The geometry of the 2D drawing that represents the feature (e.g., its vertices, edges, and faces) is kept in the B-Rep database *DB_Feature2D*.

"Cross-section" vectors store pointers to the vertices of the perimetral polygons used in sweep operations to construct features. Since sweeps are currently limited to extrusions, their cross-sections can be determined by noting that the base faces of a feature remain constant throughout the extrusion, are parallel to the cross-section, and perpendicular to the extrusion direction. In addition, the cross-section must lie within the symmetry plane or container face.

The vertices defining cross-sections are initially referenced to vertices of the input drawing. However, because the input drawing is deconstructed during the CSG modeling process, these references are not permanent. Instead, they are linked to an internal "Datum Graph," which begins as a duplicate of the input drawing but evolves during deconstruction by adding supplementary vertices and edges to accommodate datums and cross-sections. This datum graph is also used for visualizing datums and profiles.

The drawing coordinates of cross-sections originate from projecting a 3D model, which makes them unsuitable for directly constructing the sweep profile. To resolve this issue, REFER calculates unskewed profile coordinates by measuring orthogonal distances from datums to cross-sections and between vertices within each cross-section. These orthogonal distances are measured along polyline paths whose segments are datum-graph edges aligned with the principal axes. The result is a "parametric profile" that can be drawn on the appropriate datum plane before being extruded to generate the feature. These profiles are only nominally parametric: in the current version of REFER, they lack the geometric constraints required for true parametrization. Instead, they are defined solely by fixed (and imperfect) node coordinates. Finally, both extrusion direction and extrusion length are stored.

In general, the functions currently implemented in REFER to manage procedural models include:

- "Clear" functions to reset stored information.
- "Size" functions to query the amount of information stored.
- "Add" and "Remove" functions to insert or delete data items, along with "Set" and "Get" variants to upload or download data items.

It is important to note that uploading and downloading data through these functions is not as efficient as direct pointer access. However, this approach ensures safe encapsulation, as all data is accessed only through controlled functions.

Certain data items are complete structures or classes with their own management functions. The most notable example is feature geometry, which is composed of vertices, edges, and faces, and is managed via three complementary B-Rep-type databases. While integrating B-Rep databases as subsidiary items in a CSG database is not optimal, it was done because the B-Rep manager was already available. This solution serves as a practical interim measure until a definitive system can be designed to manage vertices, edges, and faces specifically for CSG models.

Some final considerations for future developers:

- The queries currently implemented in REFER to extract information about procedural models are not part of the CSG database itself; they are encapsulated within the drawing deconstruction code ("CSG_Deconstruct"). This design choice reflects limited experience in creating general queries and a desire to avoid embedding overly specific queries in the database. For future improvements, it would be desirable to analyze these queries, generalize them, incorporate them into the database, and enhance them with a layer of validation to increase robustness. Some more general queries, such as those for measuring orthogonal distances or plane orientations, have been encapsulated in "CSG_Maths."
- The current implementation was designed to help users preserve the integrity of the entities defining a 3D CSG model. Future developers should maintain consistency of the stored information when modifying database-related code.
- Since the CSG model is generated by recursively deconstructing the 2D drawing, a complementary database stores the successive stages of the deconstruction. This database consists of a vector of B-Rep models, which is used exclusively to visualize the deconstruction process.

Chapter 7

Final Remarks

7.1 Introduction

REFER is a research-oriented academic software application designed to serve as a testbed for ideas. As such, it has never been intended to offer the features or robustness of a commercial application. Moreover, REFER's development has been gradual and extended over time, so even from an academic perspective, it may arguably appear somewhat outdated. Nevertheless, REFER integrates a wide range of methods related to Sketch-Based Modeling (SBM), making it an excellent platform to support a hands-on exploration of SBM techniques.

We suggest three modes of use for the application:

- As a *demonstration* tool of the individual methods that constitute the SBM process, suitable for introductory courses in SBM.
- As an *exploration* tool to identify and investigate the limitations of the current methods, and to experiment with improvements or alternative approaches.
- For advanced users, as an *analysis* tool for studying the limitations of the SBM methodology implemented in REFER and identifying open problems and the associated challenges, which often remain difficult to resolve.

Our vision for future development is outlined in the *Future Developments* section, where we also briefly discuss our perspective on integrating AI into the SBM workflow.

7.2 REFER as a Demonstration Tool

REFER can be used as a *demonstration* tool to illustrate the operation of the various methods that compose the SBM process, particularly in an introductory course on SBM. Both the application and this handbook are structured to guide novice users through an accessible first experience. For this purpose, users can simply follow the recommended reconstruction flow, which is preconfigured by default. The many available visualization tools are specifically designed to help users interpret the results of each stage of the reconstruction process. Their use is strongly encouraged.

By exploring this handbook, readers will encounter a range of alternative workflows and configurations, which will enable experimentation. More importantly, users are encouraged to make mistakes—discovering which flows are less effective and identifying opportunities to refine or redirect the process.

REFER is highly interactive and intuitive. It can be self-taught with the support of this handbook. However, our goal in distributing REFER is to support instructors with experience in SBM who can mentor new researchers and help them build expertise in this field of study. We hope that inductive reasoning and curiosity will allow REFER's users to learn from errors and uncover new possibilities through exploration.

7.3 REFER as an Exploration Tool

REFER can also serve as an exploration tool for identifying the limitations of the implemented methods and investigating possible improvements or alternative approaches. To support this purpose, a comprehensive set of examples is provided. These examples are organized to facilitate the systematic exploration of the application's various capabilities and limitations.

In addition to the classic examples linked to research works published by the group over time, the examples are arranged in task-specific folders, with names that highlight the characteristics being tested. Some belong to series (e.g., example01, example02,...), typically representing small variations of the same geometric shape that behave differently due to distinct levels or types of sketching imperfections. For instance, the example shown in Figure 7.1 illustrates two versions of an L-bracket with a rectangular divider. Under the default tuning, the application fails to detect the

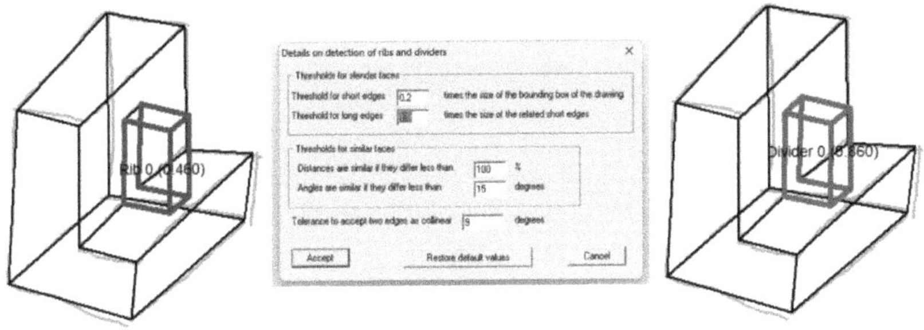

Figure 7.1. Detection of a rectangular divider in the same L-shaped bracket with different sketch imperfection thresholds.

divider's upper slender face, incorrectly classifying the feature as an incomplete rib. However, a minor adjustment of the threshold (from 2.0 to 1.9) leads to the correct identification of both the slender face and the divider.

Some examples are intentionally labeled with terms such as "failure," to indicate cases known to produce incorrect results. These do not necessarily cause runtime errors but may halt the execution of a task or yield no output. In some instances, such errors can be resolved by adjusting the default configuration parameters. Users are therefore encouraged to explore how parameter changes influence application behavior. By experimenting with the parameter configuration dialogs (or even examining the code) users will uncover one of the most significant weaknesses of the SBM method implemented in REFER: its strong dependence not only on fine (and non-universal) parameter tuning, but also on the specific order in which conditions are analyzed, since execution flow itself influences the outcome.

Because the examples are simplified fragments of more complex parts, users will also recognize that this weakness is not accidental but stems from necessity. Engineers can represent an enormous variety of shapes through sketching, and the subtle differences introduced by individual sketching styles make the problem extraordinarily challenging. Addressing this problem requires finding a psychologically plausible balance among a vast number of possible, often contradictory, interpretations (something trained engineers seem to achieve almost effortlessly).

The key takeaway from these examples is that future approaches should aim to move beyond reliance on heuristic methods. Nevertheless, the examples also demonstrate that heuristics, despite their strong dependence on parameter tuning, are capable of handling complex information in subtle and, at times, effective ways. Consequently, replacement approaches should be capable of the following:

- Extracting as much information as possible from sketches;
- Interpreting that information at both local and global levels;
- Selecting the interpretations that are most plausible for human perception.

In this regard, the parameters in REFER, which control heuristic behavior, may serve as a useful guide for how future applications might be configured or trained.

7.4 REFER as an Analysis Tool

REFER provides a representative framework that reflects the current state-of-the-art in SBM (Camba *et al.*, 2022). It serves as a valuable platform for exploring the failures and limitations of existing approaches. Throughout this handbook, various strategies for analyzing sketches and reconstructing the corresponding 3D models have been presented. In this context, REFER is capable of successfully converting somewhat imperfect sketches of certain—relatively simple—shapes into CAD models. However, a general and robust method for interpreting engineering sketches and automatically generating CAD models remains an open challenge.

In reality, we remain far from replicating the human ability to interpret engineering sketches. Yet, as Dreyfus (1992) pointed out, "negative results, provided one recognizes them as such, can be interesting." The failures encountered, especially those shared across similarly complex problems, can be instrumental in identifying common underlying causes.

Historically, SBM has been regarded as a form of cognitive simulation rooted in information processing. Dreyfus (1992) helped illuminate the broader context of this perspective: cognitive simulation problems often begin with "early success attained by working on those aspects where an heuristically guided counting out is feasible," only to encounter a turning point where "complexity is such that a global awareness would be necessary to avoid an overwhelming exponential growth of possibilities to be counted—where counted means thought out by brute-force enumeration." According to Dreyfus (1992), the limitations of cognitive simulation in domains such as SBM arise from several core challenges:

(1) Fringe consciousness is typically reduced to heuristically guided search. Fringe consciousness, as described by William James (and cited by Dreyfus (1992)), refers to the vague or peripheral awareness we experience—such as noticing a clock has stopped ticking without being consciously aware of the ticking beforehand. This kind of subtle awareness is often replicated in software using rigid heuristic rules encoded as *if-then-else* statements, failing to capture the nuanced nature of human perception.
(2) Ambiguity tolerance is replaced by context-free, overly precise approximations. Human cognition handles ambiguity by leveraging context, allowing us to operate effectively even when information is incomplete or unclear. In contrast, both classical and AI-based SBM systems lack robust context-awareness and typically attempt to resolve ambiguity by converting uncertain input into a fixed, precise form—an approach that often leads to failure in real-world scenarios.
(3) Essential/inessential discrimination is poorly mimicked by trial-and-error search. While simple tasks can be approached by exhaustively exploring possibilities, complex problems like SBM involve a combinatorial explosion of options. Humans rely on insight to isolate the core structure of a problem, but this kind of intuitive understanding resists formalization into programmable rules.
(4) Perspicuous grouping is difficult to replicate computationally. As Dreyfus notes, humans can group or classify objects as similar without explicitly identifying them as members of a type or class defined by formal traits. Machines, on the other hand, typically require predefined templates or catalogs to detect patterns. While recent systems can mimic some aspects of human perception, they struggle with transformations, distortions, and incomplete information—common characteristics of engineering sketches. Successful grouping would require a combination of context sensitivity, ambiguity tolerance, and the ability to focus on the "essence" of the pattern.

REFER exemplifies these fundamental limitations of heuristics-based systems attempting to emulate human sketch interpretation. The procedures used to identify relevant heuristics were themselves imperfect, often based on laborious experiments, interviews, and significant trial-and-error. Moreover, the resulting implementations typically include many tuning parameters that limit generalizability. Recent efforts in machine learning attempt to replicate human-like behavior by training on large datasets of sketches whose intended 3D models (i.e., ground truth) are known. However, due to the scarcity of real-world examples, training data is often synthetically generated using tools such as *Drawing Distort*. While these synthetic examples may resemble authentic engineering sketches to some degree, they frequently lack the subtle imperfections and context-specific cues that define true hand-drawn sketches.

REFER implements a form of low-level semantic processing to identify and assign meaning to the atomic components of a sketch. The system is capable of vectorizing a sketch and detecting certain individual cues. However, not all cues that humans can intuitively recognize are detectable by the system. Fortunately, the challenge, while difficult, is finite; research in visual perception offers valuable guidance by identifying the most relevant categories of cues (Hoffmann, 2000). Still, formalizing this problem computationally remains complex, as robust methods for integrating local cues into a coherent global interpretation are not yet available.

Engineering sketches serve as communication mechanisms through which humans disambiguate local elements to convey a unified global message. Computers are still far from replicating this ability. One promising approach involves assigning weights to cues based on their perceived importance. Yet, striking the right balance to simulate the nuanced human semantic process where some cues are ignored, others prioritized, and all are evaluated within context remains an open problem.

The closest approximation so far involves computing figures of merit to estimate the likelihood that a given cue conveys meaningful information. These merit scores aim to emulate human perceptual judgment, but they are typically derived from heuristics that may or may not be well-founded. While machine learning methods may help automate the computation of such figures, the underlying process still relies on heuristically defined features and deductive reasoning. Consequently, current systems fall short of capturing the full depth and adaptability of human sketch interpretation.

REFER can successfully operate within a somewhat constrained search space, such as the graph representation of a sketch—allowing it to identify candidate faces—or among sets of mutually related cues that guide the *inflation* process to generate a 3D model from a 2D sketch. However, the definition of this search space is often based on naive assumptions, and the search itself is typically guided by objective functions configured using heuristic criteria:

A general approach for constraining the search space based on the prior detection of the essential shape represented in a sketch is still lacking. Some specific cases, such as *Axonometric Inflation*, have been addressed to a limited extent.

Although the heuristic criteria that guide the objective function may adapt to reflect characteristics inferred from detected cues, this adaptation does not result from true essential/non-essential discrimination.

The automatic distinction between essential and non-essential elements remains outside the scope of REFER. This limitation reflects a broader challenge in any computer application, which generally require that the core problem be explicitly defined and restructured before processing can begin. In the absence of automatic and insightful problem restructuring, the most viable short-term solution involves transitioning from fully automated systems to more interactive tools—tools that rely on human users to recognize and isolate the essential components of the problem.

REFER also performs high-level semantic processing to assign meaning to patterns and features depicted in sketches. It can identify a variety of higher-order patterns, such as bilateral symmetry, and uses this capability to support an original Constructive Solid Geometry (CSG) strategy based on "divide and conquer" principles. In this approach, the sketch is deconstructed by detecting the relevant CSG features that define the depicted shape.

This pattern and feature recognition relies on catalogs that are inherently limited. Moreover, bottom-up information processing—based solely on local cues—has not proven sufficient to reliably detect individual features and the datums that define their mutual spatial relationships. Additionally, the system is highly sensitive to imperfections in the sketch: even minor distortions can interact in complex ways, amplifying each other's effects and significantly degrading the quality and coherence of the final geometric reconstruction.

To evaluate the current state of "non-AI SBM" in relation to its ultimate goal, we draw on Dreyfus's analysis of pattern recognition.

According to Dreyfus (1992), humans can successfully recognize patterns under four increasingly challenging conditions:

(1) The pattern may be transformed or incomplete.
(2) The number of traits (or cues) required for recognition may be unmanageable.
(3) The traits may be context dependent.
(4) Traits may be replaced by "complicated networks of overlapping similarities."

Dreyfus (1992) concluded that for automatic systems to achieve similar capabilities, they must be able to:

(1) Distinguish the essential features of a specific instance of a pattern.
(2) Make use of cues that operate on the fringes of consciousness.
(3) Consider contextual relationships (i.e., shift from atomistic to holistic approaches).
(4) Situate an individual instance relative to typical cases.

Finally, REFER does not currently implement any Artificial Intelligence (AI) techniques—a significant limitation according to some authors, who claim that many of the problems described above can be effectively addressed using AI-based approaches. The presumed superiority and inevitability of AI in SBM, however, remains debatable, as discussed in our treatment of the complex relationship between SBM and AI. While AI methods appear capable of replicating, and in some cases surpassing, the results obtained by traditional SBM approaches, their effectiveness remains constrained to a narrow class of shapes described through specific types of sketches. These sketches typically fall within strict limits of imprecision—limits that are considerably more stringent than those typically tolerated by humans.

Our critique of AI in general, and Machine Learning (ML) in particular, reflects a healthy skepticism about the capacity of ML to fully resolve the general problem of SBM. This does not imply that ML cannot provide meaningful contributions to existing techniques. Our position, summarized from (Camba *et al.*, 2022), is as follows: "The current limitations of SBM can be summarized by the fact that the thoughts we convey through sketches remain non-computational because our current notions of computation and information representation are not sufficiently rich to capture them. Machine Learning (ML) is a promising approach to overcome this limitation, but the use of low-quality CAD models to train an ML algorithm will inevitably result in intelligent systems that mimic the low-quality models that, unfortunately, are commonplace in industry. Indeed, the use of high-quality models is critical for training an ML system. We cannot limit ourselves to generating low-quality CAD models from sketches, as high-quality and semantically rich models are paramount. To this end, we suggest an alternative development of the SBM paradigm that focuses on capturing richer semantic information (design features) and training ML algorithms to interpret engineering sketches as quality CAD models."

REFER has always been guided by the ambition of contributing to a general theory of SBM—a structured framework of principles that explains the phenomenon and supports the derivation of reliable responses to specific modeling scenarios. Neural Networks, in principle, may seem to bypass the need for an explicit theory of a domain. They operate as black boxes, trained on large datasets of input-output pairs, and can produce models that can be generalized to similar unseen inputs. However, at least two major issues must be addressed:

(1) There is no universal Neural Network architecture; a suitable network must be carefully selected and adapted. In supervised learning, the programmer effectively imparts domain intelligence through training data and architecture design.
(2) The scope of the model's output is inherently limited by the scope of its training data. As a result, solutions that perform well within narrowly defined domains may fail to generalize to more complex or diverse sketch interpretations.

REFER can serve as a valuable testbed for future research and development in AI-based SBM by enabling experimentation and providing foundational insights:

(1) Experience gained from "non-AI SBM" can inform the selection and configuration of neural networks for sketch interpretation tasks.
(2) Insights from traditional SBM methods can help guide the curation of high-quality, representative training datasets—minimizing the risk of narrowing the model's generalization capabilities and ensuring the inclusion of complex and semantically rich sketch cases. In this context, "big data" should be understood not merely as large quantities of data, but as better data and deeper data understanding.

7.5 Future Developments

This handbook describes an SBM application based on "traditional" methods—a combination of heuristics and symbolic AI approaches. It is no surprise that recent advances in machine learning, mentioned throughout the handbook as extensions or ongoing developments, are producing promising results in many areas. A simplistic interpretation of this trend might suggest that traditional methods are a thing of the past, while AI represents the future. However, we argue that these approaches are best understood as complementary, each capable of enriching the other. One way to express this relationship is to say that classical approaches focus on understanding "why," whereas machine learning emphasizes determining "when."

In classical methods, the programmer attempts to understand the underlying mechanisms of human perception—why something should be interpreted in a certain way—and then replicate that reasoning through algorithms. In contrast, ML-based methods collect large sets of examples in which a particular phenomenon occurs and train a model to recognize when that phenomenon is present in new, unseen cases. Classical approaches typically move from the general to the specific: if one understands the "why," then one should be able to predict the "when." However, in practice, this reasoning relies heavily on the programmer's own intuitions about perception and interpretation. As a result, classical approaches are inherently incomplete. Engineering sketches involve countless particularities and edge cases, making it virtually impossible to account for all of them through hand-crafted rules alone.

REFER is intended to help both readers of this handbook and users of the application understand the multiple (and often subtle) interactions that make the artificial perception of engineering drawings such a complex challenge. This is, in fact, the primary goal of this handbook. A deeper reading—or perhaps a reading between the lines—as well as an analysis of the application's code structure, reveal a more fundamental insight: that despite the many technical details, the core of the application lies in storing the input 2D geometric data in the most efficient manner to support two key objectives. First, to enable a wide range of queries that extract both geometric and perceptual information implicitly present in the drawing. Second, to use the results of these queries to perform successive transformations on the 2D input, gradually refining it toward the 3D geometry we aim to reveal. In short, the process can be described as a query-driven transformation from 2D input formats to 3D output representations. We are convinced that the accumulated experience, both in identifying the most effective 2D and 3D formats and in recognizing which cues are most informative, is highly valuable for anyone exploring sketch-based modeling using artificial intelligence techniques.

Current Artificial Intelligence systems use algorithms and mathematical models to process large amounts of data and make decisions based on patterns learned through machine learning—the ability of machines to improve performance autonomously, without being explicitly programmed for every task. One key advantage of ML is its capacity to predict "when" something will occur, reducing reliance on ad hoc algorithms filled with manually tuned parameters based on intuition.

However, training an effective ML model requires a large, diverse, and unbiased dataset, which is not always achieved. Some datasets are limited in size, overly repetitive, or exhibit hidden biases. In many cases, researchers generate their own sketch datasets—for instance, by converting 3D CAD models or 2D CAD drawings into sketch-like inputs. While this strategy easily produces large volumes of examples and provides clear ground truth from the source models, it also introduces the inherent biases of the algorithms used in the transformation process.

The challenge is present both when the input consists of pictorial drawings, as in the case examined here, and when it is provided as a set of orthographic views. Nonetheless, the latter has received considerably more attention in the literature, and notable advances have been reported in this domain (Zhang *et al.*, 2023, 2025a, 2025b). Recent studies have demonstrated the use of AI-based methods to retrieve 3D CAD models from databases using queries derived from a minimal set of orthographic views—informally described as "finding parts that match these views." Despite these promising results, persistent limitations in dataset quality continue to be identified (Mahajan *et al.*, 2025).

In this regard, classical approaches can be seen as operating top-down, while machine learning takes a bottom-up approach. We view both as complementary. Top-down strategies help identify the most appropriate frameworks and decomposition of the problem, while bottom-up methods are better suited for implementing adaptive, robust solutions that are less dependent on hand-tuned parameters. In other words, striving to understand the "why" allows us to decompose complex challenges into manageable subtasks (a divide-and-conquer strategy), while maintaining awareness of how each subtask fits into the larger system. In parallel, learning the "when" enables better tuning of individual components, resulting in more general and efficient solutions.

This dual approach can benefit from the knowledge accumulated in REFER regarding both the sequence and layered structure of the process used to derive CAD models of mechanical parts from freehand sketches. Rather than attempting to train an AI system to generate CAD models directly from sketches in one step, we propose training specialized models for individual stages of the process. For example, an AI system can be trained to vectorize a sketch—this is the first stage of the process, described in the *Reconstruction* chapter. The system can then be trained to refine the vectorized drawings, and so forth. As part of the vectorization stage, we propose training a system to evaluate the merit of each stroke as a potential representation of a line segment or an elliptical arc, and proceed similarly for subsequent stages.

A large number of training examples, along with their corresponding solutions, must be generated to support the development of AI models. To this end, we propose using the REFER application itself as a platform for generating and curating data. Initially, existing examples can be leveraged, followed by the gradual inclusion of new samples until a representative and unbiased dataset is achieved.

To expand the dataset further, we revisit the idea of automatically converting existing drawings into sketches. However, we emphasize the importance of using high-quality models or drawings as input. Moreover, the resulting sketches should be varied to avoid overly repetitive patterns. The *Drawing Distort* tool, introduced in Chapter 2, exemplifies the kind of capabilities needed to introduce controlled variability into the dataset.

As a first step, we hypothesize that a sufficiently representative sample of REFER-drawn sketches, along with the corresponding outputs produced by REFER's own fitting algorithms, can be used to train an AI system to recognize and adjust ellipse segments and arcs. These AI-generated adjustments can then be directly compared with REFER's outputs.

The resulting model could be seamlessly integrated into REFER as an alternative fitting method, which would allow for a straightforward performance comparison: AI-based results can be evaluated against the "ground truth" provided by REFER's existing algorithms—already validated through comparisons with human perception and confirmed via user interviews conducted during prior validation efforts. Ideally, this process should be replicable across the entire vectorization pipeline, including later stages such as vertex determination, refinement of parallelisms, and so on. This progression may ultimately result in an AI-driven system that covers the full sequence and all semantic layers involved in converting a 2D sketch of a mechanical part into a 3D CAD model. It is important to stress that poorly constructed CAD models should not serve as ground truth for the sketches derived from them. The principle remains: "garbage in, garbage out." We recommend foundational references (Camba *et al.*, 2016, 2021; Company *et al.*, 2015a, 2025) to introduce readers to the multifaceted challenges of CAD model quality.

In summary, while the notion of "quality" is complex, high-quality parametric CAD models in an engineering design context can be broadly defined as models that are free from geometric and topological errors, can be modified without triggering regeneration failures, and accurately convey the underlying design intent.

Large Language Models (LLMs) have consistently demonstrated promising performance across a wide range of domains, although "expressing some tasks in text alone is prohibitively complex" (Li *et al.*, 2023). Fortunately, recent developments in Vision-Language Models (VLMs) appear well-suited for various tasks in Engineering Design.

The intersection of LLMs and parametric CAD modeling was recently reviewed by Zhou and Camba (2025), who argue that LLMs have shown varying degrees of success in improving productivity in parametric CAD environments—particularly by automating tasks such as sketch generation, refinement, and 3D modeling through feature recognition. However, they also highlight several persistent challenges, including understanding design intent and context, handling complex geometries, optimizing model training with diverse and representative datasets, developing consistent benchmarking frameworks, and ensuring interoperability, scalability, and security within CAD systems. The importance of expanding the applicability of these techniques across industrial settings was also emphasized.

Recent work on communicative agents (Li *et al.*, 2023) suggests that multi-agent systems powered by LLMs may be particularly effective at decomposing complex problems into a sequence of manageable subtasks, a useful capability for engineering queries that require multi-step or sequential reasoning. Another promising direction integrates traditional SBM expertise with contemporary AI techniques, particularly for reconstructing 3D models from orthographic multi-view projections. For example, Mitic *et al.* (2025) proposed a method for 3D reconstruction from 2D engineering drawings in DXF format by employing a genetic algorithm. Their approach represents the 2D drawing as a symmetric adjacency matrix, which serves as the foundational data structure for the algorithm, allowing it to optimize the reconstruction process.

Bibliography

Alevizos, P. (2013). Pictures as boolean formulas: A method for recognizing polyhedral scenes. *Journal of Mathematical Imaging and Vision*, 46, pp. 74–102. https://doi.org/10.1007/s10851-012-0371-x.

Alexis, A., Suguru, S. and Masayuki, N. (2007). CrossSketch: Freeform surface modeling with details. *SBIM '07 Proceedings of the 4th Eurographics Workshop on Sketch-Based Interfaces and Modeling*. https://doi.org/10.1145/1384429.1384442.

American Psychological Association (n.d.). Dictionary of Psychology. Perceptual cues. https://dictionary.apa.org/perceptual-cues.

Andreu, J. V. (2000). Desarrollo, implementación y prueba de un algoritmo de reconstrucción de objetos a partir de una representación axonométrica, utilizando técnicas de optimización. Facultad de Informática. Univ. Politecnica de Valencia (in Spanish).

Attneave, F. (1954). Some informational aspects of visual perception, *Psychological Review*, 61(3), pp. 183–193. https://doi.org/10.1037/H0054663.

Attneave, F. and Frost, R. (1969). The determination of perceived tridimensional orientation by minimum criteria. *Perception & Psychophysics*, 6(6B), pp. 391–396. https://doi.org/10.3758/BF03212797.

Avola, D., Bottoni, P., Dafinei, A. and Labella, A. (2011). FcBD: An agent-based architecture to support sketch recognition interfaces. *Proceedings of the 17th International Conference on Distributed Multimedia Systems*, DMS 2011, pp. 295–300.

Azariadis, P., Kyratzi, S. and Sapidis, N. S. (2013). A hybrid-optimization method for assessing the realizability of wireframe sketches. *3D Research*, 4(1), pp. 1–11. https://doi.org/10.1007/3DRes.01(2013)3.

Babic, B., Nesic, N. and Miljkovic, Z. (2008) A review of automated feature recognition with rule-based pattern recognition. *Computers in Industry*, 59(4), pp. 321–337. https://doi.org/10.1016/j.compind.2007.09.001.

Barnard, S. T. (1983). Interpreting perspective images. *Artificial Intelligence*, 21(4), pp. 435–462. https://doi.org/10.1016/S0004-3702(83)80021-6.

Bartolo, A., Camilleri, K. P., Fabri, S. G. and Borg, J. C. (2008). Line tracking algorithm for scribbled drawings. *3rd International Symposium on Communications, Control and Signal Processing*, pp. 554–559. https://doi.org/10.1109/ISCCSP.2008.4537287.

Bartolo, A., Camilleri, K. P., Fabri, S. G., Borg, J. C. and Farrugia, P. J. (2007). Scribbles to vectors: Preparation of scribble drawings for CAD interpretation. *SBIM'07: Proceedings of the 4th Eurographics Workshop on Sketch-Based Interfaces and Modeling*, pp. 123–130. https://doi.org/10.2312/SBM/SBM07/123-130.

Bhattacharjee, S. and Chaudhuri, P. (2020). A survey on sketch based content creation: from the desktop to virtual and augmented reality. *Computer Graphics Forum* 39(2), pp. 757–780. https://doi.org/10.1111/cgf.14024.

Biederman, I. (1987). Recognition-by-components: A theory of human image understanding. *Psychological Review*, 94(2), pp. 115–147. https://psycnet.apa.org/doi/10.1037/0033-295X.94.2.115.

Bohachevsky, I. O., Johnson, M. E. and Stein, M. L. (1986). Generalized simulated annealing for function optimization. *Technometrics*, 28(3), pp. 209–217. DOI: 10.1080/00401706.1986.10488128. https://doi.org/10.2307/1269076.

Bonnici, A., Akman, A., Calleja, G., Camilleri, K. P., Fehling, P., Ferreira, A., Hermuth, F., Israel, J. H., Landwehr, T., Liu, J., Panfield, N. M. J., Sezgin, M. and Rosin, P. L. (2019). Sketch-based interaction and modeling: Where do we stand? *Artificial Intelligence for Engineering Design, Analysis and Manufacturing*, 33, pp. 370–388.

Bonnici, A. and Camilleri, K. P. (2015). Vectorisation of sketched drawings using co-occurring sample circles. In Azzopardi, G., Petkov, N. (eds.) *Computer Analysis of Images and Patterns. CAIP 2015. Lecture Notes in Computer Science*, 9256. https://doi.org/10.1007/978-3-319-23192-1_58.

Bonnici A. and Camilleri, K. P. (2023). *Interactive Sketch-Based Interfaces and Modelling for Design. River Publishers, Series in Document Engineering.* ISBN: 978-87-7022-770-4.

Booker, P. J. (1979). *A History of Engineering Drawing*, Northgate Publishing Co., London. ISBN: 978-0852984246.

Bourke, P. (1989). Triangulate. Efficient Triangulation Algorithm Suitable for Terrain Modelling. Gilles Dumoulin's C++ implementation of Paul Bourke's Delaunay triangulation algorithm. http://paulbourke.net/papers/triangulate.

Brown, E. and Wang, P. S. (1996). Three-dimensional object recovery from two-dimensional images: A new approach. In *Intelligent Robots and Computer Vision XV: Algorithms, Techniques, Active Vision, and Materials Handling*, 2904, pp. 138–147. SPIE. https://doi.org/10.1117/12.256269.

Camba J. D. and Company, P. (2020). *Foundations of Computer Graphics: A User-Centered Approach.* eAcademic Books. ISBN 978-0-9962452-1-0.

Camba, J. D., Company, P. and Naya, F. (2022). Sketch-based modeling in mechanical engineering design: Current status and opportunities. *Computer-Aided Design*, 150, p. 103283. https://doi.org/10.1016/j.cad.2022.103283.

Camba, J. D., Contero, M. and Company, P. (2016). Parametric CAD modeling: An analysis of strategies for design reusability, *Computer-Aided Design*, 74, pp. 18–31. https://doi.org/10.1016/j.cad.2016.01.003.

Camba, J. D, Contero, M., Company, P. and Hartman, N. (2021). The cost of change in parametric modeling: A roadmap. *Computer-Aided Design and Applications*, 18(3), pp. 634–643. https://doi.org/0.14733/cadaps.2021.634-643.

Cardoso, R. C. and Ferrando A. (2021). A review of agent-based programming for multi-agent systems. *Computers*, 10(2), p. 16. https://doi.org/10.3390/computers10020016.

Chansri, N. and Koomsap, P. (2012). Automatic single-line drawing creation from a paper-based overtraced freehand sketch. *International Journal of Advanced Manufacturing Technology*, 59(1–4), pp. 221–242. https://doi.org/10.1007/s00170-011-3487-z.

Chatterjee, P. C. and Bhowmick P. (2025). Turn or twist? Verify locally to vectorize globally. *Computer-Aided Design*, 183, p. 103861. https://doi.org/10.1016/j.cad.2025.103861.

Chen P. and Xie, S. (1996). Freehand drawing system using a fuzzy logic concept. *Computer-Aided Design*, 28(2), pp. 77–89. https://doi.org/10.1016/0010-4485(95)00026-7.

Chernov, N., Huang, Q. and Ma, H. (2014). Fitting quadratic curves to data points. *British Journal of Mathematics and Computer Science*, 4, pp. 33–60.

Chinesta, F., Cueto, E., Abisset-Chavanne, E., Duval, J. L. and Khaldi, F. E. (2020). Virtual, digital and hybrid twins: A new paradigm in data-based engineering and engineered data. *Archives of Computational Methods in Engineering* 27, pp. 105–134. https://doi.org/10.1007/s11831-018-9301-4.

Chinesta, F., El Khaldi, F. and Cueto, E. (2023). Hybrid twin: An intimate alliance of knowledge and data. In Crespi, N., Drobot, A. T., Minerva, R. (eds.) *The Digital Twin*. Springer, Cham. https://doi.org/10.1007/978-3-031-21343-4_11.

Clowes, M. (1971). On seeing things. *Artificial Intelligence*, 2, 1, pp. 79–116. https://doi.org/10.1016/0004-3702(71)90005-1.

Company P., Aleixos N., Naya F., Varley P. A. C., Contero M. and Fernandez-Pacheco D. G. (2008). A new sketch-based computer aided engineering pre-processor. *Proceedings of the Sixth International Conference on Engineering Computational Technology*. Civil comp Ltd. Paper-149. (ISBN 978-1-905088-24-9).

Company, P., Ashley, P. and Varley, C. (2010). A method for reconstructing sketched polyhedral shapes with rounds and fillets. In *International Symposium on Smart Graphics*, pp. 152–155. Berlin, Heidelberg: Springer. https://doi.org/10.1007/978-3-642-13544-6_14.

Company, P., Camba, J. D., Patalano, S. and Lanzotti, A. (2023). On the adoption of the STEP neutral format for procedural mechanical CAD models. *Advances on Mechanics, Design Engineering and Manufacturing IV. JCM 2022. Lecture Notes in Mechanical Engineering*, pp. 677–685. https://doi.org/10.1007/978-3-031-15928-2_59.

Company, P., Conesa, J. and Aleixos, N. (2001). Axonometric inflation in line drawings reconstruction. Regeo Technical Report 01/2001.

Company, P., Contero, M., Camba, J. Aranburu, A. and Justel, D. (2025). A hierarchical multilevel modeling framework for the creation of high-quality solid models in parametric feature-based CAD systems. *Computer-Aided Design and Applications*, 22(2), pp. 229–244. https://doi.org/10.14733/cadaps.2025.229-2442025.

Company, P., Contero, M., Conesa, J. and Piquer, A. (2004). An optimisation-based reconstruction engine for 3D modelling by sketching. *Computers & Graphics*, 28(6), pp. 955–979. https://doi.org/10.1016/j.cag.2004.08.007.

Company, P., Contero, M., Naya, F. and Aleixos, N. (2006). A study of usability of sketching tools aimed at supporting prescriptive sketches. *Proceedings of the 4th Eurographics Workshop on Sketch-Based Interfaces and Modeling*, pp. 139–146.

Company, P., Contero, M., Otey, J. and Plumed, R. (2015a). Approach for developing coordinated rubrics to convey quality criteria in MCAD training. *Computer-Aided Design*, 63, pp. 101–117. https://doi.org/10.1016/j.cad.2014.10.001.

Company, P., Contero, M., Varley, P. A. C., Aleixos, N. and Naya, F. (2009). Computer-aided sketching as a tool to promote innovation in the new product development process. *Computers in Industry*, 60(8), pp. 592–603. https://doi.org/10.1016/j.compind.2009.05.018.

Company, P., Plumed, R. and Varley, P. A. (2015b). A fast approach for perceptually-based fitting strokes into elliptical arcs. *The Visual Computer*, 31, pp. 775–785. https://doi.org/10.1007/s00371-015-1099-6.

Company, P., Plumed, R. and Varley, P. (2015c). Study on perceptually based fitting elliptic arcs. Technical Report Ref. 09/2015, Regeo. Geometric Reconstruction Group. http://www.regeo.uji.es.

Company, P., Plumed, R., Varley, P. A. C. and Camba, J. D. (2019). Algorithmic perception of vertices in sketched drawings of polyhedral shapes. *ACM Transactions on Applied Perception*, 16(3), Article No. 18, pp. 1–19. https://doi.org/10.1145/3345507.

Company, P. and Varley, P. A. C. (2009). Operating modes in actual versus virtual paper-and-pencil design scenarios. *2009 Intelligent User Interfaces (IUI) Workshop on Sketch Recognition.*

Company, P. and Varley, P. A. C. (2010). A method for reconstructing sketched polyhedral shapes with rounds and fillets. *Lecture Notes in Computer Science, Smart Graphics 2010 Proceedings*, 6133, pp. 152–155. https://doi.org/10.1007/978-3-642-13544-6_14.

Company, P., Varley, P. A. C., Piquer, A., Vergara, M. and Sanchez-Rubio, J. (2010). Human perception in segmentation of sketches. *GREC 2009, LNCS* 6020, pp. 106–117. https://doi.org/10.1007/978-3-642-13728-0_10.

Company, P., Varley, P. A. C. and Plumed, R. (2011). Sketch input of engineering euclidean solid models. *2011 International Workshop on Visual Languages and Computing. In conjunction with the 17th International Conference on Distributed Multimedia Systems, Proc. of DMS 2011*, pp. 262–267.

Company, P., Varley, P. A. and Plumed, R. (2014). An algorithm for grouping lines which converge to vanishing points in perspective sketches of polyhedra. In *Graphics Recognition. Current Trends and Challenges: 10th International Workshop, GREC 2013*, Bethlehem, PA, USA, August 20–21, 2013, Revised Selected Papers 10, pp. 77–95. Springer, Berlin Heidelberg. https://doi.org/10.1007/978-3-662-44854-0_7.

Company P., Varley, P. A. C. and Plumed, R. (2017). Perimeter detection in sketched drawings of polyhedral shapes. *Eurographics, 2017 Italian Chapter Conference, STAG: Smart Tools and Applications in Graphics*. ISBN 978-3-03868-048-2. pp. 23–29. https://doi.org/10.2312/stag.20171223.

Company, P., Varley, P. A. C., Plumed, R. and Martin, R. (2012). Perceiving ribs in single-view wireframe sketches of polyhedral shapes. In *Advances in Visual Computing: 8th International Symposium, ISVC 2012*, Rethymnon, Crete, Greece, July 16–18, 2012, Revised Selected Papers, Part II, 8, pp. 557–567. Springer, Berlin Heidelberg. https://doi.org/10.1007/978-3-642-33191-6_55.

Conesa, J. (2001). Reconstrucción Geométrica de sólidos utilizando técnicas de optimización PhD Dissertation. Universidad Politecnica de Cartagena (in Spanish).

Conesa, J., Company, P. and Gomis, J. M. (1999). Initial modeling strategies for geometrical reconstruction optimization-based approaches. *Proc. 11th ADM Int. Conf. on Design Tools and Methods in Industrial Engineering*. Volume A. ISBN 88-900081-9-9, pp. 161–171.

Conesa, J., Contero, M. and Company, P. (2000). Comportamiento de los algoritmos de optimización en la reconstrucción geométrica de sólidos. *Anales de Ingeniería Mecánica*, 13, 1, pp. 425–430 (in Spanish).

Conesa, J. F. and Mula, F. J. (2016). An agent-based paradigm for the reconstruction of conical perspectives. *Expert Systems with Applications*, 55, pp. 1–20. https://doi.org/10.1016/j.eswa.2016.02.003.

Cook, M. T. and Agah, A. (2009). A survey of sketch-based 3-D modeling techniques. *Interacting with Computers*, 21, pp. 201–211. https://doi.org/10.1016/j.intcom.2009.05.004.

Cooper, M. (2008). *Line Drawing Interpretation*. Springer Science & Business Media, ISBN: 978-1-84800-228-9.

Courter, S. M. and Brewer, J. A. (1986). Automated conversion of curvilinear wire-frame models to surface boundary models; a topological approach. *Int. Conf. on Computer Graphics and Interactive Techniques*, pp. 171–178. https://doi.org/10.1145/15922.15905.

Curry, H. B. (1944). The method of steepest descent for non-linear minimization problems. *Quarterly of Applied Mathematics*, 2(3), pp. 258–261. ISSN 0033-569X. https://doi.org/10.1090/qam/10667.

Davis, T. (1996). Code to find the equation of a conic. http://mathforum.org/library/drmath/view/51735.html. Accessed Nov 2013.

De Bonet, J. S. (1995). Reconstructing rectangular polyhedra from hand-drawn wireframe sketches. MIT Final project.

Dekkers, A. and Aarts, E. (1991). Global optimization and simulated annealing. *Mathematical Programming*, 50, pp. 367–393. https://doi.org/10.1007/BF01594945.

Deloach, S. A., Wood, M. F. and Sparkman, C. H. (2001). Multiagent systems engineering. *International Journal of Software Engineering and Knowledge Engineering*, 11(3), pp. 231–258. https://doi.org/10.1142/S0218194001000542.

Dori, D. and Liu, W. (1999). Sparse pixel vectorization: An algorithm and its performance evaluation. *IEEE Transactions on Pattern Analysis and Machine Intelligence*, 21(3), pp. 202–215. https://doi.org/10.1109/34.754586.

Dori, D. and Tombre, K. (1995). From engineering drawings to 3D CAD models: Are we ready now? *Computer Aided Design*, 27(4), pp. 243–254. https://doi.org/10.1016/0010-4485(95)91134-7.

Dreyfus, H. L. (1992). *What Computers Still Can't Do: A Critique of Artificial Reason*. The MIT Press, Second Edition. ISBN 978-0262540674.

Duda, R. and Hart, P. (1972). Use of the Hough transformation to detect lines and curves in pictures. *Communications of the ACM*, 15(1), pp. 11–15. https://doi.org/10.1145/361237.361242.

Durbin, J. (1959). Efficient estimation of parameters in moving average models. *Biometrica*, 46, pp. 306–316. https://doi.org/10.1093/biomet/46.3-4.306.

Eckel, B. (2000). *Thinking in* C^{++}, Second Edition, Prentice Hall, Upper Saddle River, New Jersey.

Eggli, L., Hsu, C. Y., Bruederlin, B. D. and Elber, G. (1997). Inferring 3D models from freehand sketches and constraints. *Computer-Aided Design*, 29(2), pp. 101–112. https://doi.org/10.1016/S0010-4485(96)00039-5.

Egiazarian, V., Voynov, O., Artemov, A., Volkhonskiy, D., Safin, A., Taktasheva, M., Zorin, D. and Burnaev, E. (2020). Deep vectorization of technical drawings. *European Conference on Computer Vision (ECCV 2020)*, pp. 582–598. https://doi.org/10.1007/978-3-030-58601-0_35.

Falcidieno, B. and Giannini, F. (1989). Automatic recognition and representation of shape-based features in a geometric modeling system. *Computer Vision, Graphics and Image Processing*, 48, pp. 93–123. https://doi.org/10.1016/0734-189X(89)90106-0.

Fauvreau, J. D., Lafarge, F. and Bousseau, A. (2016). Fidelity vs. simplicity: A global approach to line drawing vectorization, *ACM Transactions on Graphics*, 35(4), Article No.: 120, pp. 1–10. https://doi.org/10.1145/2897824.2925946.

Fernandez-Pacheco, D. G., Albert, F., Aleixos, N. and Conesa, J. (2012). A new paradigm based on agents applied to free-hand sketch recognition. *Expert Systems with Applications*, 39(8), pp. 7181–7195. https://doi.org/10.1016/j.eswa.2012.01.063.

Fitzgibbon, A., Pilu, M. and Fisher, R. B. (1999). Direct least square fitting of ellipses. *IEEE Transactions on Pattern Analysis and Machine Intelligence*, 21(5), pp. 476–480. https://doi.org/10.1109/34.765658.

Fonseca, M. J. and Jorge, J. (2001). Experimental evaluation of an on-line scribble recognizer. *Pattern Recognition Letters*, 22(12), pp. 1311–1319. https://doi.org/10.1016/S0167-8655(01)00076-9.

Fonseca, M. J., Pimentel, C. and Jorge, J. A. (2002). CALI: An online scribble recognizer for calligraphic interfaces. In *AAAI Spring Symposium on Sketch Understanding*, pp. 51–58. Menlo Park: AAAI Press.

Ganin, Y., Bartunov, S., Li, Y., Keller, E. and Saliceti, S. (2021). Computer-aided design as language. *Advances in Neural Information Processing Systems*, 34, pp. 5885–5897.

Gao, C., Wang, X., Yu, Q., Sheng, L., Zhang, J. and Han, X. (2024). 3D reconstruction from a single sketch via view-dependent depth sampling. *IEEE Transactions on Pattern Analysis and Machine Intelligence*, 46(12), pp. 9661–9676, https://doi.org/10.1109/TPAMI.2024.3424404.

Gibson, J. A. (1986). *The Ecological Approach to Visual Perception*. Psychology Press. Taylor & Francis Group.

Gomis, J. M. and Company, P. (1995). La proyeccion en el proceso de representacion y en los sistemas de representacion. *Anales de Ingenieria Grafica* (ISSN 1137-7704), 4(1), pp. 20–31. (In Spanish).

González-Lluch, C., Company, P., Contero, M., Camba, J. D. and Plumed, R. (2017). A survey on 3D CAD model quality assurance and testing tools. *Computer-Aided Design*, 83, pp. 64–79. https://doi.org/10.1016/j.cad.2016.10.003.

Governi, L., Furferi, R., Palai, M. and Volpe, Y. (2013). 3D geometry reconstruction from orthographic views: A method based on 3D image processing and data fitting. *Computers in Industry*, 64(9), pp. 1290–1300. https://doi.org/10.1016/j.compind.2013.02.003.

Gryaditskaya, Y., Hahnlein, F., Liu, C., Sheffer, A. and Bousseau, A. (2020). Lifting freehand concept sketches into 3D. *ACM Transactions on Graphics (Proc. SIGGRAPH Asia)*, 39(6), Article 167, pp. 1–16. https://doi.org/10.1145/3414685.3417851.

Guzman, A. (1968). *Computer Recognition of Three-Dimensional Objects in a Visual Scene*. Tech Rept MAC-TR-59, M.I.T., Cambridge USA.

Halir, R. and Flusser, J. (1998). Numerically stable direct least squares fitting of ellipses. In *Proceedings of 6th International Conference in Central Europe on Computer Graphics and Visualization* (WSCG'98), pp. 125–132.

Han, J. H. (1996). Survey of Feature Research. Technical Report IRIS-96-346, University of Southern California Institute for Robotics and Intelligent Systems.

Han, J. H. (1997). On multiple interpretations. In Hoffmann, C. and Bronsvort, W. (eds) *Proceedings: Fourth Symposium on Solid Modeling and Applications*, pp. 311–321. ACM Press.

Han, Z., Ma, B., Liu, Y.-S. and Zwicker, M. (2020). Reconstructing 3D shapes from multiple sketches using direct shape optimization. *IEEE Transactions on Image Processing*, 29, 9184255, pp. 8721–8734. https://doi.org/10.1109/TIP.2020.3018865.

Hähnlein, F. (2022). *Binary Optimization for the Analysis and Synthesis of Concept Sketches*, PhD Dissertation, Université Cote d'Azur.

Hähnlein, F., Gryaditskaya, Y., Sheffer, A. and Bousseau, A. (2022). Symmetry-driven 3D reconstruction from concept sketches. *SIGGRAPH'22 Conference Proceedings*. Article No. 19, pp. 1–8. https://doi.org/10.1145/3528233.3530723.

Heath L. S., Paripati, P. K. and Roach, J. W. (1993). Representing polyhedra: Faces are better than vertices. *Computational Geometry*, 3(6), pp. 327–351. https://doi.org/10.1016/0925-7721(93)90006-R.

Hertzmann, A. (2024). Toward a theory of perspective perception in pictures. *Journal of Vision*, 24(4), 23, pp. 1–41. https://doi.org/10.1167/jov.24.4.23.

Hilaire, X. and Tombre, K. (2006). Robust and accurate vectorization of line drawings. *IEEE Transactions on Pattern Analysis and Machine Intelligence*, 28(6), pp. 890–904. https://doi.org/10.1109/TPAMI.2006.127.

Hintze, A. (2016). *Understanding the Four Types of Artificial Intelligence. Consulté le*, 10(11), 2021. https://doi.org/10.64628/AAI.ar4dfgrcg.

Hochberg, J. and Brooks, V. (1960). The psychophysics of form: Reversible perspective drawings of spatial objects. *American Journal of Psychology*, 73(3), pp. 337–354. https://doi.org/10.2307/1420172.

Hoffmann, C. M. (1989). *Geometric and Solid Modeling*. Morgan Kaufmann Publishers Inc., San Francisco, CA, United States. ISBN 1558600671.

Hoffmann, C. M. and Arinyo, R. J. (1992). Erep an editable high-level representation for geometric design and analysis. Purdue University, Purdue e-Pubs. Report Number: 92-055. https://docs.lib.purdue.edu/cstech/976.

Hoffmann, D. (2000). *Visual Intelligence. How we Create What we See*, WW Norton & Company, New York.

Hohenberg, F. (1956). *Konstruktive Geometrie fur Techniker*. Wien, Springer.

Huffman, D. (1971). *Impossible Objects as Nonsense Sentences. Machine Intelligence*. Edinburgh University Press, pp. 295–323.

Igarashi, T., Matsuoka, S., Kawachiya, S. and Tanaka, H. (1997). Interactive beautification: A technique for rapid geometric design. *Proceedings of UIST'97*, pp. 105–114. https://doi.org/10.1145/263407.263525.

Igarashi, T., Matsuoka, S. and Tanaka, H. (1999). Teddy: A sketching interface for 3D freeform design. *ACM SIGGRAPH99 Conference Proc*, pp. 409–416. https://doi.org/10.1145/311535.311602.

Isenberg, T., Freudenberg, B., Halper, N., Schlechtweg, S. and Strothotte, T. (2003). A developer's guide to silhouette algorithms for polygonal models. *IEEE Computer Graphics and Applications*, 23(4), pp. 28–37. https://doi.org/10.1109/MCG.2003.1210862.

Issa, R. (2019). *Essential Mathematics for Computation Design. RhinoDeveloper.* 4th Edition. https://www.rhino3d.com/download/rhino/6/essentialmathematics/.

Ito, T., Kaneko, T., Tanaka, Y. and Saga, S. (2022). An interactive sketch-based CAD interface realizing geometrical and topological editing across multiple objects based on fuzzy logic. *Computers & Graphics*, 103, pp. 153–167. https://doi.org/10.1016/j.cag.2022.02.007.

Jain, A. K. (2010). Data clustering: 50 years beyond K-means. *Pattern Recognition Letters*, 31(8), pp. 651–666, ISSN 0167-8655.

Jarret, R. and Su, P. (2002). *Building Tablet PC Applications*. Microsoft Press, Redmond, Washington.

Jennings, N. R., Sycara, K. and Wooldridge, M. (1998). A roadmap of agent research and development. *Autonomous Agents and Multi-Agent Systems*, 1(1), pp. 7–38. https://doi.org/10.1023/A:1010090405266.

Jones, R. M. (2000). *Introduction to MFC Programming with Visual C^{++}*, Prentice Hall, Upper Saddle River, New Jersey.

Jorge, J. A. and Fonseca, M. J. (1999). A simple approach to recognize geometric shapes interactively. *GREC 1999. Lecture Notes in Computer Science*, vol. 1941. Springer, Berlin, Heidelberg. https://doi.org/10.1007/3-540-40953-X_23.

Kanatani, K. (1986). The constraints on images of rectangular polyhedra. *IEEE Transactions on Pattern Analysis and Machine Intelligence; PAMI*, 8(4), pp. 456–463. https://doi.org/10.1109/TPAMI.1986.4767809.

Kanatani, K. (1990). *Group-Theoretical Methods. Image Understanding, Number 20 in Springer Series in Information Sciences.*

Kang, D. J., Masry, M. and Lipson, H. (2004). Reconstruction of a 3D object from main axis system. *AAAI Fall Symposium Series: Making Pen-Based Interaction Intelligent and Natural.*

Kettner L. and Welzl, E. (1997). Contour edge analysis for polyhedron projections. Tech. rep., Geometric modeling: Theory and practice, Springer. *Proc. International Conference on Theory and Practice of Geometric Modeling.* Blaubeuren, Germany. https://doi.org/10.3929/ethz-a-006651848.

Kim, Y. and Bae, S. H. (2016). SketchingWithHands: 3D sketching handheld products with first-person hand posture. *Proceedings of the 29th Annual Symposium on User Interface Software and Technology UIST 2016.* 16 October 2016, pp. 797–808; Code 124327. https://doi.org/10.1145/2984511.2984567.

Kirkpatrick, S., Gelatt, C. D. and Vecchi, M. P. (1982). Optimization by simulated annealing, Research Report RC 9355, IBM (Yorktown Heights, NY, 1982).

Koch, S., Matveev, A., Jiang, Z., Williams, F., Artemov, A., Burnaev, E., Alexa, M., Zorin, D. and Panozzo, D. (2019). ABC: A big CAD model dataset for geometric deep learning. *Proceedings of the IEEE/CVF Conference on Computer Vision and Pattern Recognition*, pp. 9601–9611. https://doi.org/10.1109/CVPR.2019.00983.

Kotagiri, S. (2010). C# implementation for fitting an ellipse for a set of points. Srikanth Kotagiri's Blog. https://blog.sdvc.io/posts/fit-ellipse/ Accessed June 2025.

Kyratzi, S. and Azariadis, P. (2015). Geometric definition of the hidden part of a line drawing in a sketch-to-solid methodology. *Computer-Aided Design and Applications*, 12(3), pp. 355–365, https://doi.org/10.1080/16864360.2014.981466.

Ku, D. C., Qin, S. F. and Wright, D. K. (2006). Interpretation of overtracing freehand sketching for geometric shapes. *WSCG'2006, Conference proceedings* ISBN 80-86943-03-8.

Lamb, D. and Bandopadhay, A. (1990). Interpreting a 3D object from a rough 2D line drawing. *Proceedings of the First IEEE Conference on Visualization: Visualization '90*, pp. 59–66. https://doi.org/10.1109/VISUAL.1990.146365.

Larson, E. J. (2021). *The Myth of Artificial Intelligence: Why Computers Can't Think the Way We Do.* The Belknap Press of Harvard University Press, Cambridge, Massachusetts • London, England. ISBN 978-0674983519.

Leclerc, Y. and Fischler, M. (1992). An optimization-based approach to the interpretation of single line drawing as 3D wire frames. *International Journal of Computer Vision*, 9(2), pp. 113–136. https://doi.org/10.1007/BF00129683.

LeCun, Y., Bengio, Y. and Hinton, G. (2015). Deep learning. *Nature*, 521, pp. 436–444.

Lee, Y. T. and Fang, F. (2011). 3D reconstruction of polyhedral objects from single parallel projections using cubic corner. *Computer-Aided Design*, 43(8), pp. 1025–1034. https://doi.org/10.1016/j.cad.2011.03.008.

Li, C., Pan, H., Bousseau, A. and Mitra, N. J. (2022). Free2CAD: Parsing freehand drawings into CAD commands. *ACM Transactions on Graphics*, 41, pp. 1–16. https://doi.org/10.1145/3528223.3530133.

Li, G., Hammoud, H. A. A., Itani, H., Khizbullin, D. and Ghanem, B. (2023). CAMEL: Communicative Agents for "Mind" exploration of large language model society. *Advances in Neural Information Processing Systems*, 36, pp. 51991–52008. https://doi.org/10.48550/arXiv.2303.17760.

Lim, S., Qin, S. F., Prieto, D. and Shackleton, J. (2004). A study of sketching behaviour to support free-form surface modeling from on-line sketching. *Design Studies*, 25, pp. 393–413. https://doi.org/10.1016/j.destud.2003.10.008.

Lipson, H. and Shpitalni, M. (1996). Optimization-based reconstruction of a 3D object from a single freehand line drawing. *Computer-Aided Design*, 28(8), pp. 651–663. https://doi.org/10.1016/0010-4485(95)00081-X.

Liu, C. and Bessmeltsev, M. (2025). State-of-the-art report in sketch processing. *Computer Graphics Forum*, e70079. https://doi.org/10.1111/cgf.70079.

Liu, Y. and Lee, Y. T. (2010). 3D reconstruction of freeform shapes from 2D line drawings. *VRCAI '10 Proceedings of the 9th ACM SIGGRAPH Conference on Virtual-Reality Continuum and its Applications in Industry*. https://doi.org/10.1145/1900179.1900244.

Long, A. C., Landay, J. A., Rowe, L. A. and Michiels, J. (2000). Visual similarity of pen gestures. *Proceedings of Human Factors in Computer Systems (SIGCHI)*, pp. 360–367. https://doi.org/10.1145/332040.332458.

Luck, M., McBurney, P., Shehory, O. and Willmott, S. (2005). *Agent Technology: Computing as Interaction. A Roadmap to Agent Based Computing*. Agentlink. ISBN 085432-845-9.

Lun, Z., Gadelha, M., Kalogerakis, E., Maji, S. and Wang, R. (2017). 3D shape reconstruction from sketches via multi-view convolutional networks. *2017 International Conference on 3D Vision (3DV)*, pp. 67–77.

Lundy, M. and Mees, A. (1986). Convergence of an annealing algorithm, *Mathematical Programming*, 34(1), pp. 111–124. https://doi.org/10.1007/BF01582166.

Magee, M. J. and Aggarwal, J. K. (1984). Determining vanishing points from perspective images. *Computer Vision, Graphics and Image Processing*, 26(2), pp. 256–267. https://doi.org/10.1016/0734-189X(84)90188-9.

Mahajan, S. N., Karthik, K. M., and Muthuganapathy, R. (2025). OrthoCAD-322K: A cross-modal approach for retrieving 3D CAD models from orthographic views using a graph-based framework on a developed large-scale dataset. *Computers & Graphics*, 132, p. 104357, https://doi.org/10.1016/j.cag.2025.104357.

Marill, T. (1991). Emulating the human interpretation of line-drawings as three-dimensional objects. *International Journal of Computer Vision*, 6(2), pp. 147–161. https://doi.org/10.1007/BF00128154.

Masood, A. and Sarfraz, M. (2007). Corner detection by sliding rectangles along planar curves. *Computers & Graphics*, 31, pp. 440–448. https://doi.org/10.1016/j.cag.2007.01.021.

McLean, G. F. and Kotturi, D. (1995). Vanishing point detection by line clustering. *IEEE Transactions on Pattern Analysis and Machine Intelligence*, 17(11), pp. 1090–1095. https://doi.org/10.1109/34.473236.

Meeran, S. and Pratts, M. J. (1993). Automated feature recognition from 2D drawings. *Computer Aided Design*, 25(1), pp. 7–17. https://doi.org/10.1016/0010-4485(93)90061-R.

Meeran, S. and Taib, J. M. (1999). A generic approach to recognising not-interacting, nested and interacting features from 2D drawings. *Computer-Aided Design*, 31(14), pp. 891–910. https://doi.org/10.1016/S0010-4485(99)00082-2.

Miao, Y., Hu, F., Zhang, X., Chen, J. and Pajarola, R. (2015). SymmSketch: Creating symmetric 3D free-form shapes from 2D sketches. *Computational Visual Media*, 655, 1(1), pp. 3–16. https://doi.org/10.1007/s41095-015-0002-8.

Miles, V., Giani, S. and Vogt, O. (2023). Approaching STEP file analysis as a language processing task: A robust and scale-invariant solution for machining feature recognition. *Journal of Computational and Applied Mathematics*, 427, 115166. https://doi.org/10.1016/j.cam.2023.115166.

Minnameier, G. (2017). *Forms of Abduction and an Inferential Taxonomy*. In Magnani, L. and Bertolotti, T. (eds.) Springer Handbook of Model-Based Science.

Mitić, P., Kočović, V., Mišić, M., Stefanović, M., Đorđević, A., Pantić, M. and Projović, D. (2025). A novel approach in 3D model reconstruction from engineering drawings based on symmetric adjacency matrices using DXF files and genetic algorithm. *Symmetry*, 17(5), p. 771. https://doi.org/10.3390/sym17050771.

Moreno, C. (1998a). Efficient 2-D Geometric Operations, part 1. *C/C++ Users Journal*, 16(11), pp. 25–36. https://dl.acm.org/doi/10.5555/297521.297525.

Moreno, C. (1998b). Efficient 2-D Geometric Operations, part 2. *C/C++ Users Journal*, 16(12), pp. 21–35. https://dl.acm.org/doi/10.5555/306490.306492.

Moreno, C. (1998c). Efficient 2-D Geometric Operations (C^{++} code). Code available at https://mochima.com/articles/cuj_geometry_article/cuj_geometry_article.html.

Nagendra, I. V. and Gujar, U. G. (1988). 3-D objects from 2-D orthographic views — A survey. *Computers & Graphics*, 12(1), pp. 111–114. https://doi.org/10.1016/0097-8493(88)90015-5.

Naya, F., Conesa, J., Contero, M., Company, P. and Jorge, J. (2003). Smart sketch system for 3D reconstruction based modeling. In *Smart Graphics: Third International Symposium on Smart Graphics, SG 2003*. Heidelberg, Germany, July 2–4, Proceedings 3, pp. 58–68. Springer, Berlin, Heidelberg. https://doi.org/10.1007/3-540-37620-8_6.

Naya F., Contero, M., Aleixos, N. and Company, P. (2007). ParSketch: A sketch-based interface for 2D parametric geometry editor. *Lecture Notes in Computer Science. HCII 2007*. Volume 4551, pp. 115–124. https://doi.org/10.1007/978-3-540-73107-8_13.

Naya, F., Contero, M., Aleixos, N., Jorge, J. A. and Company, P. (2008). Sketch-based interfaces for parametric modelling. In Pavlidis, I., (ed.), *Human Computer Interaction*, InTech, pp. 44–55. https://doi.org/10.5772/6310.

Neto, D. M., Oliveira, M. C., Menendes, L. F. and Alves, J. L. (2013). Nagata patch interpolation using surface normal vectors evaluated from the IGES file. *Finite Elements in Analysis and Design*, 72, pp. 35–46. https://doi.org/10.1016/j.finel.2013.03.004.

Olsen, L., Samavati, F. F., Costa-Sousa M. and Jorge J. A. (2009). Sketch-based modelling: A survey. *Computers & Graphics*, 33(1), pp. 85–103. https://doi.org/10.1016/j.cag.2008.09.013.

Onkar, P., and Sen, D. (2010). Functional segmentation of strokes for product sketch understanding. *International Journal of Shape Modeling*, 16(1 and 2), pp. 9–38. https://doi.org/10.1142/S0218654310001262.

Palmer, S. E. (1999). *Vision Science, Photons to Phenomenology*. Cambridge, MA: The MIT Press.

Paton, K. (1969). An algorithm for finding a fundamental set of cycles of a graph. *Communications of the ACM*, 12(9), pp. 514–518. https://doi.org/10.1145/363219.363232.

Pereira, J., Jorge, J., Branco, V. and Nunes, F. (2000). Towards calligraphic interfaces: Sketching 3D scenes with gestures and context icons. *WSCG'2000 Conference Proc*. Skala V. Ed.

Perkins, D. N. (1968). Cubic Corners, Quarterly Progress Report, 89, pp. 207–214, MIT Research Laboratory of Electronics.

Perkins, D. N. (1971). Cubic Corners, Oblique Views of Pictures, the Perception of Line Drawings of Simple Space Forms. Geometry and the Perception of Pictures: Three Studies. Harvard Project Zero. Technical Report No. 56.

Piquer, A. (2003). Percepción artificial de dibujo lineales. PhD Dissertation. Universitat Jaume I (in Spanish).

Piquer, A., Martin, R. and Company, P. (2003). Skewed mirror symmetry for depth estimation in 3D line-drawings. In *International Workshop on Graphics Recognition*, pp. 142–153. Berlin, Heidelberg: Springer. https://doi.org/10.1007/978-3-540-25977-0_13.

Piquer, A., Martin, R. R. and Company, P. (2004). Skewed mirror symmetry for depth estimation in 3D line-drawings. *Lecture Notes in Computer Science. GREC 2003 Post-proceedings*, 3088. pp. 138–149. https://doi.org/10.1007/b99011.

Plumed, R. (2014). Sobre la percepción de indicios en bocetos de ingeniería y su aplicación al modelado basado en bocetos. PhD Dissertation. Universitat Jaume I (in Spanish).

Plumed, R., Company, P., Piquer, A. and Varley, P. A. C. (2010). Do engineers use convergence to a vanishing point when sketching? *Proc. International Symposium on Distributed Computing and Artificial Intelligence*, pp. 241–250. https://doi.org/10.1007/978-3-642-14883-5_32.

Plumed, R., Company, P. and Varley, P. A. C. (2013a). Metrics of human perception of vanishing points in perspective sketches. *WSCG 2013, 21st WSCG Conference on Computer Graphics, Visualization and Computer Vision*, pp. 59–68.

Plumed, R., Company, P. and Varley, P. A. C. (2013b). Human-like recognition of straight lines in sketched strokes. CEIG 2013, *Spanish Computer Graphics Conference*, pp. 56–66.

Plumed, R., Company, P. and Varley, P. A. C. (2015). A new approach for perceptually-based fitting strokes into straight segments. *CEIG 2015, XXV Spanish Computer Graphics Conference*, pp. 81–89.

Plumed, R., Company, P. and Varley, P. A. (2016). Detecting mirror symmetry in single-view wireframe sketches of polyhedral shapes. *Computers & Graphics*, 59, pp. 1–12. https://doi.org/10.1016/j.cag.2016.05.013.

Plumed, R., Company, P., Varley, P. A. C. and Martin, R. R. (2013c). From Sketches to CAM Models: Perceiving pockets and steps in single-view wireframe sketches of polyhedral shapes. *PerFab2013, UbiComp'13 Adjunct Proceedings of the 2013 ACM Conference on Pervasive and Ubiquitous Computing Adjunct Publication*, September 2013, Zurich, Switzerland, pp. 951–958, ISBN: 978-1-4503-2215-7. https://doi.org/10.1145/2494091.2499207.

Plumed, R., Company, P., Varley, P. and Martin, R. (2014). CSG feature trees from engineering sketches of polyhedral shapes. In *Eurographics 2014*, short papers. 2014, pp. 33–36.

Plumed, R., Varley, P. and Company, P. (2013d). Features and design intent in engineering sketches. *Intelligent Computer Graphics 2012*, pp. 77–106. https://doi.org/10.1007/978-3-642-31745-3_5.

Plumed, R., Varley, P. A. C., Company, P. and Martin, R. R. (2022). Extracting datums to reconstruct CSG models from 2D engineering sketches of polyhedral shapes. *Computers & Graphics*, 102, pp. 349–359. https://doi.org/10.1016/j.cag.2021.10.013.

Pop, M., Duncan, C., Barequet, G., Goodrich, M., Huang, W. and Kumar, S. (2001). Efficient perspective-accurate silhouette computation and applications. In *Proc. 17th Ann. ACM Symp. on Computational Geometry*, pp. 60–68. https://doi.org/10.1145/378583.378618.

Pusch, R., Samavati, F., Nasri, A. and Wyvill, B. (2007). Improving the sketch based interface. *The Visual Computer*, 23, pp. 955–962. https://doi.org/10.1007/s00371-007-0160-5.

Qin, S. (2005). Intelligent classification of sketch strokes. *EEE EUROCON2005-Computer as a Tool*, 1374–1377. https://doi.org/10.1109/EURCON.2005.1630216.

Qin, S., Wright, D. and Jordanov, I. (2001). On-line segmentation of freehand sketches by knowledge-based nonlinear thresholding operations. *Pattern Recognition*, 34, pp. 1885–1893. https://doi.org/10.1016/S0031-3203(00)00122-9.

Qin, S. F., Wright, D. K. and Jordanov, I. N. (2000). From on-line sketching to 2D and 3D geometry: A system based on fuzzy knowledge. *Computer-Aided Design*, 32(14), pp. 851–866. https://doi.org/10.1016/S0010-4485(00)00078-6.

Qin, Z., Jia, J., Li, T.-T. and Lu, J. (2007). Extracting 2d projection contour from 3d model using ring-relationship-based method. *Information Technology Journal*, 6, pp. 914–918. https://doi.org/10.3923/itj.2007.914.918.

Rai, R., Tiwari, M. K., Ivanov, D. and Dolgui, A. (2021). Machine learning in manufacturing and industry 4.0 applications. *International Journal of Production Research*, 59(16), pp. 4773–4778. https://doi.org/10.1080/00207543.2021.1956675.

Rattarangsi, A. and Chin, R. T. (1992). Scale-based detection of corners of planar curves. *IEEE Transactions on Pattern Analysis and Machine Intelligence*, 14(4), pp. 430–449. https://doi.org/10.1109/34.126805.

Reisfeld, E. and Sharf, A. (2023). OneSketch: Learning high-level shape features from simple sketches. *The Visual Computer*, 39, pp. 2811–2822. https://doi.org/10.1007/s00371-022-02494-2.

Roberts, L. (1965). *Optical and Electro-Optical Information Processing*. The MIT Press, Cambridge, Massachusetts and London, England, Chapter 9: Machine Perception of three-dimensional solids.

Roberts, L. G. (1963). *Machine Perception of Three-Dimensional Solids*. Ph.D. Dissertation, MIT Dep. of Electrical Engineering.

Rosin, P. L. (1993). Ellipse fitting by accumulating five-point fits. *Pattern Recognition Letters*, 14, pp. 661–669. https://doi.org/10.1016/0167-8655(93)90052-F.

Roth-Koch, S. and Westkaemper, E. (2010). The implementation of a sketch-based virtual product development. *Prod. Eng. Res. Devel.* 4, pp. 175–183. https://doi.org/10.1007/s11740-010-0222-9.

Rother, C. (2002). A new approach to vanishing point detection in architectural environments. *Image and Vision Computing*, 20(9–10), pp. 647–655. https://doi.org/10.1016/S0262-8856(02)00054-9.

Rubine, D. (1992). Combining gestures and direct manipulation. *Proceedings ACM CHI'92 Conference Human Factors in Computing Systems*, pp. 659–660.

Saga, S. and Makino, H. (1993). Fuzzy spline interpolation and its application to online freehand curve identification. *Second IEEE International Conference on Fuzzy Systems*, 2, pp. 1183–1190. https://doi.org/10.1109/FUZZY.1993.327560.

Salomon, D. (2011). *The Computer Graphics Manual*. Volume 1. Springer. ISBN 978-0-85729-885-0.

Salomons, O. W., Van Houten, F. J. A. M. and Kais H. J. J. (1993). Review of research in feature-based design. *Journal of Manufacturing Systems*, 12(2), pp. 113–132. https://doi.org/10.1016/0278-6125(93)90012-I.

Sanfilippo, E. M. and Borgo, S. (2016). What are features? An ontology-based review of the literature. *Computer-Aided Design*, 80, pp. 9–18, https://doi.org/10.1016/j.cad.2016.07.001.

Sarfraz, M., Swati Z. N. K. and Zaman, S. (2009). A novel corner detector approach using sliding ellipses. *Sixth International Conference on Computer Graphics, Imaging and Visualization*. https://doi.org/10.1109/CGIV.2009.33.

Saund, E. and Moran, T. (1994). A perceptually-supported sketch editor. *Proceedings UIST (1994)*, pp. 175–184. https://doi.org/10.1145/192426.192494.

Schildt, H. (1997). Windows NT 4 Programming from the Ground Up. McGraw-Hill Osborne. ISBN-13: 978-0078825736.

Schmidhuber, J. (2015). Deep learning in neural networks: An overview. *Neural Networks*, 61, pp. 85–117. https://doi.org/10.1016/j.neunet.2014.09.003.

Schmidt, J. M. (2013). A simple test on 2-vertex- and 2-edge-connectivity. *Information Processing Letters*, 113(7), pp. 241–244. https://doi.org/10.1016/j.ipl.2013.01.016.

Schweikardt, E. and Gross, M. D. (2000). Digital Clay: Deriving digital models from freehand sketches. *Automation in Construction*, 9, pp. 107–115. https://doi.org/10.1016/S0926-5805(99)00052-7.

Seff, A., Ovadia, Y., Zhou, W. and Adams, R. P. (2020). Sketchgraphs: A large-scale dataset for modeling relational geometry in Computer-Aided Design. arXiv preprint: arXiv:2007.08506.

Seff, A., Zhou, W., Richardson, N. and Adams, R. P. (2022). Vitruvion: A generative model of parametric CAD sketches. arXiv preprint: arXiv:2109.14124. DOI: 10.48550/arXiv.2109.1412.

Shahin, T. M. M. (2008). Feature-based design – An overview. *Computer-Aided Design and Applications*, 5(5), pp. 639–653. https://doi.org/10.3722/cadaps.2008.639-653.

Shpitalni, M. and Lipson, H. (1996). Identification of faces in a 2D line drawing projection of a wireframe object. *IEEE Transactions on Pattern Analysis and Machine Intelligence*, 18(10), pp. 1000–1012. https://doi.org/10.1109/34.541409.

Shpitalni, M. and Lipson, H. (1997). Classification of sketch strokes and corner detection using conic sections and adaptive clustering. *Journal of Mechanical Design, Transactions of the ASME*, 119(1), pp. 131–135. https://doi.org/10.1115/1.2828775.

Simo-Serra, E., Iizuka, S., Sasaki, K. and Ishikawa, H. (2016). Learning to simplify: Fully convolutional networks for rough sketch cleanup. *ACM Transactions on Graphics*, 35(4), a121. http://dx.doi.org/10.1145/2897824.2925972.

Stone, P. (1998). *Layered Learning in Multi-Agent Systems*. PhD Dissertation. Carnegie-Mellon University, School of Computer Science, Pittsburgh, PA.

Stroud, I. and Nagy, H. (2011). *Solid Modelling and CAD Systems. How to Survive a CAD System*. Springer, London, https://doi.org/10.1007/978-0-85729-259-9.

Sugihara, K. (1984). Interpretation of an axonometric projection of a polyhedron. *Computers & Graphics*, 8(4), pp. 391–400. https://doi.org/10.1016/0097-8493(84)90037-2.

Sugihara, K. (1986). *Machine Interpretation of Line Drawings*. The MIT Press, Cambridge, Massachusetts. ISBN: 978-0262192545.

Suh, Y. S. (2007). Reconstructing polyhedral CAD models by recognizing extrusion features from single-view drawings. In *International Design Engineering Technical Conferences and Computers and Information in Engineering Conference*, 48035, pp. 197–206. https://doi.org/10.1115/DETC2007-35186.

Sun, Y., Li, J., Xu, Z., Zhang, J., Liu, X., Zhang, D. and Lu, G. (2025). Sketch2Seq: Reconstruct CAD models from feature-based sketch segmentation. *IEEE Transactions on Visualization and Computer Graphics*, pp. 1–14. https://doi.org/10.1109/TVCG.2025.3566544.

Sunil, V. B., Agarwal, R. and Pande, S. S. (2010). An approach to recognize interacting features from B-Rep CAD models of prismatic machined parts using a hybrid (graph and rule based) technique. *Computers in Industry*, 61(7), pp. 686–701. https://doi.org/10.1016/j.compind.2010.03.011.

Suwa, M. and Tversky, B. (1997). What do architects and students perceive in their design sketches? A protocol analysis. *Design Studies*, 18(4), pp. 385–403. https://doi.org/10.1016/S0142-694X(97)00008-2.

Swati, Z. N. K., Zaman, S. and Sarfraz, M. (2009). A novel corner detector approach using sliding two ellipses and one rectangle. In *Proceedings of the 7th International Conference on Frontiers of Information Technology*, pp. 1–5.

Sycara, K. P. (1998). Multiagent systems. *AI Magazine*, 19(2), pp. 79–92. https://doi.org/10.1609/aimag.v19i2.1370.

Szpak, Z., Chojnacki, W. and Van den Hengel, A. (2012). Guaranteed ellipse fitting with the Sampson distance. In *Proceedings of 12th European Conference on Computer Vision (ECCV). LNCS*, 7576, pp. 87–100. https://doi.org/10.1007/978-3-642-33715-4_7.

Tanaka, M., Asano, T. and Higashino, C. (2020). Isometric conversion of mechanical sketches into 3D models. *Computer-Aided Design and Applications*, 18(4), pp. 772–785. https://doi.org/10.14733/cadaps.2021.772-785.

Tanaka, M., Asano, T. and Higashino, C. (2022). Abstraction of sketch features for predicting hidden shapes of sketches for the automatic conversion into 3D models. *Computer-Aided Design and Applications*, 19(5), pp. 977–987. https://doi.org/10.14733/cadaps.2022.977-987.

Tanaka, M., Asano, T., Higashino, C. and Ohtsuki, M. (2024). Handling chains, springs, and screws for automatic conversion of mechanical sketches into 3D models. *Computer-Aided Design and Applications*, 21(1), pp. 29–38. https://doi.org/10.14733/cadaps.2024.29-38.

Tanaka, M. and Kaneeda, T. (2015). Feature extraction from sketches of objects. *Computer-Aided Design and Applications*, 12(3), pp. 300–309. https://doi.org/10.4304/jait.5.3.74-78.

Tanaka, M., Takamiya, Y., Tsubota, N., Asanuma, S. and Iwama, K. (2017). Reconstruction of solid models from sketches including curves with inductive learning technique. *Computer-Aided Design and Applications*, 14, pp. 1–10. https://doi.org/10.1080/16864360.2016.1273580.

Tardif, J. P. (2009). Non-iterative approach for fast and accurate vanishing point detection. *12th International Conference on Computer Vision, IEEE*, 1250–1257. https://doi.org/10.1109/ICCV.2009.5459328.

Turner, A., Chapmann, D. and Penn, A. (2000). Sketching space. *Computers & Graphics*, 24, pp. 869–879. https://doi.org/10.1016/S0097-8493(00)00089-3.

Tyan, L. W. and Devarajan, V. (1998). Automatic identification of non-intersecting machining features form 2D CAD input. *Computer Aided Design*, 30, pp. 357–366. https://doi.org/10.1016/S0010-4485(97)00084-5.

Van Dijk, D. (1995). New insights in computer-aided conceptual design. *Design Studies*, 16(1), pp. 62–80. https://doi.org/10.1016/0142-694X(95)90647-X.

Vandenbrande, J. H. and Requicha, A. A. G. (1993). Spatial reasoning for the automatic recognition of machinable features in solid models. *IEEE Transactions on Pattern Analysis and Machine Intelligence*, 15(12), pp. 1269–1285. https://doi.org/10.1109/34.250845.

Varley, P. A. C. (2002). Automatic Creation of Boundary-Representation Models from Single Line Drawings. PhD Dissertation. Department of Computer Science. University of Wales College of Cardiff.

Varley, P. A. and Company, P. (2010). A new algorithm for finding faces in wireframes. *Computer-Aided Design*, 42(4), pp. 279–309. https://doi.org/10.1016/j.cad.2009.11.008.

Varley, P. and Martin, R. (2000a). A system for constructing boundary representation solid models from a two-dimensional sketch-frontal geometry and sketch categorisation. *In Proc. GMP 2000, IEEE.* https://doi.org/10.1109/GMAP.2000.838235.

Varley, P. and Martin, R. (2000b). A system for constructing boundary representation solid models from a two-dimensional sketch-topology of hidden parts. *In Proc. GMP 2000, IEEE.*

Varley, P., Suzuki, H. and Martin, R. R. (2004b). Making the most of using depth reasoning to label line drawings of engineering objects. *ACM Symposium on Solid Modeling and Applications*, pp. 191–202. https://doi.org/10.2312/sm.20041390.

Varley, P. A. C. and Martin, R. R. (2000c). Constructing boundary representation solid models from a two-dimensional sketch frontal geometry and sketch categorisation. *Proceedings of the First UK-Korea Workshop on Geometric Modeling and Computer Graphics*, Kyung Moon Publishers, pp. 113–128.

Varley, P. A. C., Martin, R. R. and Suzuki, H. (2004). Can machines interpret line drawings? *SBM'04, Proceedings of the First Eurographics Conference on Sketch-Based Interfaces and Modeling*, pp. 107–116. https://doi.org/10.2312/SBM/SBM04/107-116.

Varley, P. A. C., Martin, R. R. and Suzuki, H. (2005). Frontal geometry from sketches of engineering objects: Is line labelling necessary? *Computer-Aided Design*, 37, pp. 1285–1307. https://doi.org/10.1016/j.cad.2005.01.002.

Vaseghi, S. (2008). *Advanced Digital Signal Processing and Noise Reduction*, 4th ed. Wiley, The Atrium, Southern Gate, Chichester, West Sussex, UK. https://doi.org/10.1002/0470094966.

Verma, A. K. and Rajotia, S. (2010). A review of machining feature recognition methodologies. *International Journal of Computer Integrated Manufacturing*, 23(4), pp. 353–368. https://doi.org/10.1080/09511921003642121.

Waltz, D. (1972). *Generating Semantic Descriptions from Drawings of Scenes with Shadows*. Tech. rep., Artificial Intelligence Laboratory, Cambridge, MA, USA.

Wang, S., Qin, S. and Gao, M. (2012). Overtraced strokes to single line drawing: Preparation of sketch for 3D interpretation. *18th International Conference on Automation and Computing (ICAC)*, Loughborough, UK, 2012, pp. 1–6.

Wang, S., Qin, S. and Gao, M. (2014). New grouping and fitting methods for interactive overtraced sketches. *The Visual Computer*, 30, pp. 285–297. https://doi.org/10.1007/s00371-013-0844-y.

Wang, S., Wang, G., Gao, M. and Yu, S. (2013). Using fuzzy hybrid features to classify strokes in interactive sketches. *Advances in Mechanical Engineering*, 2013, Article ID 259152. https://doi.org/10.1155/2013/259152.

Wang, S., Wang, S., Li, Y. and Zhang, Q. (2015b). Grouping of multiple overtraced strokes in interactive freehand sketches. In *2015 14th*

International Conference on Computer-Aided Design and Computer Graphics (CAD/Graphics), pp. 232–233. IEEE. https://doi.org/10.1109/CADGRAPHICS.2015.27.

Wang, W. and Grinstein, G. (1993). A survey of 3D solid reconstruction from 2D projection line drawing. *Computer Graphics Forum*, 12(2), pp. 137–158. https://doi.org/10.1111/1467-8659.1220137.

Wang, Y., Chen, Y., Liu, J. and Tang, X. (2009). 3D reconstruction of curved objects from single 2D line drawings. *IEEE Conference on Computer Vision and Pattern Recognition CVPR 2009*, pp. 1834–1841. https://doi.org/10.1109/CVPR.2009.5206841.

Watanabe, T., Yoshikawa, T., Ito, T., Miwa, Y., Shibata, T. and Saga, S. (2020). An infinite-resolution grid snapping technique based on fuzzy theory. *Applied Soft Computing*, 89, p. 106112. https://doi.org/10.1016/j.asoc.2020.106112.

Wenyin, L. and Dori, D. (1999). From raster to vectors: Extracting visual information from line drawings. *Pattern Analysis and Applications*, 2(1), pp. 10–21. https://doi.org/10.1007/s100440050010.

Wolin, A., Eoff, B. and Hammond, T. (2008) Shortstraw: A simple and effective corner finder for polylines. *SBIM 2008*, pp. 33–40. https://doi.org/10.2312/SBM/SBM08/033-040.

Wright, L. (1983). *Perspective in Perspective*. Routledge & Kegan Paul, London.

Xin, M., Sharlin, E. and Costa-Sousa, M. (2008). Napkin sketch – Handheld mixed reality 3D sketching. VRST '08: *Proceedings of the 2008 ACM Symposium on Virtual Reality Software and Technology*, pp. 223–226. http://dx.doi.org/10.1145/1450579.1450627.

Xiong, Y. and LaViola Jr, J. J. (2010). A shortstraw-based algorithm for corner finding in sketch-based interfaces. *Computers & Graphics*, 34(5), pp. 513–527. https://doi.org/10.1016/j.cag.2010.06.008.

Xu, B., Chang, W., Sheffer, A., Bousseau, A., McCrae, J. and Singh, K. (2014). True2Form: 3D curve networks from 2D sketches via selective regularization. *ACM Transactions on Graphics*, 33(4), 131, pp. 1–13. https://doi.org/10.1145/2601097.2601128.

Xuetao, Y., Wonka, P. and Razdan, A. (2009). Generating 3D building models from architectural drawings: A survey. *IEEE Computer Graphics and Applications*, 29(1), pp. 20–30. https://doi.org/10.1109/MCG.2009.9.

Yang, H. and Zhang, H. (2017). Automatic 3D reconstruction of a polyhedral object from a single line drawing under perspective projection. *Computers and Graphics*, 65, pp. 45–59. https://doi.org/10.1016/j.cag.2017.04.003.

Yin, J., Liu, C., Lin, R., Vining, N., Rhodin, H. and Sheffer, A. (2022). Detecting viewer-perceived intended vector sketch connectivity. *ACM Transactions on Graphics*, 41(4), Article No. 87, pp. 1–11. http://dx.doi.org/10.1145/3528223.3530097.

Yu, J., Kulkarni, S. R. and Poor, H. V. (2013). Robust ellipse and spheroid fitting. *Pattern Recognition Letters*, 33, pp. 492–499. https://doi.org/10.1016/j.patrec.2011.11.025.

Yuan, S., Tsui, L. Y. and Jie, S. (2008). Regularity selection for effective 3D object reconstruction from a single line drawing, *Pattern Recognition Letters*, 29(10), pp. 1486–1495. https://doi.org/10.1016/j.patrec.2008.03.003.

Zeleznik, R. C., Herndon, K. P. and Hughes, J. F. (1996). SKETCH: An interface for sketching 3D scenes. *SIGGRAPH'96 Conference Proceedings*, pp. 163–170. http://dx.doi.org/10.1145/3596711.3596720.

Zhang, C., Pinquié, R., Polette, A., Carasi, G., De Charnace, H., and Pernot, J. P. (2023). Automatic 3D CAD models reconstruction from 2D orthographic drawings. *Computers & Graphics*, 114, pp. 179–189. https://doi.org/10.1016/j.cag.2023.05.021.

Zhang, C., Polette, A., Pinquié, R., Carasi, G., De Charnace, H., and Pernot, J. P. (2025a). eCAD-Net: Editable parametric CAD models reconstruction from dumb B-Rep models using deep neural networks. *Computer-Aided Design*, 178, p. 103806. https://doi.org/10.1016/j.cad.2024.103806.

Zhang, C., Polette, A., Pinquié, R., Iida, M., De Charnace, H., and Pernot, J. P. (2025b). Reinforcement learning-based parametric CAD models reconstruction from 2D orthographic drawings. *Computer-Aided Design*, 188, pp. 103–925. https://doi.org/10.1016/j.cad.2025.103925.

Zhang, X., Song, J., Dai, G. and Lyu, M. R. (2006). Extraction of line segments and circular arcs from freehand strokes based on segmental homogeneity features. *IEEE Transactions on Systems, Man, and Cybernetics, Part B (Cybernetics)*, 36(2), pp. 300–311. https://doi.org/10.1109/TSMCB.2005.857288.

Zhang, Z., Jaiswal, P. and Rai, R. (2018). FeatureNet: Machining feature recognition based on 3D Convolution Neural Network, *Computer-Aided Design*, 101, pp. 12–22. https://doi.org/10.1016/j.cad.2018.03.006.

Zhang, Z., Liu, X., Li, C. Wu, H. and Wen, Z. (2022). Vectorizing line drawings of arbitrary thickness via boundary-based topology reconstruction. *Computer Graphics Forum*, 41(2), pp. 433–445. https://doi.org/10.1111/cgf.14485.

Zhou, J. and Camba, J. D. (2025). The status, evolution, and future challenges of multimodal large language models (LLMs) in parametric CAD. *Expert Systems with Applications*, 282, p. 127520. https://doi.org/10.1016/j.eswa.2025.127520.

Index

www.ingramcontent.com/pod-product-compliance
Lightning Source LLC
Chambersburg PA
CBHW040144210626
46794CB00002BA/10
* 9 7 8 9 8 1 9 8 2 5 6 5 3 *